The IMA Volumes
in Mathematics
and its Applications

Volume 50

Series Editors
Avner Friedman   Willard Miller, Jr.

Institute for Mathematics and
its Applications
IMA

The **Institute for Mathematics and its Applications** was established by a grant from the National Science Foundation to the University of Minnesota in 1982. The IMA seeks to encourage the development and study of fresh mathematical concepts and questions of concern to the other sciences by bringing together mathematicians and scientists from diverse fields in an atmosphere that will stimulate discussion and collaboration.

The IMA Volumes are intended to involve the broader scientific community in this process.

Avner Friedman, Director
Willard Miller, Jr., Associate Director

* * * * * * * * * *

IMA ANNUAL PROGRAMS

1982–1983	Statistical and Continuum Approaches to Phase Transition
1983–1984	Mathematical Models for the Economics of Decentralized Resource Allocation
1984–1985	Continuum Physics and Partial Differential Equations
1985–1986	Stochastic Differential Equations and Their Applications
1986–1987	Scientific Computation
1987–1988	Applied Combinatorics
1988–1989	Nonlinear Waves
1989–1990	Dynamical Systems and Their Applications
1990–1991	Phase Transitions and Free Boundaries
1991–1992	Applied Linear Algebra
1992–1993	Control Theory and its Applications
1993–1994	Emerging Applications of Probability

IMA SUMMER PROGRAMS

1987	Robotics
1988	Signal Processing
1989	Robustness, Diagnostics, Computing and Graphics in Statistics
1990	Radar and Sonar
1990	Time Series
1991	Semiconductors
1992	Environmental Studies: Mathematical, Computational, and Statistical Analysis

* * * * * * * * * *

SPRINGER LECTURE NOTES FROM THE IMA:

The Mathematics and Physics of Disordered Media

Editors: Barry Hughes and Barry Ninham
(Lecture Notes in Math., Volume 1035, 1983)

Orienting Polymers

Editor: J.L. Ericksen
(Lecture Notes in Math., Volume 1063, 1984)

New Perspectives in Thermodynamics

Editor: James Serrin
(Springer-Verlag, 1986)

Models of Economic Dynamics

Editor: Hugo Sonnenschein
(Lecture Notes in Econ., Volume 264, 1986)

Richard A. Brualdi   Shmuel Friedland
Victor Klee
Editors

Combinatorial and Graph-Theoretical Problems in Linear Algebra

With 18 Illustrations

Springer-Verlag
New York Berlin Heidelberg London Paris
Tokyo Hong Kong Barcelona Budapest

Richard A. Brualdi
Department of Mathematics
University of Wisconsin
Madison, WI 53706 USA

Shmuel Friedland
Department of Mathematical Statistics
   and Computer Science
University of Illinois at Chicago
Chicago, IL 60680-4348 USA

Victor Klee
Department of Mathematics
University of Washington
Seattle, WA 98195-0001 USA

Series Editors:
Avner Friedman
Willard Miller, Jr.
Institute for Mathematics and its
   Applications
University of Minnesota
Minneapolis, MN 55455 USA

Mathematics Subject Classifications (1991): 05, 15

Library of Congress Cataloging-in-Publication Data
Combinatorial and graph-theoretical problems in linear algebra /
   Richard A. Brualdi, Shmuel Friedland, Victor Klee, editors.
      p.   cm. — (The IMA volumes in mathematics and its
   applications ; v. 50)
      Includes bibliographical references.
      ISBN 0-387-94086-3
      1. Algebras, Linear — Congresses.   2. Graph theory — Congresses.
   3. Combinatorial analysis — Congresses.   I. Brualdi, Richard A.
   II. Friedland, Shmuel.   III. Klee, Victor.   IV. Series.
   QA184.C64   1993
   512′.5 — dc20   93-2099

Printed on acid-free paper.

Production managed by Hal Henglein; manufacturing supervised by Vincent R. Scelta.
Camera-ready copy prepared by the IMA.
Printed and bound by Edwards Brothers, Inc., Ann Arbor, MI.
Printed in the United States of America.

9 8 7 6 5 4 3 2 1

ISBN 0-387-94086-3 Springer-Verlag New York Berlin Heidelberg
ISBN 3-540-94086-3 Springer-Verlag Berlin Heidelberg New York

The IMA Volumes
in Mathematics and its Applications

Current Volumes:

FOREWORD

This IMA Volume in Mathematics and its Applications

COMBINATORIAL AND GRAPH-THEORETICAL PROBLEMS IN LINEAR ALGEBRA

is based on the proceedings of a workshop that was an integral part of the 1991-92 IMA program on "Applied Linear Algebra." We are grateful to Richard Brualdi, George Cybenko, Alan George, Gene Golub, Mitchell Luskin, and Paul Van Dooren for planning and implementing the year-long program.

We especially thank Richard Brualdi, Shmuel Friedland, and Victor Klee for organizing this workshop and editing the proceedings.

The financial support of the National Science Foundation made the workshop possible.

Avner Friedman

Willard Miller, Jr.

PREFACE

The 1991-1992 program of the Institute for Mathematics and its Applications (IMA) was Applied Linear Algebra. As part of this program, a workshop on *Combinatorial and Graph-theoretical Problems in Linear Algebra* was held on November 11-15, 1991. The purpose of the workshop was to bring together in an informal setting the diverse group of people who work on problems in linear algebra and matrix theory in which combinatorial or graph-theoretic analysis is a major component. Many of the participants of the workshop enjoyed the hospitality of the IMA for the entire fall quarter, in which the emphasis was discrete matrix analysis. The stimulating and congenial environment that they helped create will undoubtedly remain in our memory for a long time to come. The papers presented in this volume include eleven of the fourteen invited talks as well as four other papers offered by other invited participants. Invited presentations by Fan Chung (on Laplacians of graphs and hypergraphs), Alex Lubotzky (on Ramanujan diagrams) and Hans Schneider (on spectral radii of certain iteration matrices and cycle means of graphs) are regrettably not included in this volume. We hope that this volume offers a glimpse of the increasing role of combinatorial structure in matrix analysis and conversely of the powerful tool that linear algebra provides in combinatorics and graph theory.

We are grateful to the IMA Director, Avner Friedman, and the Associate Director, Willard Miller, Jr., for their enthusiastic support of the workshop, and to the wonderful IMA staff for the cheerful and effective way in which they manage to juggle the requests of so many people at the same time. We thank Ms. K. Smith, Ms. P. Brick, and Mr. S. Skogerboe for their obvious 'TEX-pertise.'

Richard A. Brualdi
Shmuel Friedland
Victor Klee

CONTENTS

SYMBOLIC DYNAMICS AND MATRICES

MIKE BOYLE*

Contents

---

*University of Maryland at College Park

Introduction. The main purpose of this article is to give some overview of matrix problems and results in symbolic dynamics. The basic connection is that a nonnegative integral matrix A defines a topological dynamical system known as a shift of finite type. Questions about these systems are often equivalent to questions about "persistent" or "asymptotic" aspects of nonnegative matrices. Conversely, tools of symbolic dynamics can be used to address some of these questions. At the very least, the ideas of conjugacy, shift equivalence and strong shift equivalence give viewpoints on nonnegative matrices and directed graphs which are at some point inevitable and basic (although accessible, and even elementary).

My motivation for this article was to try to communicate some of this to matrix theorists. The earlier sections are more descriptive. The later sections move more to current frontiers and are oriented more to presenting open problems.

Trying to stay close to matrices, I've neglected huge parts of symbolic dynamics. Also even some matrix matters get short shrift. I've barely mentioned state splitting (3.4) and resolving maps (Sec. 10), which are important for constructions [AM,BMT,A2] and applications [ACH,AFKM,MSW]. Marker methods are a key to some inverse spectral results, but to avoid a nontrivial excursion into nonmatrix matters I just recommend some references (Sec.8). For more general background, one can dig into [DGS],[BMT],[PT1],[P2] and their references. Unfortunately, at present there is no single book which gives an appropriate introduction (although one by Adler, Lind and Marcus could appear any year now).

This article is a followup to the talk I was invited to give at the November 1991 I.M.A. Workshop on Combinatorial and Graph-theoretic problems in Linear Algebra. The talk was on joint work with David Handelman [BH1,2] on solving inverse spectral (and other) problems for nonnegative matrices, using tools from symbolic dynamics. My warm thanks go to the organizers Richard Brualdi, Shmuel Friedland and Victor Klee, for the interdisciplinary stimulation and good company.

1. SHIFTS OF FINITE TYPE

1.1 Subshifts. For the purposes of this paper, a topological dynamical system will be a continuous map T from a compact metric space X into itself. We can represent this as (X, T) or just T. Except in Section 9, T is a homeomorphism.

The system which is the full shift on n symbols (know more succinctly as the n-shift) is defined as follows. We endow a finite set of n elements–say, $\{0, 1, ..., n-1\}$– with the discrete topology. (This finite set is often called the alphabet.) We let X be the countable product of this set, indexed by $\mathbf{Z}$. We think of an element of X as a doubly infinite sequence $x = ...x_{-1}x_0x_1...$ where each x_i is one of the n elements. X is given the product topology and thus becomes a compact zero dimensional metrizable space. A metric compatible with the topology is given by (for x not equal to y)

$$\text{dist}(x, y) = 1/(k+1), \quad \text{where } k = \min\{|i| : x_i \neq y_i\}.$$

That is, two sequences are close if they agree in a large stretch of coordinates around the zero coordinate.

A finite sequence of elements of the alphabet is called a word. If W is a word of length $j - i + 1$, then the set of sequences x such that $x_i...x_j = W$ is called a cylinder set. The cylinder sets are closed and open, and they give a basis for the product topology on X.

There is a natural shift map S sending X into X, defined by shifting the index set by one: $(Sx)_i = x_{i+1}$. It is easy to see that S is bijective, S sends cylinders to cylinders, and thus S is a homeomorphism. The full shift on n symbols is the system (X, S).

A subshift is a subsystem of some full shift on n symbols. This means that it is a homeomorphism obtained by restriction of the shift to some compact set Y invariant under the shift and its inverse. The complement of Y is open and is thus a union of cylinder sets. Because Y is shift invariant, it follows that there is a (countable) list of words such that Y is precisely the set of all sequences y such that for every word W on the list, for every $i \leq j$, W is not equal to $y_i...y_j$. If Y is a set which may be obtained by forbidding a finite list of words, then the subshift is called a subshift of finite type, or just a shift of finite type (SFT). For example, we get an SFT by restricting the two-shift to the set Y of sequences in which the word 00 never occurs.

1.2 Vertex shifts. We will define vertex shifts, which are examples of shifts of finite type. For some n, let A be an $n \times n$ zero-one matrix. We think of A as the adjacency matrix of a directed graph with n vertices; the vertices index the rows and the columns, and $A(i, j)$ is the number of edges from vertex i to vertex j. Let Y be the space of doubly infinite sequences y such that for every k in Z, $A(y_k, y_{k+1}) = 1$. We think of Y as the space of doubly infinite walks through the graph, where the walks/itineraries are presented by recording the vertices traversed. The restriction of the shift to Y is a shift of finite type: a sufficient list of forbidden words is the set of words ij such that there is no arc from i to j.

Notation: throughout this paper, "graph" means "directed graph".

1.3 Edge shifts. Again let A be an adjacency matrix for a directed graph, but now allow multiple edges: so, the entries of A are nonnegative integers. Let the set of edges be the alphabet. Let Y be the set of sequences y such that for all k, the terminal vertex of y_k is the initial vertex of y_{k+1}. Again, we can think of Y as the space of doubly infinite walks through the graph, now presented by the edges traversed. The shift map restricted to Y is an edge shift and it is a shift of finite type: a sufficient list of forbidden words is the set of edge pairs ij which do not satisfy the head-to-tail rule.

In the sequel, unless otherwise indicated an SFT defined by a matrix A is intended to be the edge shift defined by A. We denote this SFT by S_A.

1.4 Codes. Suppose (X, S) and (Y, T) are subshifts. A map f from X to Y is called a code if it is continuous and intertwines the shifts, i.e. $fS = Tf$. We think of a code as a homomorphism of these dynamical systems.

Now suppose F is a function from words of length $2n + 1$ which occur in S-sequences into some finite set A. Then the rule

$$(fx)_i = F(x_{i-n}...x_{i+n}), \quad \text{for all } i \text{ in } \mathbf{Z},$$

defines a code f, called a block code, into the full shift on the alphabet A. This block code defines a code from S into any subshift T which contains the image of f. The "Curtis-Hedlund-Lyndon Theorem" asserts that every code is given by a block code. The argument is easy: given f, one obtains F and n above for $i = 0$ as a consequence of uniform continuity, and the formula for all i follows because f intertwines the shifts.

If a code f is surjective, then it is called a quotient or factor map. If it is injective, then it is called an embedding. If it is injective and surjective, then it is an isomorphism or conjugacy of subshifts. This notion of isomorphism is our fundamental equivalence relation.

To expand on this a little, think of a code f from S_A to S_B as a map of infinite paths in graphs. If we think of x_i as our location on this path at time i, then we think of $(fx)_i$ as our location on the image path at time i. The rule F above determines that location $(fx)_i$, knowing the location x_i with memory of the last n locations and anticipation of the next n locations. The same rule F works for any i. If f is an isomorphism, then in a strong sense the structure of these infinite path spaces is essentially the same.

1.5 Higher block presentations. Let S be a subshift. Suppose n is a positive integer and j, k are nonnegative integers such that $j + k + 1 = n$. Given this we will define a code f with domain the subshift S by the rule

$$(fx)_i = x_{i-j}...x_{i+k}, \quad i \text{ in } \mathbf{Z}.$$

So, for the output sequence fx, the symbol $(fx)_i$ is the word of length n in x in the coordinates $i - j$ through $i + k$. The shift map on the set of all output sequences is a subshift T, and f is an isomorphism from S to T. (f is a block code and it is clearly one-to-one.) The system T doesn't depend on the choice of j (although the map does). T is called the n-block presentation of S.

An easy exercise is to construct a one-block isomorphism between the n-block presentation of S and the subshift obtained by passing to the 2-block presentation $n-1$ times.

For an important example, let S be the edge shift defined by a matrix A. Let G be the graph with adjacency matrix A. A symbol in the alphabet of the two block presentation is a word uv, where u and v are edges and the terminal vertex of u equals the initial vertex of v (i.e. uv is a path of length 2 in G). We can define a new graph G' whose vertices are the edges of G, and where there is an arc from u to v if the terminal vertex of u equals the initial vertex of v. If we give such an arc the name uv, then we see that the two-block presentation of S_A is the SFT presented by the matrix which is the adjacency matrix of G'. That is, for SFT's defined by matrices, passing to the two-block presentation amounts to passing from the defining graph to its edge graph (remember, all our graphs are directed—the edge graph in this category is the directed graph we've just described, it rarely has a symmetric adjacency matrix).

Similarly, we can think of the n-block presentation of S_A as given by a graph $G(n)$ whose vertices are the paths of length $n-1$ in G. Here (for $n > 2$) there is an edge from vertex $a(1)...a(n-1)$ to a vertex $b(1)...b(n-1)$ iff $a(2)...a(n-1) = b(1)...b(n-2)$. (For $n = 2$, there is an edge from vertex $a(1)$ to vertex $b(1)$ iff the terminal vertex of $a(1)$ equals the initial vertex of $b(1)$.) Note, if the original subshift contains infinitely many points, then as n goes to infinity the size of the adjacency matrix for $G(n)$ must go to infinity. In particular matrices of very different size may define isomorphic SFT's.

1.6 One-block codes. Suppose f is a code from S to T. Then there is some n such that for all x, $(fx)_0$ is determined by the word $x_{-n}...x_n$ of length $2n + 1$. Define an isomorphism from S to its $(2n+1)$-block presentation S' as in §1.5, using $j = k = n$. Let h be the inverse of this isomorphism. Then fh is a code from S' to S_B, and fh is a one-block map. Often, given a code from S, by passing to a higher block presentation in this way we can assume that the code is just a one-block map.

For example, if there is a map from S_A to S_B, then there is a one-block map from S_C to S_B, where C is a matrix giving some higher-block presentation of A. That is, by passing to an iterated edge graph H of the graph with adjacency matrix A (H has adjacency matrix C), there is a graph homomorphism from H to the graph with adjacency matrix B which (applied at every edge along a path) gives a map from the set of infinite A-paths to the set of infinite B-paths.

1.7 Isomorphic SFT's. Any SFT (Y, S) is isomorphic to a vertex shift. To show this, let W be a finite list of words such that Y is the set of all sequences (from some full shift) in which no word of W occurs. Let $n+1$ be the length of the longest word in W (the SFT is then "n-step"). Without loss of generality, assume $n > 0$. Let V be the set of all words of length n which occur in sequences in Y. Let V be the vertex set of a directed graph. In this graph, there is an edge from

$u = u_1...u_n$ to $v = v_1...v_n$ if $uv_n = u_1v$ and this word of length $n + 1$ occurs in a sequence in Y. This graph defines a vertex SFT (X, T). An isomorphism f from Y to X is given by the rule $(fy)_i = y_i...y_{i+n-1}$.

Also, any SFT is isomorphic to an edge shift, because the two-block presentation of a vertex shift is an edge shift.

Even if one is only interested in SFT's defined from graphs, it is useful to consider general n-step SFT's, because working with these gives access to topological and combinatorial arguments which can in turn yield results about the graphically defined SFT's. The vertex shifts are sometimes more simple to work with than the edge shifts. The edge shifts are very useful. One reason is conciseness: an edge shift presented by a small matrix (perhaps though with large entries) may be presentable as a vertex shift only by a large matrix. Also, the set of zero-one matrices (the matrices which define vertex shifts) is not closed under various operations under which the set of nonnegative integer matrices is closed. Working only with zero-one matrices rules out some very useful matrix arguments (e.g. [F2]) and interpretations. For one of these, first a little preparation.

If S is a subshift, then we let S^n denote the homeomorphism obtained by iterating S n times. The homeomorphism S^n is isomorphic to a subshift T whose alphabet is the set of S-words of length n. An isomorphism from S^n to T is given by the map f which sends a point x to the point y such that for all k in $\mathbf{Z}$,

$$y_k = x_{kn}...x_{(k+1)n-1}.$$

Now, let an edge shift S be defined by a matrix A. Then the subshift S^n is conjugate to the edge shift defined by A^n. The number of edges from vertex i to vertex j in the directed graph with adjacency matrix A^n is just the (i, j) entry of the matrix A^n. This is also the number of paths of length n from i to j in the directed graph with adjacency matrix A. We can use bijections of these edges and paths to replace symbols y_k of the construction of the previous paragraph with edges in the directed graph defined by A^n. Then that construction provides the claimed isomorphism.

1.8 Topological Markov shifts. An SFT is also called a topological Markov shift, or topological Markov chain. This terminology is appropriate because an SFT can be viewed as the topological support of a finite-state stochastic Markov process, and also as the topological analogue of such a process. (This viewpoint was advanced in the seminal 1964 paper of Parry [P1].)

Roughly speaking, in a Markov process the past and future are independent if conditioned on the present (or more generally if conditioned on some finite time interval). We can say precisely why an SFT is a topological analogue of this. Suppose the SFT is n-step (given by forbidding a certain list of words of length at most $n + 1$). Also suppose that x and y are points (doubly infinite sequences) in the SFT such that $x_0...x_{(n-1)} = y_0...y_{n-1}$. Then it follows that the doubly infinite sequence z defined by

$$z_i = x_i \quad \text{if } i < n$$
$$= y_i \quad \text{if } i \geq 0$$

must also be a point in the SFT. That is: the possibilities for the future (sequences in positive coordinates) and the past (sequences in negative coordinates) are independent conditioned on the near-present (i.e., the word in a certain finite set of coordinates).

1.9 Applications of SFT's. For completeness I'll mention in the most cursory fashion two ways in which SFT's appear in a natural and useful way.

First, imagining very long tapes of zeros and ones, consider infinite strings of zeros and ones (i.e., points in the full shift on two symbols). It is natural to think of a block code as a machine which takes an input tape and recodes it, and to suppose that somehow the study of block codes may be relevant to efficiently encoding and decoding data. This turns out to be the case [ACH,MSW], in fact I understand that some constructions arising from symbolic dynamics have actually been built into IBM hardware.

Second, imagine a homeomorphism (or diffeomorphism) h on some space X. One way to study h is by symbolic dynamics. Crudely, cut X into n pieces. Name the pieces $1, 2, ..., n$. To any given point x in X there is associated a sequence y in the full shift on n symbols, where y is defined by setting y_i to be the piece containing $h^i(x)$, for each integer i. This gives some set of sequences y. The sequence associated to $h(x)$ is the shift of y. This establishes some relation between the topological dynamics of the shift space and the dynamics of h. Sometimes a relation of this sort is very useful (for example for analyzing h-invariant measures or periodic points), when the shift space which arises is a shift of finite type [Bow1,2].

A variation on the last theme going back to Hadamard is the study of geodesic flows with symbolic dynamics [AF].

2. MATRIX INVARIANTS

Throughout this section A will represent a matrix with integral entries. Unless otherwise indicated, we also assume that A is nondegenerate (every row has a nonzero entry and every column has a nonzero entry) and that the entries are nonnegative. (This is because if A were nonnegative with ith row or column zero, then A would define the same SFT as would the principal submatrix obtained by deleting row i and column i — it is only the "nondegenerate core" of A which carries information about the SFT defined by A.) We let S_A denote the shift of finite type defined by A.

By a matrix invariant of A we will mean something determined by A which is the same for matrices A and B which determine isomorphic shifts of finite type. Some of these matrix invariants correspond to dynamically important invariants of the associated shift. The matrix invariants usually have an algebraic flavor, often being defined for (not necessarily nonnegative) integral matrices. Then one has an associated realization problem: which of the algebraic invariants can be realized by nonnegative matrices? We'll list the most important of these invariants and discuss the corresponding dynamical properties.

2.1 Mixing properties. The nonnegative matrix A is primitive if some power has all entries greater than zero. A is irreducible if for every position (i,j) there is $n > 0$ such that $A^n(i,j) > 0$. Otherwise A is reducible.

The associated SFT S is mixing if and only if A is primitive. It has a dense forward orbit if and only if A is irreducible. The most important class to understand is the class of mixing SFT's. Then one understands other SFT's by how they are built up from the mixing SFT's. This is analogous to the situation with nonnegative matrices, which one understands by first understanding the primitive case. (Caveat: often the general case of a problem for SFT's follows very easily from the mixing case, but sometimes the generalization is quite difficult.) In the sequel we will sometimes make the simplifying assumption that A is primitive. Sometimes this is only for simplicity and sometimes we are avoiding serious difficulties.

2.2 Entropy. The premier numerical invariant of a dynamical system S is its (topological) entropy, $h(S)$. For a subshift S,

$$h(S) = \limsup_n \frac{\log(\#W_n(S))}{n}$$

where $W_n(S)$ is the set of words of length n occurring in sequences of S. That is, the entropy is the exponential growth rate of the S-words. For a full shift on n symbols, the entropy is $\log(n)$. For an SFT defined by a matrix A, the entropy is the log of the spectral radius of A. This is easy to prove because the number of words of length n is the sum of the entries of A^n.

What numbers can be entropies of mixing SFT's? Equivalently, what numbers can be spectral radii of primitive integral matrices? This was settled by Lind [L]: a number is the spectral radius of a primitive integral matrix if and only if it is a Perron number. A Perron number is a positive real number which is an algebraic integer which is strictly greater than the modulus of any other root of its minimal polynomial.

2.3 Periodic points. The periodic points of a topological dynamical system are often involved in its dynamics in a crucial way. Let $\text{Fix}(S)$ denote the set of fixed points of a map S, i.e. the set of points x such that $Sx = x$. Suppose that for every positive integer n, the set $\text{Fix}(S^n)$ is finite. (This will be true for any subshift S, for which a fixed point of S^n is a periodic sequence of period dividing n.) Then the sequence $\#\text{Fix}(S^n)$ contains all the information one has from restricting S to its periodic points and forgetting the topology. The favored choice in dynamics for compiling this information is the (Artin-Mazur) zeta function of S,

$$\zeta_S(z) = \exp \sum_{n=1}^{\infty} \frac{\#Fix(S^n)}{n} z^n.$$

(We will postpone to Section 4 some justification for this choice.) The zeta function can be considered as a formal power series whenever $\text{Fix}(S^n)$ is finite for all n. If also

$$\limsup_n [\#Fix(S^n)]^{1/n} = a < \infty$$

then the zeta function is defined as an analytic function on the open disc of radius $1/a$ around the origin in the complex plane. For subshifts, such a number a always exists, not larger than the cardinality of the alphabet. For an SFT defined by a matrix A, a is the spectral radius of A.

If S is an SFT defined by a matrix A, then the number of fixed points of S is simply the trace of A: a fixed point is a sequence consisting of one edge repeated forever. Similarly, for all positive integers n

$$\#Fix(S^n) = tr(A^n).$$

From this one can compute

$$\zeta_S(z) = \exp \sum_{n=1}^{\infty} \frac{tr(A^n)}{n} z^n = \left[\prod (1 - az) \right]^{-1} = [\det(I - zA)]^{-1}$$

where the product is over the eigenvalues a of A, repeated according to multiplicity. (So the inverse zeta function of an SFT is a polynomial with integral coefficients and constant term 1.) The first equality follows from the definition of the zeta function and the previous equation. The last two equalities hold for any square matrix A with real entries, as we now argue. The last equality follows from dividing the equation

$$\det(zI - A) = \prod_a (z - a)$$

by z^k (where A is k by k) and then replacing $1/z$ with z. The second equality follows from three facts:

(1) $tr(A^n) = \sum_a a^n$,

(2) for any complex number a,

$$\exp \left(\sum_{n=1}^{\infty} \frac{(az)^n}{n} \right) = 1/(1 - az)$$

(to see this take the derivative of the log of each side),

(3) $\exp(\sum_{n=1}^{\infty} \sum_a \frac{(az)^n}{n}) = \prod_a \exp(\sum_{n=1}^{\infty} \frac{(az)^n}{n})$.

Problem. What are the nonzero spectra of primitive integral matrices?

In dynamical language, this problem is: what are the zeta functions of mixing shifts of finite type? (Clearly the nonzero part of the spectrum of a matrix A—counting multiplicities, always—determines $\det(I - zA)$, and vice versa.) This is a difficult problem, but there is very strong evidence supporting a conjectural answer of appealing simplicity, which we discuss in Section 8 below.

2.4 Isomorphism. Matrices A and B are elementary strong shift equivalent over a semiring S if there are matrices U, V with entries from S such that $A = UV$ and $B = VU$. The matrices U, V need not be square. If the semiring is not specified, then it is understood to be the nonnegative integers. A and B are strong shift equivalent over S if they are linked by a finite chain of elementary strong shift equivalences— that is, strong shift equivalence is the transitive closure of elementary strong shift equivalence.

Two shifts of finite type S_A and S_B are isomorphic if and only if the matrices A and B are strong shift equivalent [W1]. It is not trivial to prove isomorphism gives strong shift equivalence, so we refer to [W1] or [PT1,Sec.V.3] for this direction. However the other direction is easy. Suppose $A = UV$, $B = VU$. Let G_A, G_B be the directed graphs with adjacency matrices A, B and let these graphs have no vertices in common. Let U be the adjacency matrix for a set of arcs with initial vertices in G_A and terminal vertices in G_B; similarly V describes arcs from G_B to G_A. Let lower case letters a, b, u, v represent arcs corresponding to A, B, U, V. From the matrix equations we may choose bijections of arcs and paths of length 2,

$$\{a\} \longleftrightarrow \{uv\}, \qquad \{b\} \longleftrightarrow \{vu\}$$

respecting initial and terminal vertex. We view a point of S_A as an infinite path $\ldots a_{-1} a_0 a_1 \ldots$ of edges a_i and apply the first bijection at each a_i to get the following picture.

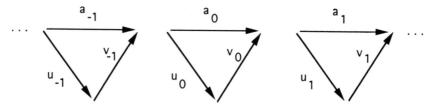

(Here, for example, $u_0 v_0$ is the path corresponding to a_0.) We apply the second bijection to give a correspondence $v_i u_{i+1} \longleftrightarrow b_i$ and get a larger picture:

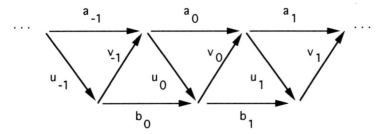

This picture gives a rule which sends a point of S_A to a point of S_B. It is easy to check that the rule defines an isomorphism of the two SFT's.

Strong shift equivalence and shift equivalence (below) were introduced by Williams [W1]. These are crucial ideas in the subject.

2.5 Eventual isomorphism. Two matrices A and B are shift equivalent over a semiring S if there are matrices U, V over S and a positive integer ℓ (called the lag) such that the following equations hold.

$$A^\ell = UV \qquad B^\ell = VU$$
$$AU = UB \qquad BV = VA.$$

Again, S is the nonnegative integers if the semiring is not specified.

Two systems S and T are eventually isomorphic if S^n and T^n are isomorphic for all but finitely many n. If a pair U, V give a shift equivalence of lag ℓ from A to B, then A^ℓ and B^ℓ are strong shift equivalent. Also the pair $A^n U, V$ gives a shift equivalence of lag $l + n$ from A to B. Consequently, if A and B are shift equivalent, then the SFT's S_A and S_B are eventually isomorphic. The converse also holds because the shift equivalence of A^p and B^p implies the shift equivalence of A and B if p is a sufficiently large prime [KR1].

Regardless of the semiring S, strong shift equivalent matrices must be shift equivalent. This follows from manipulating a chain of ℓ elementary strong shift equivalences to produce a lag ℓ shift equivalence.

At first glance, shift equivalence may appear to be a more obscure and complicated equivalence relation than strong shift equivalence. In fact, it is just the opposite. (We will say more about this below.) Williams introduced the idea of shift equivalence with the intent of reducing strong shift equivalence to a manageable equivalence relation on matrices. He conjectured that shift equivalence implies strong shift equivalence for matrices over $\mathbf{Z}_+$ [W1]. This conjecture was finally refuted in the reducible case. This important result was proved by Kim and Roush [KR7] using fundamental advances by themselves and Wagoner [KRW]. However, the crucial irreducible case remains open.

Problem. Suppose A and B are irreducible matrices shift equivalent over $\mathbf{Z}_+$. Must they be strong shift equivalent over $\mathbf{Z}_+$?

It is not hard to show that the answer is yes to this question if and only if it is yes to the question for primitive matrices.

The classification of reducible SFT's will not follow easily from the classification of irreducible SFT's. However, work in progress by Kim and Roush indicates that one will be able to classify shift equivalent reducible SFT's in terms of the classification of irreducible SFT's and the range of the dimension representation [KRW] on their automorphism groups.

2.6 Flow equivalence. Two homeomorphisms are flow equivalent if they are cross sections of the same flow. If the matrices A and B are irreducible and neither is a permutation matrix, then S_A and S_B are flow equivalent if and only if

(i) $\det(I - A) = \det(I - B)$, and

(ii) the cokernels of $I - A$ and $I - B$ are isomorphic.

(Here, for example, if A is $n \times n$ then the cokernel of $I - A$ is the abelian group which is $\mathbf{Z}^n$ modulo the image of $I - A$.) This was ultimately proved by Franks [F2] following the earlier work of Bowen&Franks [BF] and Parry&Sullivan [PS].

The classification up to flow equivalence of SFT's defined by reducible matrices is difficult and interesting. Cuntz [C] has introduced algebraic invariants in a special case from the viewpoint of associated C^*-algebras. Danrun Huang, beginning from the work of Franks and Cuntz, has gone much further on this problem [Hu]. Huang's work is very much in the vein of discerning the right invariants and then showing they are complete by way of matrix constructions realizing prescribed algebraic actions subject to positivity constraints.

We won't be concerned with flow equivalence in this paper, except to flesh out certain algebraic patterns below.

2.7 Relations

For matrices over $\mathbf{Z}_+$:

strong shift equivalence $\Rightarrow$ shift equivalence $\Rightarrow$

same zeta function $\Rightarrow$ same entropy.

Also, for irreducible matrices at least,

shift equivalence $\Rightarrow$ same flow equivalence class.

None of these implications can be reversed.

3. SHIFT EQUIVALENCE AND STRONG SHIFT EQUIVALENCE

In this section we will try to explain what it means for two integral matrices to be shift equivalent over $\mathbf{Z}$, and we will indicate some tools one derives from the ideas of strong shift equivalence.

3.1 Shift equivalence and the dimension group. Suppose A is a $n \times n$ integral matrix. Then A acts on $\mathbf{Z}^n$ and from this we can form the direct limit group $G(A)$, on which A induces an automorphism A'. This gives a pair $(G(A), A')$. Two integral matrices A and B are shift equivalent over $\mathbf{Z}$ if and only if there is a group isomorphism f from $G(A)$ to $G(B)$ such that $fA' = B'f$.

This was pointed out by Krieger [Kr1,2] who also did something much deeper, giving a K-style construction of this group from certain compact subsets of the shift space. We will skip the (easy) proof of the previous paragraph and any explanation of the topological theory (see [BMT] for more).

We remark that $G(A)$ is called a dimension group (and therefore $(G(A), A')$ has been called the "dimension pair") for historical reasons. Krieger's topological construction was adapted from K-theoretic constructions in operator algebras–in fact, $G(A)$ is K_0 of an associated C^*-algebra [CuK2]. It is natural to think of K_0 of a ring R as a "dimension group", because K_0 is concerned with (stable) isomorphism classes of finitely generated projective R-modules and for R a field the isomorphism class of such a module is given by its (vector space) dimension.

We will give a concrete description of $(G(A), A')$. Of course A acts on $\mathbf{Q}^n$. Let V_A be the rational subspace of $\mathbf{Q}^n$ which is the image of A^n. (So, if A is not nilpotent, then V_A is the largest invariant subspace on which the action of A is nonsingular.) Now we can give the following presentation for $G(A)$:

$$G(A) = \{v \text{ in } V : \text{for some } k > 0, vA^k \text{ is in } \mathbf{Z}^n\}.$$

With this presentation, the automorphism A' is just multiplication by A. (We have arbitrarily chosen the action here to be on row vectors. The choice does matter, as a matrix need not be shift equivalent to its transpose [PT1]. However matrices A and B are (strong) shift equivalent if and only if their transposes are.)

For example, if $A = [2]$, then in this presentation $G(A)$ is the dyadic rationals–all rational numbers p/q where p and q are integers and q is a power of 2. If $|\det A| = 1$, then $G(A)$ is just $\mathbf{Z}^n$.

Note: if $(G(A), A')$ and $(G(B), B')$ are isomorphic, then the actions obtained by tensoring with $\mathbf{Q}$ are isomorphic. In other words, the restrictions of A and B to V_A and V_B are isomorphic as linear transformations of rational vector spaces. (In particular these restrictions have the same characteristic polynomial, which is just the characteristic polynomial of A divided by the appropriate power of the indeterminate. Equivalently, $\det(I - tA) = \det(I - tB)$.) Another way to say this is that A and B have the same Jordan form away from zero (i.e., the nonnilpotent parts of the Jordan forms of A and B are the same modulo conjugation by a permutation matrix).

An example of two matrices with the same characteristic polynomial which are not shift equivalent over $\mathbf{Z}$ is given by the pair

$$A = \begin{pmatrix} 1 & 2 \\ 2 & 3 \end{pmatrix} \qquad B = \begin{pmatrix} 1 & 1 \\ 4 & 3 \end{pmatrix}.$$

To see this easily, suppose there is a shift equivalence. Note that because $|det A| = 1$, we must have $|det U| = 1$ (where U is the matrix in the defining equations for shift equivalence). If $AU = UB$, then

$$U^{-1}(A - I)U = (B - I).$$

Since every matrix on the left side is integral and $A - I$ is divisible by 2, the matrix on the right side must have every entry divisible by 2, a contradiction. (For an alternate proof, note that $\operatorname{cok}(I - A) \cong \mathbf{Z}/2 \oplus \mathbf{Z}/2$ and $\operatorname{cok}(I - B) \cong \mathbf{Z}/4$, so that A and B do not even define flow equivalent SFT's.)

The general solution to the decision problem for shift equivalence is very difficult [KR1,3]. But there are many classes of tractable examples. For example, if A has a single nonzero eigenvalue k, then A is shift equivalent over $\mathbf{Z}$ to the one by one matrix $[k]$. If $p(t)$ is the minimal polynomial of an algebraic integer λ, then the shift equivalence classes over $\mathbf{Z}$ of integral matrices with characteristic polynomial $p(t)$ are in bijective correspondence with the ideal classes of the ring $\mathbf{Z}[1/\lambda]$ ([BMT]). Such algebraic aspects of shift equivalence turn out to correspond to coding relations among corresponding shifts of finite type ([BMT], [KMT], [As2]).

3.2 (Strong) Shift equivalence over $\mathbf{Z}_+$. If A and B are shift equivalent over $\mathbf{Z}$, then they are strong shift equivalent over $\mathbf{Z}$. The same is true for matrices over any principal ideal domain [E2],[W2] or Dedekind domain [BH2].

It is not known whether primitive matrices shift equivalent over $\mathbf{Z}_+$ must be strong shift equivalent over $\mathbf{Z}_+$. This is still unknown if $\mathbf{Z}$ is replaced by the rationals $\mathbf{Q}$, or even the reals $\mathbf{R}$! So the order requirement complicates the situation for strong shift equivalence drastically.

For shift equivalence, there are also complications but they are fewer. The best news is that two primitive matrices are shift equivalent over $\mathbf{Z}_+$ if they are shift equivalent over $\mathbf{Z}$. (This is still true by the same proof if $\mathbf{Z}$ is replaced by any unital subring of the reals.) This was first proved geometrically by Parry and Williams [PW], also see [KR1]. (Caveat: irreducible matrices may be shift equivalent over $\mathbf{Z}$ but not over $\mathbf{Z}_+$, as an example of Kaplansky and myself shows [B2].)

It is easy to sketch a proof of this. Suppose A and B are primitive and a pair U, V gives a shift equivalence of lag ℓ ($AU = UB$ etc.). The idea is, for large n the matrices $A^n U$, $V A^n$ are positive (possibly after replacing U, V with $-U, -V$) and they give a shift equivalence of lag $2n + \ell$. To understand positivity, remember that the Perron Theorem implies that for large n, A^n is "approximately" $(a^n)RL$ where R is a positive right eigenvector, L is a positive left eigenvector, $LR = 1$, and a is the spectral radius of A. Here "approximately" means that the error is growing at an exponentially smaller rate. Thus it suffices to show that LU and VR are positive. Because the Perron eigenvalue is a simple root of the characteristic polynomial of B, the eigenvectors LU and VR for B must be multiples of the corresponding (positive) eigenvectors for B. Because $LUVR = LA^\ell R$ is positive, either both of these multiples are negative (then we replace U, V with $-U, -V$) or both are positive (and we are done).

One puts the requirement of nonnegativity into the dimension group context as follows [Kr1]. An ordered group is a group G with a set G_+ (called the positive set) such that G_+ is closed under addition and every element of G is a difference of elements of G_+. An isomorphism of ordered groups is a group isomorphism mapping the positive set of the domain group onto the positive set of the range group. Now one just adds to the "dimension group" structure an order structure on the group to reflect shift equivalence over $\mathbf{Z}_+$ rather than $\mathbf{Z}$. With this structure, the isomorphism A' above is an isomorphism of the ordered group $(G(A), G_+(A))$. The "dimension pair" now becomes a "dimension triple" (G, G_+, A').

We'll describe this in terms of the explicit presentation described in §3.1. Given a nonnegative matrix A, define the positive set

$$G_+(A) = \{v \text{ in } G(A) : \text{for some } k > 0, vA^k \text{ has all entries nonnegative}\}.$$

Multiplication by A induces an ordered-group automorphism A' on $(G(A), G_+(A))$. Now matrices A, B over $\mathbf{Z}_+$ are shift equivalent if and only if there is an isomorphism of their ordered groups intertwining A' and B'.

For more on dimension groups, see [E1].

3.3 The dimension module of an SFT. Here we make explicit a reformulation of the dimension data, which will seem trivial (but correct) from a homological viewpoint [Br]. This reformulation is by no means due to me (see [Wa1, pp.92,120]).

To an SFT defined by a matrix A, we associated a "dimension pair" $(G(A), A')$. The action of A' gives an action of the group $\mathbf{Z}$ on $G(A)$. Whenever a group H acts on an abelian group K by automorphisms of K, the group K acquires a $\mathbf{Z}H$-module structure, where $\mathbf{Z}H$ is the integral group ring of H [Br]. An isomorphism of such H-actions is equivalent to a $\mathbf{Z}H$-module isomorphism. So instead of referring to the "dimension pair" $(G(A), A')$ we can just refer to the dimension module. Here we mean the $\mathbf{ZZ}$-module $G(A)$. The ring $\mathbf{ZZ}$ is isomorphic to the ring $\mathbf{Z}[t, t^{-1}]$ of integral Laurent polynomials in one variable.

Since "dimension pair" and "dimension module" carry the same information, to some extent passing from the latter to the former is just a matter of cleaner terminology. But it is also a matter of a better functorial setup— "thinking right" as the group cohomologists say. We'll see more of this in Sections 5 and 7.

We encode the order information of the "dimension triple" by making the $\mathbf{ZZ}$-module an ordered module in the natural way. First, $\mathbf{ZZ}$ is an ordered ring in a natural way, with the semiring $\mathbf{Z}_+\mathbf{Z}$ (formal nonnegative integral combinations of the set $\mathbf{Z}$) the positive set. (If we think of $\mathbf{ZZ}$ as the Laurent polynomials $\mathbf{Z}[t, t^{-1}]$, then the positive set is $\mathbf{Z}_+[t, t^{-1}]$, the Laurent polynomials whose coefficients are nonnegative integers.) Now an ordered $\mathbf{ZZ}$-module is an ordered group (G, G_+) which is a $\mathbf{ZZ}$-module such that $\mathbf{Z}_+\mathbf{Z}$ sends G_+ into itself. It is easy to check that this data is equivalent to the data in the dimension triple.

"Dimension module" was used in a different sense by Tuncel [T]—we reconcile our viewpoints in Section 6.

3.4 Strong shift equivalence and state-splitting. Let A be an $n \times n$ matrix. Let A' be an $(n+1) \times n$ matrix related to A as follows: row i of A is the sum of row i and row $(n+1)$ of A', otherwise the rows of A and A' are equal. Now define an $(n+1) \times (n+1)$ matrix B related to A' as follows: column $(n+1)$ of B equals column i of A', and the first n columns of A' equal those of B. Then there is an elementary strong shift equivalence (U, V) between A and B with $A' = V$. For example,

$$A = \begin{pmatrix} 0 & 9 \\ 8 & 5 \end{pmatrix} \quad A' = \begin{pmatrix} 0 & 9 \\ 6 & 1 \\ 2 & 4 \end{pmatrix} \quad B = \begin{pmatrix} 0 & 9 & 9 \\ 6 & 1 & 1 \\ 2 & 4 & 4 \end{pmatrix} \quad U = \begin{pmatrix} 1 & 0 & 0 \\ 0 & 1 & 1 \end{pmatrix}$$

In this case, or in the case where the roles of row and column are reversed, we say that B is obtained from A by an elementary state-splitting (the state i is split into two new states) and A is obtained from B by an elementary amalgamation. One of the fundamental tools introduced in Williams' paper [W1] is the following: if two SFT's S_A and S_B are topologically conjugate, then there is a finite sequence of state splittings and amalgamations which begins with A and ends with B. (In fact this can be chosen to be a finite sequence of row splittings followed by a finite sequence of column amalgamations [P2].)

3.5 The Masking Lemma. As one application of Williams' theorem that strong shift equivalence of nonnegative integral matrices is equivalent to isomorphism of the SFT's they represent, we state a case of Nasu's Masking Lemma. (This is more or less his original statement [N], but his argument gives a much more general result [BH1, App.1].)

THEOREM (NASU). *Suppose A and B are square nonnegative integral matrices, and there is a subsystem of S_A which is conjugate to S_B. Then there is a matrix A' which defines an SFT conjugate to S_A, such that A' contains B as a principal submatrix.*

I have no idea how one would prove this from scratch. With Williams' theorem, the basic idea is very simple. First one passes from A to a higher block presentation, which has as a principal submatrix a matrix B' defining an SFT conjugate to S_B. Williams' result gives a strong shift equivalence from B to B'. Now one simply experiments with 2×2 block forms and sees that each elementary strong shift equivalence along this chain can be extended. For details, see [N] or [BH1].

3.6 Algebraic topology. Wagoner [Wa1-5] has introduced ideas of algebraic topology into the study of shift equivalence, strong shift equivalence and other matters beyond the scope of this survey. Essentially, these let one make arguments and constructions by way of topological objects constructed as quotient spaces of certain infinite simplices from the relations of (strong) shift equivalence. (Incidentally in this way natural new questions arise not easily visible from the matrix situation.)

This setup has actually led to new constructions and results on matrices over Z_+. In particular, at present results from this context are a crucial part of the work [KRW] on which the Kim-Roush counterexample [KR7] to Williams' conjecture rests.

4. Zeta Functions and Spectra

Recall the zeta function of §2.2,

$$\zeta_S(z) = \exp\left(\sum_{n=1}^{\infty} \frac{\#\mathrm{Fix}(S^n)}{n} z^n\right).$$

At first glance this may seem an unnecessarily tortured way to encode the information of the periodic points. We'll consider some justification for this, particularly from the viewpoint of matrices and shifts of finite type.

4.1 "Higher mathematics". The zeta function comes to us from more exalted zeta functions in algebraic geometry and number theory. The zeta function was introduced by Artin and Mazur [AM]. They used its natural relationship to certain algebraic geometric systems to obtain constraints on the periodic points of large sets of diffeomorphisms by proving rationality of the zeta function. The potential connections with such systems, and other algebraically defined systems, are one reason to use the zeta function to count periodic points. Algebraic topology also appears as a powerful tool for analyzing zeta functions even of systems which do not arise from algebra [F2,Fri1].

4.2 Rationality constraints. Given a system S, consider the sequence $f_n = \#\mathrm{Fix}(S^n)$, $n = 1, 2, \dots$. Assume each f_n is finite. We can capture the information in this sequence in a generating function or a zeta function,

$$g(z) = \sum_{n=1}^{\infty} f_n z^n$$

$$\zeta(z) = \exp\left(\sum_{n=1}^{\infty} \frac{f_n}{n} z^n\right).$$

These functions do carry the same information. Also, if the zeta function is a ratio of polynomials, then so is the generating function (it is the derivative of the log of the zeta function, multiplied by z). But the converse is false.

The generating function is rational if and only if the sequence f_n eventually satisfies some recursion relation. It turns out [BowL] that the zeta function is rational if and only if there are integral matrices C, D such that for all n,

$$f_n = \mathrm{tr}\, C^n - \mathrm{tr}\, D^n.$$

It is sometimes the case in dynamics that one can prove the rationality of the zeta function for interesting systems, precisely by finding such matrices [F1,Fri1,Fri2]. The rationality of the zeta function then sharply and transparently captures this constraint.

4.3 Product formula. The zeta function can be written as a (usually infinite) product,

$$\zeta_S(z) = \prod [1 - z^n]^{-1}$$

where there is a term $[1 - z^n]^{-1}$ for each S-orbit of cardinality n. The numbers $\#\mathrm{Fix}(S^n)$ are good for algebra, but the numbers of orbits of cardinality n are more fundamental dynamically. The product formula relates them in a nice way.

4.4 Why not the characteristic polynomial? If A is a matrix then its characteristic polynomial $\chi_A(z) = \det(zI - A)$ is almost the zeta function of $S = S_A$; if A is $n \times n$, then

$$\zeta_S^{-1}(z) = \det(I - zA) = z^n \det(z^{-1}I - A) = z^n \chi_A(z^{-1}).$$

However, the characteristic polynomial contains extraneous, noninvariant information; only the nonzero spectrum matters for the traces of powers of A. Also, sometimes working with $\det(I - zA)$ rather than $\det(zI - A)$, one avoids extraneous but nontrivial complications of sign (e.g., §4.5). Still, sometimes for practicality one works not with the zeta function but with the characteristic polynomial of A away from zero—this is the unique polynomial with nonzero constant term which can be obtained from the characteristic polynomial by dividing by a power of the indeterminate.

4.5 Cycles. Let A (for simplicity) be a nonnegative integer matrix. One can check that $\det(I - zA)$ is a sum of terms $(-1)^m z^k$, where there is one term for each set C of pairwise disjoint simple cycles in the directed graph with adjacency matrix A, k is the sum of the lengths of the cycles in C, and m is the number of cycles in C. Here a cycle is simple if it visits no vertex twice, and two cycles are disjoint if they have no vertex in common. (Lee Neuwirth informed me this viewpoint on $\det(I - zA)$ has been rediscovered many times–see [CDS]. I learned it from [W3] and [Ara]; these authors consider matrices with more general entries, corresponding to graphs labelled by polynomials—sometimes commuting, sometimes not—where the viewpoint is especially useful and the terms acquire coefficients which are products of the labels along the cycles.) So there is some nice connection between the structure of simple cycles and the zeta function.

4.6 Matrices with polynomial entries. We discuss this in the next section.

5. Graphs via Polynomial Matrices

5.1 Introduction. Recall, a $n \times n$ matrix over $\mathbb{Z}_+$ can be considered as the adjacency matrix of a directed graph with n vertices. Using such matrices to represent SFT's (as edge shifts) allows a more concise presentation of SFT's than one has using only zero-one matrices (for vertex shifts), and gives access to additional arguments.

There is a still more general way to present a directed graph (hence an SFT), by using matrices with entries in $t\mathbb{Z}_+[t]$ (polynomials in one variable t, with nonnegative integer coefficients, with every term divisible by t—i.e., the only constant term allowed is zero). This allows still more concise presentations, additional access to matrix arguments and algebra, and a concordance of formal patterns which make it a convincing candidate for the "right" general way to present a directed graph (or SFT).

The idea (like so many others) can already be found in rough basic form in Shannon [Sh]. The computation of the zeta function from a polynomial matrix presentation was worked out in [BGMY]. More recently, the idea has been used to develop more sophisticated and manageable state-splitting arguments for analyzing Markov shifts [MT1] and constructing codes [AFKM, HMS]. Exciting work in preparation by Kim, Roush and Wagoner, which we will not preempt with further discussion here, develops very interesting and useful new constructions of conjugacies by methods which appeal in a fundamental way to this polynomial matrix setting.

I thank Hang Kim and Fred Roush for suggesting to me that these polynomial matrices may also be important for studying the inverse spectral problem and related problems for nonnegative matrices.

5.2 The construction. The basic idea is extremely simple. Let A be an $n \times n$ matrix over $t\mathbf{Z}_+[t]$. From A we will construct a directed graph. Its vertex set will include a set of n vertices (say $1, 2, ..., n$) which index the rows and the columns of $A(t)$. If for example, $A(1,2) = 3t^2 + t^4$, then there will be three paths of length 2 and one path of length 4 from vertex 1 to vertex 2. At each interior vertex on one of these paths (a path of length k has $k - 1$ interior vertices), there is just one incoming edge and one outgoing edge. These interior vertices are disjoint from $1, 2, ..., n$. This recipe produces a graph which can have many more than n vertices (hence the conciseness of the presentation). For example, the matrix

$$A(t) = \begin{bmatrix} 0, & t^2 \\ 2t^3, & t + t^3 \end{bmatrix}$$

produces the directed graph

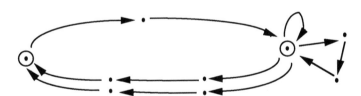

Note as in [BGMY] (fondly referred to as the "bigamy" paper) that the distinguished set of n vertices (corresponding to the indices of the rows and the columns of A) is a "rome": any sufficiently long path in the graph hits the rome. Given a rome in a directed graph, one can reverse the procedure and produce a presenting matrix A over $t\mathbf{Z}_+[t]$, where A is $n \times n$ if the rome has cardinality n. If B is a matrix over $\mathbf{Z}_+$ which is the adjacency matrix of a directed graph, then the matrix tB is one matrix over $t\mathbf{Z}_+[t]$ which presents the graph in the new formalism.

Another viewpoint is to think of A as giving a directed graph G' with labelled edges. The number of edges from i to j is the (i, j) entry of A evaluated at $t = 1$. An edge is labelled by a power of t. The power corresponds to the length of a path.

5.3 Spectra. Suppose B is a matrix over $\mathbf{Z}_+$ and A is a matrix over $t\mathbf{Z}_+[t]$, and A and B are presentations of the same directed graph G. We will prove that

(*) $$det(I - tB) = det(I - A).$$

(Of course by inverting we get the zeta function of the associated SFT.) We can picture the argument in terms of the graph G' described in §5.2. We choose, if possible, some arc from i to j labelled by t^{k+1}, with $k > 0$; then we delete this arc, add a vertex i', add an arc labelled t from i to i', and add an arc labelled t^k from i' to j. It is clear that a finite sequence of such moves produces the graph G, with every edge labelled by t. So we are done if we show the invariance of $\det(I - A)$ under one such elementary move.

This is a simple computation. With i, i', j and A as above, we may assume $i' = 1, i = 2$. Let A' be the matrix derived from A by the elementary move above. Adding t times row 1 of $(I - A')$ to row 2 does not change the determinant. The resulting matrix M has determinant equal to $\det(I - A)$, because the upper left entry is 1, every other entry in the first column is zero, and the lower right diagonal block is $I - A$.

Below, by the zeta function of a matrix (over $\mathbf{Z}$ or $t\mathbf{Z}_+[t]$) we will mean the inverse of the quantity (*) above. By its spectral radius we will mean the spectral radius of B in (*)—i.e., $1/a$, where a is the smallest positive root of (*).

5.4 Shift equivalence and flow equivalence

In this part let B be the adjacency matrix for a graph which is also presented by a matrix A over $t\mathbf{Z}_+[t]$. Let B have size N. Let $L = \mathbf{Z}[t, t^{-1}]$ be the ring of Laurent polynomials over $\mathbf{Z}$. Let L^N represent the N-fold direct sum.

The matrix $I - tB$ maps L^N into itself. It is well known [Wa1, pp.92,120] that the group cokernel$(I - tB)$ is isomorphic to the dimension group $G(B)$ of §3.1. It is not hard to check that an isomorphism is determined by the map which sends an element $[t^n e_i]$ of cok$(I-tB)$ (where e_i represents the usual canonical basis vector) to the vector $e_i(B')^{-n}$ (where B' is the isomorphism of $G(B)$ in §3.1). Moreover, if we let cok$(I - tB)$ be an L-module in the obvious way, then under this correspondence the action of t^{-1} on cok$(I - tB)$ corresponds to the action of B' on $G(B)$. So, the L-module cok$(I - tB)$ is a version of the dimension module of §3.3. (Similarly, we could use the module cok$(I - t^{-1}B)$; then the action of t would correspond to the action of B.) For integral matrices C and D, there is a module isomorphism between cok$(I - tC)$ and cok$(I - tD)$ if and only if C and D are shift equivalent over $\mathbf{Z}$.

To summarize: the L-module cok$(I - tB)$ is the shift-equivalence-over-$\mathbf{Z}$ class. Also $\det(I - tB)$ is the (inverse of the) zeta function.

Obviously we would like these facts to hold for A in place of tB. They do! (This was observed independently by, at least, Wagoner and myself.) One proof that cok$(I - tB)$ and cok$(I - A)$ give isomorphic modules follows the pattern of the proof of §5.3 for $\det(I - A)$. Again check each elementary step. Note that the matrix M in the proof of §5.3 has first column with just the first entry (which is 1) nonzero. Adding t^k times the first column of M to column j of M does not change the isomorphism class of the cokernel and it produces a matrix which is the direct sum of 1 and $I - A$. This is the proof.

Now also assume that B is primitive and nontrivial (i.e., B is not [1]). We will sketch how shift equivalence and flow equivalence are nicely unified in this polynomial setting. Recall, $\det(I - B)$ and $\operatorname{cok}(I - B)$ are complete invariants of flow equivalence for S_B within the class of irreducible SFT's. Obviously we can get these from $\det(I - tB)$ and $\operatorname{cok}(I - tB)$ by setting $t = 1$. Can we recover these invariants from $\det(I - A)$ and $\operatorname{cok}(I - A)$ by setting $t = 1$? Yes.

A satisfying way to express this is the following. Given A, form the L-module $\operatorname{cok}(I - A)$. This is the shift equivalence class and one invariant of it is $\det(I - A)$. Now, in the language of group cohomology [Br], apply the coinvariants functor to this $\mathbf{Z}[t, t^{-1}]$ module to get a $\mathbf{Z}$-module. This $\mathbf{Z}$-module (abelian group) $\operatorname{cok}(I - A(1))$ is the Bowen–Franks group. This group and $\det(I - A(1))$ are complete invariants of flow equivalence. The latter invariant is obtained by applying the coinvariants functor to $\det(I - A)$.

In other words, at least in this primitive case we get the flow equivalence class by applying the coinvariants functor to the shift equivalence class. It will be interesting to see how well this viewpoint carries over to more general situations.

5.5 Powers. Let B be a nondegenerate adjacency matrix for a graph which is also presented by a matrix A over $t\mathbf{Z}_+[t]$. Also, let n be a positive integer greater than 1. Recall (§1.7) that the matrix B^n over $\mathbf{Z}_+$ presents an SFT isomorphic to $(S_B)^n$.

In contrast, the SFT T_n presented by the polynomial matrix A^n will never be isomorphic to $(S_B)^n$. It turns out that the SFT T_1 is a quotient of T_n by a map which is everywhere n-to-1! (In particular, A and A^n present SFT's of equal entropy.) To see this, let A have size k, and let $1, 2, ..., k$ represent the vertices comprising the natural rome in the graph G_n presented by A^n. Similarly, let $1, 2, ..., k$ represent the vertices comprising the natural rome in the graph G_1 presented by A. An entry $A^n(i, j)$ represents the paths in G_n from vertex i which end at vertex j and whose interior vertices do not intersect the rome in G_n. This entry also represents the paths from i to j in G_1 whose interior vertices hit the rome in G_1 exactly $n - 1$ times. This correspondence gives us a map from paths in G_n to paths in G_1. It is easy to check this map determines a code from T_n to T_1 which is everywhere n-to-1.

The inverse zeta function of T_n is $\det(I - A^n)$; this polynomial is divisible by the polynomial $\det(I - A)$, which is the inverse zeta function of S_B. The ratio is $\det(I + A + ... + A^{n-1})$. It is possible for the latter polynomial to be trivial (so $\det(I - A) = \det(I - A^n)$). An example for $n = 2$ (i.e., $\det(I + A) = 1$) is given by

$$A = \begin{bmatrix} t^2, & t \\ t, & 0 \end{bmatrix}.$$

For completeness, we record an example A which satisfies $\det(I + A) = 1$ and which also defines a mixing SFT:

$$A = \begin{bmatrix} t^2 & t & 0 & 0 & 0 & 0 \\ t & 0 & 0 & 0 & t & 0 \\ 0 & 0 & t^2 & t & 0 & 0 \\ 0 & 0 & t & 0 & t & 0 \\ 0 & 0 & 0 & t^2 & 0 & t \\ 0 & t & 0 & 0 & 0 & 0 \end{bmatrix}.$$

Finally, we indicate briefly how the shift equivalence data for $(S_B)^n$ may be recovered from the polynomial matrix A. In this polynomial setting, as in §5.4 we think of the shift equivalence class of $(S_B)^n$ as the isomorphism class of a certain ordered L-module. We derive a representative M of this class from the ordered L-module $\mathrm{cok}(I - A)$ as follows. As an ordered abelian group, we let M equal $\mathrm{cok}(I - A)$. We define the action of L on M by defining the action of t on M to be the action of t^n on the original module $\mathrm{cok}(I - A)$.

5.6 Small presentations. If a matrix is $n \times n$, then we say it has size n. What is the smallest size of a matrix with a given nonzero spectrum? With a given shift equivalence class? These are difficult questions with unhappy answers.

For example, consider the 4-tuple $(\sqrt{2}, i, -i, \epsilon)$, where ϵ is small and positive. This will be the nonzero spectrum of a primitive real matrix, but as ϵ goes to zero the minimum size of such a matrix goes to infinity [BH1, Ap.3]. The problem of determining the minimum size at which a primitive matrix can realize a given nonzero spectrum is notoriously difficult [BH1, Ap.3].

For an example over the integers, consider the 3-tuple (5,1,1). There is an infinite collection of primitive integral matrices which have nonzero spectrum (5,1,1) but which are pairwise not shift equivalent. (This follows from the general results of [BH2]. It is also given by a barehanded construction in [B2,Example 3.4], which works for tuples (M,a,a) when M and a are positive integers with M > a+3 .) In particular, as one runs through the possible shift equivalence classes, the sizes of the realizing primitive matrices must go to infinity (since the entries of primitive integral matrices with bounded spectral radius and size are uniformly bounded).

It is a plausible and exciting prospect that one can obtain much more control on the size of a presenting matrix over $t\mathbf{Z}_+[t]$. As a quick example, we remark that it is easy to convert the above-mentioned primitive matrices (those constructed in [B2,Example 3.4]) into polynomial matrix presentations of size 4. That is, we get infinitely many shift equivalence classes with the same zeta function, all presented by polynomial matrices of size 4. (Of course, the degrees of the polynomials in these matrices must become unbounded.)

The striking theorems below of Handelman and Perrin (§5.8 and §5.9), and the result in §5.10, provide some general positive results for the existence of small polynomial matrices presenting graphs (or equivalently, presenting nonnegative integral matrices) with prescribed properties. They help motivate the following general problems, which basically translate intractable problems about nonnegative matrices into the polynomial matrix setting, which seems more promising.

Problem 1. Suppose C is a primitive integral matrix and A is a matrix over $tZ_+[t]$, where C is strong shift equivalent over Z_+ to the adjacency matrix for the graph presented by A. Given C, what is the minimum size possible for A?

Problems 2 & 3. Suppose an integral matrix C of size n is shift equivalent over Z to a primitive matrix. Find good bounds for the minimum size of a matrix A over $tZ_+[t]$ such that A presents a graph with primitive adjacency matrix B, where

(Problem 2) B is shift equivalent over Z to C

(Problem 3) B and C have the same nonzero spectrum.

We remark that for problems 1 and 2, the minimum number of generators for the group $\text{cok}(I - C)$ is a lower bound for the size of A. There are additional and independent constraints involving the sequence trC^n and the spectral radius of C. For example, if C has spectral radius less than 2 and $trC > k$, then the size of A is greater than k.

5.7 Sparse matrices and labelled graphs

Suppose for the moment, for concreteness and simplicity, that U is a unital subsemiring of the reals (e.g., Q or R_+). Suppose C is a square matrix whose entries lie in $tU[t]$–i.e., they are polynomials over U with zero constant term. Then we can think of C as presenting a labelled graph much as above. If C has size k, then there are k distinguished vertices in the graph. A monomial term ct^n in $C(i, j)$ contributes a path of n arcs from vertex i to vertex j, whose interior vertices connect to no additional vertices. The first arc on the path is labelled c and the others are labelled 1.

Let B be the matrix which is the adjacency matrix of this graph. So, B is $N \times N$, where N is the total number of vertices in the graph. $B(i, j)$ is zero if there is no arc from i to j, otherwise it is the label on the arc from i to j. Now B might be quite sparse, and so we can think of C as a concise presentation of B. Also, it is an easy exercise following §5.3 to check that $\det(I - C) = \det(I - tB)$. (Similarly, one can prove an analogue of §5.4.)

The point we wish to make is that matrices over polynomial rings may be a good way to present sparse matrices. In particular, one can hope for more satisfying results on the intractable problem of determining the smallest size nonnegative matrix with a given nonzero spectrum, by allowing polynomial matrix presentations.

Problems 2R & 3R. Suppose a real matrix C of size n is shift equivalent over R to a primitive matrix. Find good bounds for the minimum size of a matrix A over $tR_+[t]$ such that A presents a graph with primitive adjacency matrix B, where

(Problem 2R) B is shift equivalent over R to C

(Problem 3R) B and C have the same nonzero spectrum.

Of course, in problems 2R and 3R above, in place of R we could consider any unital subring of R.

5.8 Handelman's theorem.
Which Perron numbers λ can occur as the spectral radius of a matrix A over $tZ_+[t]$ of size one? Handelman [H6] proved that there is such an A of size 1 if and only if no conjugate of λ over the rationals is a positive

real number. The matrix A he constructs presents a graph with primitive adjacency matrix.

5.9 Perrin's theorem. One can ask, given a Perron number λ, what is the smallest matrix A over $t\mathbf{Z}_+[t]$ with spectral radius λ? Dominique Perrin has explained to me that for any Perron number λ there is a matrix A over $t\mathbf{Z}_+[t]$ which has size 2 and has spectral radius λ! This follows from his construction in [Pe]. Given λ, the nonnegative integral matrix B on page 364 of [Pe] has spectral radius λ^n, for some positive n. The desired matrix A is obtained by noticing that the labelled graph for $t^n B$ has a rome consisting of the two vertices 1 and k.

To be honest, there is a nontrivial imperfection to Perrin's striking result: the adjacency matrix for the corresponding graph will be irreducible but not necessarily primitive, and the period of the matrix (which will be the integer n above) may be large.

Problem. Can one prove Perrin's result, but with the realizing matrix A corresponding to a graph of period 1?

5.10 A theorem on extensions

Lind [L] proved that every Perron number (positive algebraic integer with modulus strictly greater than that of any conjugate) is the spectral radius of a primitive integer matrix (the converse is obvious from the Perron theorem). He did not bound the size of such a matrix. Using the polynomial matrix presentation, we'll produce a realizing matrix whose size is the degree of the Perron number. Basically the proof just lifts a corollary [BMT 5.14] of work of Lind and Handelman into the polynomial matrix setting.

As in [BMT], we say a matrix is IEP if it is square, every entry is an integer, and it is eventually positive (i.e., all sufficiently large positive powers of the matrix have every entry strictly positive).

LEMMA. *Let B be an IEP matrix of size m. Then there is a primitive matrix C of size m over $t\mathbf{Z}_+[t]$ with the same spectral radius as B such that the dimension module for B is a quotient of that for C.*

Proof. Following the lines of Lind's proof [L] (but avoiding most of the difficulties via the IEP hypothesis), one can construct a primitive matrix A with spectral radius equal to that of B such that the dimension module for B is a quotient of that for A. This was done in [BMT, 5.14]. The matrix A constructed there is the adjacency matrix of a $(\mathbf{Z}_+)$-labelled graph in which m vertices (denoted there as $v(i,n)$, $1 \le i \le \beta$; β there corresponds to m here) comprise an obvious rome. Using this rome we pass to the size m matrix presentation over $t\mathbf{Z}_+[t]$.

☐

THEOREM.

(1) *Suppose λ is a Perron number of degree n. Then there is a directed graph with primitive adjacency matrix B (over $\mathbf{Z}_+$) which can be presented by a matrix A over $t\mathbf{Z}_+[t]$, where A has size n.*

(2) *Suppose C is an $n \times n$ integral matrix whose spectral radius λ is a simple root of the characteristic polynomial which is strictly greater than the modulus of any other root. Then there is a directed graph with primitive adjacency matrix B (over $\mathbb{Z}_+$) which can be presented by a matrix A over $t\mathbb{Z}_+[t]$, such that the following hold*

- *B has spectral radius λ*
- *the dimension module for C is a quotient of the module for B*
- *if λ is irrational, then A has size n*
- *if λ is rational, then A has size at most $n + k$,*

where k is the smallest integer such that $\lambda^k \geq n + k$.

Proof. First we prove (2) using results of Handelman. If λ is irrational, then C is similar over the integers to an IEP matrix [H1]. If λ is rational, then C is shift equivalent over the integers to an IEP matrix of size at most $n + k$ (k as defined in (2)) [H2]. In either case, the lemma provides the desired matrix A.

Now we prove (1). If λ is rational, then λ is a positive integer, and we let $A = [t\lambda]$. If λ is irrational, then we let C be the companion matrix of the minimal polynomial of λ and appeal to (2). This finishes the proof.

$\square$

6. MORE WILLIAMS' PROBLEMS

6.1 Introduction. The equations defining (strong) shift equivalence can be used to define (strong) shift equivalence for morphisms in any category. Always, strong shift equivalence implies shift equivalence. By a Williams' problem we mean the problem of whether the converse is true.

It turns out that shift equivalence and strong shift equivalence arise in several natural ways from problems in symbolic dynamics. We'll consider them in this section. In each case we get a Williams' problem. Often there is a dynamical interpretation for strong shift equivalence and shift equivalence (isomorphism and eventual isomorphism) which gives the problem direct dynamical meaning. Usually shift equivalence turns out not to imply strong shift equivalence (but understanding the difference is fundamental).

First we recall a definition. If G is a semigroup, then the integral semigroup ring $\mathbb{Z}G$ of G is the free abelian group with generator set G, with the multiplication defined on G by the semigroup operation and then extended to $\mathbb{Z}G$ by the distributive law. We picture an element of $\mathbb{Z}G$ as a formal integral combination of elements of G. We let $\mathbb{Z}_+G$ denote nonnegative integral sums of elements of G. We make $\mathbb{Z}G$ an ordered ring by designating $\mathbb{Z}_+G$ to be the positive set. ($\mathbb{Z}_+G$ is closed under addition and multiplication and every element of $\mathbb{Z}G$ is a difference of elements of $\mathbb{Z}_+G$.) If G is a group, then $\mathbb{Z}G$ is also called the integral group ring.

If G is $\mathbb{Z}^n$, then $\mathbb{Z}G$ is isomorphic to the ring of Laurent polynomials in n variables, $\mathbb{Z}[x_1, x_1^{-1}, ..., x_n, x_n^{-1}]$.

6.2 Markov shifts and matrices of Laurent polynomials. By a Markov shift we will mean an irreducible shift of finite type (SFT) together with a shift-invariant Markov measure with full support. It turns out that the analysis of Markov shifts is intimately related to understanding matrices of integral Laurent polynomials, and in particular their shift equivalence classes.

A Markov shift can be defined by an irreducible stochastic matrix P. Let A be the zero-one matrix such that $A(i,j)$ is zero if and only if $P(i,j)$ is. Then S_A is the underlying SFT. We view P as giving labels to the edges in the graph with adjacency matrix A—that is, P is a function from arcs into the reals. P determines the Markov measure on the SFT S_A as follows. Let ℓ be the positive left eigenvector for P whose entries sum to 1. Then for any i, the measure of the set of points which see a given word $a_0...a_k$ in coordinates $i, ..., i+k$ is $\ell(v_0)P(a_0)P(a_1)...P(a_k)$, where v_0 is the initial vertex of the arc a_0.

A code between Markov shifts $S(P)$ and $S(Q)$ is a code between their underlying SFT's which sends the P-measure to the Q-measure. (For more on these codes and their relatives, we recommend [P2] and [MT1].)

It is natural to try to generalize the ideas of (strong) shift equivalence to this category by using the (strong) shift equivalence equations on stochastic matrices. To see why this fails, suppose we have $P = UV$ and $Q = VU$ with P and Q stochastic. We would like to build up some elementary isomorphism between $S(P)$ and $S(Q)$ with these equations. We could try to follow the construction of §2.3. But now we don't know how to break up UV-paths into arcs. An entry of UV can be interpreted as a sum of terms (weights) on paths. Such a weight has the form $U(i,j)V(j,k)$. But $UV(i,k)$ could be the sum of, say, several such small terms or just a few larger terms. It turns out that what one really needs to know are the path weights with multiplicities—information which is lost on multiplying the real matrices.

The solution [PT2] is to regard the entries of the stochastic matrix P as lying not in the reals $\mathbf{R}$ but in a larger ring, $\mathbf{Z}\mathbf{R}_+^*$, the integral group ring of the group $\mathbf{R}_+^*$ of positive reals under multiplication. (Parry and Tuncel [PT2] actually used an isomorphic ring with a more analytic flavor.) It turns out [PT2] that in a natural way, P and Q are strong shift equivalent over $\mathbf{Z}_+\mathbf{R}_+^*$ if and only they define isomorphic Markov shifts. Also, P and Q are shift equivalent as matrices over $\mathbf{Z}_+\mathbf{R}_+^*$ if and only if all but finitely many powers of the Markov shifts are isomorphic (i.e., they are eventually isomorphic) [MT1].

Finally there is a crucial simplification. Given P, Parry and Schmidt [PSc] showed it is possible to pass in a canonical way from $\mathbf{R}_+^*$ to a finitely generated subgroup of $\mathbf{R}_+^*$—that is, to consider only matrices over a certain finitely generated subgroup of $\mathbf{R}_+^*$. This group must be isomorphic to $\mathbf{Z}^n$ for some n. After choosing generators for the group, one simply works with matrices over $\mathbf{Z}_+\mathbf{Z}^n$—that is, matrices whose entries are Laurent polynomials in n variables with nonnegative integral coefficients. So in the end, isomorphism and eventual isomorphism of Markov shifts are determined by strong shift equivalence and shift equivalence, but now of matrices over Laurent polynomials in several variables. (For a complete explanation of these things, we recommend [MT1].)

Much of the structure for SFT's generalizes here. For example, the information carried by the periodic points is perfectly encoded in the stochastic zeta function of Parry and Tuncel [PT2] , given by the formula

$$\zeta(z) = \exp \sum_{n=1}^{\infty} \frac{tr(P^n)}{n} z^n = [\det(I - zP)]^{-1}.$$

Here the entries of P lie not in $\mathbf{R}$ but in its integral group ring (or, if one prefers, in the isomorphic ring $\mathbf{Z}[\exp]$ of Parry and Tuncel, see [PT2] or [MT1, Defn. 4.2]). The equation above makes sense at the level of formal power series. (In earlier work of Parry and Williams [PW], a "stochastic zeta function" was offered which was given by the same formula but with the entries of P still regarded as lying in $\mathbf{R}$. This gives an invariant, but one which does not capture all the desired information. We have appropriated the term "stochastic zeta function" for the Parry-Tuncel function because we regard it as the correct end product of this line of development.)

For SFT's, the dimension module (§5.3) was a certain $\mathbf{Z}[t, t^{-1}]$-module. The dimension module of a Markov shift is a $R[t, t^{-1}]$-module, where R may be taken to be the ring of Laurent polynomials in n variables (by identifying the variables with generators of the canonical subgroup mentioned earlier)—it is a version of Tuncel's dimension module [Tu]. The variable-length graphs still work—but now they are labelled by elements of $tR_+[t]$ rather than $t\mathbf{Z}_+[t]$. Recall if A is a matrix over $t\mathbf{Z}_+[t]$ presenting an SFT, then the shift-equivalence-over-$\mathbf{Z}$ data for the SFT is encoded as the isomorphism class of the $\mathbf{Z}[t, t^{-1}]$-module $\operatorname{cok}(I - A)$. In the stochastic case, the entries of A lie in $tR_+[t]$, and the shift-equivalence-over-R data is encoded as the isomorphism class of the $R[t, t^{-1}]$-module $\operatorname{cok}(I - A)$. (There is even a notion of stochastic flow equivalence, for which one invariant [P2,Ara] can be interpreted in the following way: apply the coinvariants functor to the $R[t, t^{-1}]$-module $\operatorname{cok}(I - A)$ to obtain the R-module $\operatorname{cok}(I - A(1))$, which is an invariant of stochastic flow equivalence. Here $A(1)$ denotes the matrix obtained by substituting 1 for t in A. Similarly the element $\det(I - A(1))$ is an invariant of stochastic flow equivalence [P2,Ara].)

Even in the primitive case, shift equivalence does not imply strong shift equivalence for matrices over R_+ [B3]. (A matrix over R_+ is primitive if some power has every entry nonzero.) Thus even in the primitive case we must be concerned with order in the classification of matrices up to shift equivalence over R_+. We can express the order information in the module framework by considering the $R_+[t, t^{-1}]$-module $\operatorname{cok}(I - A)$ as an ordered module. Here the positive set of $\operatorname{cok}(I - A)$ is the set of vectors which have all entries in $R_+[t]$ after multiplication by a sufficiently large power of A. Now shift equivalence of matrices over $tR_+[t]$ is equivalent to isomorphism of their ordered modules, i.e. a module isomorphism which takes one positive set onto the other.

The realization problems for polynomial matrices are vastly more difficult. Recall, Lind characterized the spectral radii of nonnegative integral matrices [L]. The analogue for irreducible matrices over R_+ is the beta function of Tuncel (see [MT1]).

To think of this in a finite way, consider an irreducible matrix A over R_+ as a matrix with entries in $Z_+[x_1, x_1^{-1}, ..., x_n, x_n^{-1}]$ and let $p(t)$ be its characteristic polynomial. Factor $p(t)$ in $Z[x_1, x_1^{-1}, ..., x_n, x_n^{-1}][t]$. There will be one factor such that for any substitution of positive reals for the variables $x_1, ..., x_n$, this factor will have the largest root. You may regard that factor as the beta function. What are the beta functions of irreducible matrices over R_+? There is not even a good conjecture at present. Progress to date has rested on a blend of techniques from algebra, geometry and analysis [H3-6,deA].

Problem. What are the beta functions of primitive matrices of Laurent polynomials?

Problem. What are the zeta functions arising from primitive matrices of Laurent polynomials? (That is, which polynomials can arise as $\det(I - tA)$ for some primitive matrix A whose entries are Laurent polynomials with nonnegative integral coefficients?)

6.3 Boolean matrices. The Boolean semiring $\mathbf{B}$ is the set $\{0, 1\}$, with addition and multiplication defined as the quotient of these operations on the nonnegative reals by the map sending 0 to 0 and sending all positive numbers to 1. So, 1 is a multiplicative identity, 0 is an additive identity, $1 + 1 = 1$, $0 \times 0 = 0$. Relations among nonnegative matrices project to relations among Boolean matrices (matrices over $\mathbf{B}$). For example, if A and B are shift equivalent nonnegative matrices, then their Boolean images are shift equivalent over $\mathbf{B}$; and if A and B are to be strong shift equivalent, then their Boolean images must be. So it makes sense to look at Williams' problem for matrices over $\mathbf{B}$, if only to check necessary conditions on Williams' problem for matrices over Z_+ (or R_+, or Q_+).

Kim and Roush have completely classified all Boolean matrices up to shift equivalence and also up to strong shift equivalence [KR5]. Their general result is lovely, but for simplicity we will just state the answer for primitive Boolean matrices (those for which some power is the matrix with every entry 1). They are all shift equivalent! And two such matrices are strong shift equivalent if and only if they have the same powers with zero trace (there are only finitely many such powers). In particular, shift equivalence does not imply strong shift equivalence in this setting.

One useful spinoff of their work is an intriguing tool for controlling the sign patterns of nonnegative matrices. As remarked in [KR5, p.154], if A and B are two nonnegative matrices whose Boolean projections are strong shift equivalent, and their entries lie in a nondiscrete unital subring S of the reals, then B is strong shift equivalent over S_+ to a matrix with the same block sign pattern as A. For example, this with the classification result [KR5] shows that a matrix over the nonnegative rationals Q_+ with positive trace is strong shift equivalent over Q_+ to a matrix with all entries strictly positive. In particular, if Δ is the nonzero spectrum of a primitive real matrix with positive trace, then Δ is the nonzero spectrum of a strictly positive real matrix [BH1, Ap.4].

6.4 Markov shifts over infinite alphabets. Wagoner [Wa1,2] has developed shift equivalence and strong shift equivalence for SFT's over infinite alphabets. Here a crucial feature is that the morphisms are required to be uniformly continuous.

6.5 Sofic shifts. There are also notions of shift equivalence and strong shift equivalence available for sofic shifts (these are the subshifts which are quotients of SFT's) [BK]. Again strong shift equivalence corresponds to isomorphism and shift equivalence corresponds to isomorphism of all large powers. The equations of (strong) shift equivalence are now applied to elements of an integral semigroup ring, where the semigroup is the semigroup under multiplication of infinite zero-one matrices with all row sums at most one and with all but finitely many entries zero. Now, instead of considering matrices over a commutative integral group ring, we are looking at the (noncommutative) integral semigroup ring of a nonabelian semigroup. Because of this noncommutativity, it is a serious problem even to define an appropriate zeta function [B4]. Nevertheless, Kim and Roush showed that shift equivalence in this setting is decidable [KR4], which matches their result in the SFT setting [KR3]. A key reduction in their proof is an analogue of the Parry-Schmidt result [PSc] for Markov chains, which allowed one to restrict to modules over smaller rings: in the sofic case, for considering shift equivalence (but not strong shift equivalence), Kim and Roush observed by appeal to work of Nasu [N] that for two given systems it suffices to consider modules over the integral semigroup ring of a certain *finite* semigroup (the zero-one matrices of a certain bounded size).

For an explanation of these ideas, we refer to [BK,Sec.1] and [KR4]. Our aim here is primarily to indicate by yet another example that the basic ideas of shift and strong shift equivalence return in various guises to describe symbolic dynamical structures.

7. GENERAL MATRICES

In this section we'll consider shift equivalence and strong shift equivalence a bit more generally. This has algebraic and order aspects. We justify the investigation by the symbolic dynamical relevance of (strong) shift equivalence in various settings, and because these relations are algebraically natural in general (e.g., $sr \sim rs$ generates a definition of $K_0(R)$ via projections in $GL(R)$).

First we consider the algebraic aspect. Let R be a unital ring (unital means that R has a multiplicative identity). We define shift equivalence and strong shift equivalence of matrices over R by the same equations we use for $R = \mathbf{Z}$. Does shift equivalence imply strong shift equivalence?

If $R = \mathbf{Z}$, the answer is yes. This was proved in old unpublished preprints by Williams and by Effros [E2],[W2]. They offer related but different arguments, both of which go through for any principal ideal domain. The answer is still yes for a Dedekind domain [BH2]. On the other hand the following is open.

Problem [BH2]. Suppose R is a unital subring of the reals which is not a Dedekind domain. If two matrices are shift equivalent over R, must they be strong shift equivalent over R? What if R has finite homological dimension?

Next we consider order. A ring R is ordered if there is a subsemiring R_+ (called the positive set) such that every element of R can be written as the difference of two elements in R_+. We have already looked at shift and strong shift equivalence for $R = \mathbf{Z}$ and $R_+ = \mathbf{Z}_+$. There Williams' problem is open in the irreducible case. Even for matrices of size 2, after all this time, it is settled (in the affirmative) only if the determinant is not less than -1 [Ba1,CuKr,W1].

Problem [Ba1]. Suppose A and B are primitive size 2 matrices over $\mathbf{Z}_+$ which are shift equivalent over $\mathbf{Z}_+$, with $\det A < -1$. Must A and B be strong shift equivalent over $\mathbf{Z}+$?

There are additional ideas [Ba2] but the problem is tough (as can be attested by the calibre of some of the mathematicians who have spent months or years on Williams' problem for matrices over $\mathbf{Z}_+$—Williams, Parry, Franks, Krieger, Marcus, Baker, Handelman, Kim&Roush....).

A sane response is to back off and consider Williams' problem for matrices over the nonnegative rationals or reals. Here Kim and Roush proved that shift equivalent matrices over $\mathbf{Q}_+$ are strong shift equivalent—if each has exactly one nonzero eigenvalue, with multiplicity one (equivalently, the inverse zeta function is $1 - at$, where a is the nonzero eigenvalue) [KR6]. (It is some indication of the difficulty of the problem that this seemingly simple case was open for so long.) They have also added to Baker's viewpoint [Ba2] the development of approximation and homotopic techniques [KR8,9].

But still: strong shift equivalence remains very poorly understood.

Problem. Does shift equivalence over R_+ imply strong shift equivalence over R_+ for strictly positive matrices over the reals?

8. Inverse Problems for Nonnegative Matrices

In [BH1,2], Handelman and I studied certain inverse problems for nonnegative matrices, using tools from symbolic dynamics. As we took pains to explain and motivate the problems there, in this section the discussion will be brief, and we refer to [BH1,2] for more.

8.1 The inverse spectral problem. Let $\Delta = (d_1, ..., d_n)$ be an n-tuple of complex numbers. An old problem asks, when is Δ the spectrum of a nonnegative real matrix of size n? (Δ is the spectrum of a matrix A if the characteristic polynomial is $\chi_A(t) = \prod_i (t - d_i)$. So Δ includes the information about multiplicities and we don't care about the order in which the d_i are listed.)

Necessary conditions on Δ are discussed in [BH1], especially in Appendix 3. The best reference to the literature on this problem is still [BePl], for a more recent discussion see [Mi]. To my knowledge the problem first appears in print in Suleimanova's 1949 paper [Su] (if we neglect the glorious work of Perron and Frobenius early in this century). It has been rather intractable. The solution is known if $n = 3$ [LL] or if $n = 4$ and the entries of Δ are real [Ke]. The problem is still open even for $n = 5$ (or $n = 4$ with some entries not real).

Thus it is natural to relax the old problem in some useful way. From the viewpoint of symbolic dynamics, one only cares about the nonzero spectrum (as in

§2.3), and it is appropriate to ask when an n-tuple Δ of nonzero complex numbers can be the nonzero part of the spectrum of some nonnegative matrix. This is also natural from just the matrix viewpoint and certainly occurred to matrix theorists working on the problem (e.g. Charles Johnson). I would guess this viewpoint was never seriously pursued because constructions which could make much use of the relaxed condition were unavailable.

Aside from motivating the pursuit of the nonzero spectrum, symbolic dynamics enters the picture by providing some tools which let one exploit the extra room provided by passing to arbitrarily large matrices subject to a given nonzero spectrum. In some cases there are matrix constructions for which codes need not be mentioned, but whose inspiration comes from coding constructions. There are also the ideas around strong shift equivalence, which provide some direct constructions, and which let one translate coding constructions into matrix results. In this vein there is especially one result, the Submatrix Theorem, which plays an essential role in [BH1,2] and for which the only proof known (so far) relies in a fundamental way on coding ideas independent of matrices. These ideas are the marker methods of the proof of Krieger's Embedding Theorem [Kr3]. Because the Embedding Theorem has been a basic and useful tool in the study of shifts of finite type (read: asymptotic theory of nonnegative integral matrices), it may be that the Submatrix Theorem will have other applications in the study of "asymptotic" aspects of more general nonnegative matrices.

8.2 Submatrix theorem. Let S be a unital subring of the reals. Given square nonnegative matrices A and B over S, with A primitive, the Submatrix Theorem produces (subject to "obvious" necessary conditions on the nonzero spectra) a matrix A' with B as a principal submatrix, where A' is strong shift equivalent over S to A. If we are interested only in invariants of strong shift equivalence (such as the nonzero spectrum), then this provides tremendous control over submatrices for constructions.

If the ring S is discrete, then it must be the integers. The necessary trace condition for the Submatrix Theorem is different in this case, and to avoid discussing it we just refer to [BH1, Thm.1.10].

Recall (§2.3) the polynomial $\det(I - tC)$ determines the nonzero spectrum of a matrix C and vice versa.

SUBMATRIX THEOREM (NON-DISCRETE CASE) [BH1]. *Let S be a dense unital subring of the reals. Suppose that A and B are nonnegative matrices with entries from S, such that A is primitive. Then there exists a primitive matrix C with entries from S such that B is a proper principal submatrix of C and $\det(I-tC) = \det(I-tA)$ if and only the following three conditions hold:*

(1) *The spectral radius of B is strictly smaller than that of A.*

(2) *For all positive integers n, $tr B^n \le tr A^n$.*

(3) *For all positive integers n and k, if $tr B^n < tr A^n$, then $tr B^{nk} < tr A^{nk}$.*

Given conditions (1),(2) and (3), C may be chosen strong shift equivalent over S_+ to A.

8.3 Spectral conjecture. Handelman and I [BH1] conjectured that certain "obvious" necessary conditions are sufficient for an n-tuple $\Delta = (d_1, ..., d_n)$ of complex numbers to be the nonzero spectrum of a primitive matrix whose entries lie in a given unital subring S of the reals. (The general case follows easily from the primitive case [BH1].) It seems to us that the supporting evidence [BH1] is fairly overwhelming. (Not that we can prove it.) For example, the conjecture is true if S is the reals, or if S is nondiscrete and $\sum d_i$ is nonzero [BH1].

Those necessary conditions on Δ are a Perron condition, a Galois condition, and a trace condition. The Perron condition is that there be a positive real number which is listed just once in Δ and which is strictly larger than the modulus of any other entry. The Galois condition is simply that the degree n polynomial $\prod(t - d_i)$ must have its coefficients in S. The trace condition is different if $S = \mathbf{Z}$, so here we will just state the trace condition when S is nondiscrete. For $k > 0$, let $t(k)$ denote the sum of the kth powers of the entries of Δ. Then the trace condition is that for all positive integers m and k, two things hold:

(1) $t(k) \geqslant 0$, and

(2) if $t(m) > 0$, then $t(mk) > 0$.

8.4 Generalizing the spectral conjecture. Again let S be a unital subring of the reals. Suppose A is a square matrix over S. The Spectral Conjecture asserts that there exists a primitive matrix B over S with the same nonzero spectrum as A if two necessary conditions, a Perron condition and a trace condition, are satisfied. (Here the Galois condition is automatically satisfied.) The Generalized Spectral Conjecture of [BH2] asserts that under these same necessary conditions, one can require the primitive matrix B to be shift equivalent over S to A. Handelman and I take this opportunity to make a further generalization: we conjecture that under these same necessary conditions, there exists a primitive matrix B strong shift equivalent over S to A.

But, remember, at this moment we do not know if two matrices shift equivalent over S must be strong shift equivalent over S (Sec.7)—i.e., it may be that this generalization is equivalent to the Generalized Spectral Conjecture of [BH2].

9. ONE-SIDED SHIFTS

Let A be a square nonnegative integral matrix. Then A defines a one-sided SFT T_A. This is defined just as the two-sided SFT S_A was, with one difference: now the sequences are $x = x_0 x_1 ...$ (that is, the coordinates are indexed only by nonnegative integers, not by all the integers). Here the shift map is still continuous and (if A is nondegenerate) surjective, but usually it is not invertible. (In fact, it is invertible only when the space of sequences is finite. When A is nondegenerate, this means that up to conjugation by a permutation matrix, A is a direct sum of permutation matrices.)

We have the same facts about codes and block codes as before, but now a block code can only depend on nonnegative coordinates. That is, if there is a code f from T_A to T_B, then there is some positive integer N and some function F from T_A

words of length N to the alphabet of T_B such that for all x and for all $i \geq 0$,

$$(fx)_i = F(x_i...x_{i+N-1}).$$

Williams classified these systems up to isomorphism [W1]. (There is an exposition of this in [BFK], which also has a good deal more about these systems.) The classification is beautifully simple, so we will describe it.

Given the matrix A, if possible choose two equal columns i and j; then add row j to row i, and then erase row j and column j. This produces a smaller matrix A_1. The matrix A_1 may also have a pair of equal columns, which we can "amalgamate" as before to get a yet smaller matrix. If B is a matrix obtained by a (finite) sequence of column amalgamations in this way from A, then we call B an amalgamation of A. If B has no pair of equal columns (i.e. cannot be further column-amalgamated), then B is called a total amalgamation of A.

Williams [W1] proved that the total amalgamation of A is independent (up to conjugation by a permutation matrix) of the choices of columns at each step, and that T_A is isomorphic to T_B if and only if the total amalgamations of A and B are the same (up to conjugation by a permutation matrix). So for one-sided SFT's the classification has a clean and simple solution.

However there is another fundamental problem which seems much harder in the one-sided case. Recall a matrix is nondegenerate if it has no zero rows or columns, and a primitive matrix is a square nonnegative matrix some power of which has every entry strictly positive.

Problem. Suppose B is a square nondegenerate nonnegative matrix and A is a primitive matrix. Give necessary and sufficient conditions under which the one-sided SFT T_B is isomorphic to a proper subsystem of the one-sided SFT T_A.

Let us immediately express this problem in purely matrix terms. With no loss of generality, we may (and should) assume that B is a total amalgamation. Then T_B is isomorphic to a proper subsystem of T_A if and only if for some matrix C presenting a higher block presentation of T_A, the matrix C has a proper principal submatrix D such that B is a total amalgamation of D.

The solution of the corresponding two-sided case is Krieger's embedding theorem [Kr3]. There, simple necessary conditions on entropy and periodic points are sufficient for the proper embedding: the entropy of S_A must strictly exceed that of S_B, and for every positive integer n the number of orbits of cardinality n for S_B cannot exceed the corresponding number for S_A.

These necessary conditions do not suffice in the one-sided case. For example, a point with k preimages must be sent to a point with at least k preimages. Similarly there are constraints counting preimages of preimages, etc.; and these are mixed with constraints on periodic points (e.g. a fixed point with 3 preimages must go to a fixed point with at least three preimages).

Krieger's theorem is proved by a marker argument, but the possible markers are much more limited and tightly defined in the one-sided case [As1]. One-sided systems are much more closely tied to their presenting matrices than are two sided systems [W1,BFK]. So it is plausible that in the one-sided case a proof of the embedding theorem may be much more closely tied to graphs and matrices (and therefore could lead to better algorithms for constructing the embedding codes which exist by Krieger's theorem in the two-sided case). There have only been a few papers on one-sided SFT's ([W1],[BFK],[As1]), so we have a situation common in symbolic dynamics: there's not too much to learn, but there is something to invent.

The problem above is posed for embedding SFT's only (rather than general subshifts), in contrast to the statement of Krieger's theorem. This is to emphasize the matrix aspect. There is almost surely no loss of generality. It is an exercise to check that an embedding of a subshift (one-sided or two) into a shift of finite type always extends to an embedding of some SFT containing the subshift. Every subshift is a nested intersection of SFT's in a well-understood way. So from the solution of the problem above, one should be able without great difficulty to give the solution to the general problem. (This is certainly the case for two-sided shifts—but in that case, at present there is no real simplification of Krieger's proof obtained by embedding only SFT's.)

10. Quotients

Let S_A and S_B be (for simplicity) mixing SFT's. When is S_B a quotient of S_A? That is, when is there a code from S_A onto S_B? It is easy to check that a necessary condition is the entropy inequality

$$h(S_A) \geq h(S_B).$$

There is a fundamental dichotomy for such maps. If $h(S_A) > h(S_B)$ and f is a quotient map from S_A onto S_B, then the points with uncountably many preimages under f comprise a residual subset of S_B. If $h(S_A) = h(S_B)$ and f is a quotient map from S_A onto S_B, then there is a positive integer N such that no point in S_B has more than N preimages under f [CP].

As it turns out, the case of strict inequality is relatively easy [B1]. If $h(S_A) > h(S_B)$, then S_B is a quotient of S_A if and only if the trivially necessary periodic point condition holds: for all $n > 0$,

(*) $$tr(A^n) > 0 \Longrightarrow tr(B^n) > 0.$$

The equal-entropy case is much more rigid, subtle and algebraic. The periodic point condition (*) is still necessary for S_B to be a quotient of S_A. But also, the dimension module of S_B must be a quotient of a closed submodule of the dimension module of S_A [KMT]. (In the terminology of §3.1, a closed submodule is given by restricting A' to a subgroup H of $G(A)$ which is the intersection of $G(A)$ with an A'-invariant subspace of V_A. A submodule given by restricting A' to some A'-invariant subgroup H is not closed if $(\mathbf{Q}H) \cap G(A)$ contains a point not in H.)

CONJECTURE. *Suppose A and B are primitive matrices over $\mathbf{Z}_+$ of equal spectral radius satisfying the periodic point condition (*) such that the dimension module of B is a quotient of a closed submodule of the dimension module for A.*

Then S_B is a quotient of S_A.

This conjecture is strongly supported by the work of Jonathan Ashley [As2]. For primitive integral matrices A and B of equal spectral radius satisfying (*), he showed that S_B is a quotient of S_A by a closing map if and only if the dimension module of S_B is a quotient or closed subsystem of the dimension module of S_A. (Closing maps are topologically conjugate to resolving maps, which can be constructed from certain matrix equations and are the most useful codes for industrial applications [ACH]. See [AM] and [BMT] for background.)

Ashley's proof begins with a construction (the Eventual Factors Theorem of [BMT]) which produces interrelated quotient maps of higher powers of the shifts. This construction is derived from matrix equations which capture the quotient relation for the dimension modules. One scheme for approaching the conjecture above is to mimic this pattern: first find matrix equations for the dimension condition, then find an "eventual" construction from these analogous to the starting result from [BMT], and then adapt Ashley's arguments for coding into and out of long periodic blocks. (Caveat: this may be a red herring.)

There are, at least, some nice matrix equations for the dimension condition. Let A and B be primitive integral matrices of equal spectral radius. Then the dimension module for B is a quotient of a closed submodule of the dimension module for A if and only if there are positive integral matrices R, S and a positive integer n such that for all positive integers k,

$$RA^k S = B^{n+k}.$$

We leave a proof to the interested reader (it may help to consult [BMT,Sec.2]).

REFERENCES

[ACH] R.ADLER, D.COPPERSMITH & M.HASSNER, *Algorithms for sliding block codes*, IEEE Trans. Info.Th., Vol. IT-29, pp. 5-22, 1983.

[AF] R.ADLER & L.FLATTO, *Geodesic flows, interval maps, and symbolic dynamics*, Bulletin (New Series) A.M.S. Vol. 25, No.2 (1991), pp.229-334.

[AFKM] R.ADLER, J.FRIEDMAN, B.KITCHENS & B.MARCUS, *State splitting for variable-length graphs*, IEEE Trans. Info. Th. Vol. IT-32, No. 1, pp. 108-113, 1986.

[Ara] P. ARAUJO, *A stochastic analogue of a theorem of Boyle's on almost flow equivalence*, Erg.Th.& Dyn.Syst. to appear.

[ArMa] M.ARTIN & B.MAZUR, *On periodic points*, Annals of Math. 81 (1969), 82-99.

[As1] J.ASHLEY, *Marker automorphisms of the one-sided d-shift*, Erg.Th.& Dyn.Syst. 10 (1990), 247-262.

[As2] J.ASHLEY, *Resolving factor maps for shifts of finite type with equal entropy*, Ergod. Th. & Dyn. Syst. 11 (1991), 219-240.

[Ba1] K.A.BAKER, *Strong shift equivalence of 2x2 matrices of non-negative integers*, Ergod.Th.& Dyn.Syst. 3 (1983), 541-558.

[Ba2] K.A.BAKER, *Strong shift equivalence and shear adjacency of nonnegative square integer matrices*, Linear Algebra Appl. 93: 131-147 (1987).

[BePl] A. BERMAN & R.PLEMMONS, *Nonnegative matrices in the mathematical sciences*, Academic Press (1979).

[BGMY] L.BLOCK, J.GUCKENHEIMER, M.MISUREWICZ & L.YOUNG, *Periodic points and topological entropy of one-dimensional maps*, Springer Lec. Notes 819, Springer, 18-34.

[Bow1] R. BOWEN, *Equilibrium States and the Ergodic Theory of Anosov Diffeomorphisms*, Springer Lec. Notes in Math. 470, Springer-Verlag (1975).

[Bow2] R. BOWEN, *On Axiom A diffeomorphisms*, CBMS Regional Conf. Ser. in Math., No. 35, Amer.Math.Soc., Providence, RI (1977).

[BowF] R. BOWEN & J. FRANKS, *Homology for zero-dimensional basic sets*, Annals of Math. 106 (1977), 73-92.

[BowL] R. BOWEN & O.E.LANFORD, *Zeta functions of restrictions of the shift transformation*, Proc. Symp. Pure Math. A.M.S. 14 (1970), 43-50.

[B1] M. BOYLE, *Lower entropy factors of sofic systems.*, Erg.Th.&Dyn.Syst. 4, (1984), 541-557.

[B2] M. BOYLE, *Shift equivalence and the Jordan form away from zero*, Erg.Th.& Dyn. Syst. 4 (1984), 367-379.

[B3] M. BOYLE, *A zeta function for homomorphisms of dynamical systems*, J. London Math. Soc. (2) 40 (1989) 355-368.

[B4] M. BOYLE, *The stochastic shift equivalence conjecture is false*, Symbolic dynamics and its applications, Contemporary Math. 135, Amer. Math. Soc. (1992), 107-110.

[BFK] M.BOYLE, B.KITCHENS & J.FRANKS, *Automorphisms of one-sided subshifts of finite type*, Ergod.Th.&Dyn.Syst. 10 (1990), 421-449.

[BH1] M. BOYLE & D. HANDELMAN, *The spectra of nonnegative matrices via symbolic dynamics*, Annals of Math. 133 (1991), 249-316.

[BH2] M. BOYLE & D. HANDELMAN, *Algebraic shift equivalence and primitive matrices*, Trans. AMS, to appear.

[BMT] M.BOYLE, B. MARCUS & P. TROW, *Resolving Maps and the Dimension Group for Shifts of Finite Type*, Memoirs AMS 377 (1987).

[BK] M.BOYLE & W. KRIEGER, *Almost Markov and shift equivalent sofic systems*, in *Dynamical Systems, Proc. Special Year in Dynamics*, 1986-87 at the University of Maryland, Springer Lec. Notes in Math 1342, Springer-Verlag.

[Br] K. BROWN, *Cohomology of Groups*, Springer GTM 87. Springer-Verlag 1982.

[CDS] D.CVETKOVIC, M.DOOB & H.SACHS, *Spectra of graphs*, Academic Press (1980).

[CP] E. COVEN & M. PAUL, *Endomorphisms of irreducible subshifts of finite type*, Math Systems Th. 8(1974), 167-175.

[Cu] J.CUNTZ, *A class of C*-algebras and topological Markov chains II: reducible chains and the Ext-functor for C*-algebras*, Inventiones Math. 63, 25-40 (1981).

[CuKr1] J.CUNTZ & W.KRIEGER, *Topological Markov chains with dicyclic dimension groups*, J.fur Reine und Angew. Math. 320 (1980), 44-51.

[CuKr2] J.CUNTZ & W.KRIEGER, *A class of C*-Algebras and topological Markov chains*, Inventiones Math. 56, 251-268 (1980).

[deA] V. DE ANGELIS, *Polynomial beta functions and positivity of polynomials*, PhD. dissertation, University of Washington, 1992.

[DGS] M. DENKER, C. GRILLENBERGER & K.SIGMUND, *Ergodic Theory on Compact Spaces*, Springer Lec. Notes in Math. 527, Springer-Verlag (1976).

[E1] E.G.EFFROS, *Dimensions and C*-Algebras*, CBMS 46 (1981), A.M.S., Providence, Rhode Island.

[E2] E.G.EFFROS, *On Williams' problem for positive matrices*, Unpublished manuscript, ca. 1981.

[F1] J. FRANKS, *Homology and Dynamical Systems*, CBMS 49 (1982), A.M.S., Providence, Rhode Island.

[F2] J. FRANKS, *Flow equivalence of subshifts of finite type.*, Erg.Th.Dyn. Syst. 4(1984), 53-66.

[Fri1] D. FRIED, *Rationality for isolated expansive sets*, Advances in Math. 65, 35-38 (1987).

[Fri2] D. FRIED, *Finitely presented dynamical systems* Erg.Th.Dyn.Syst.7(1987), 489-507.

[H1] D. HANDELMAN, *Positive matrices and dimension groups affiliated to C*-Algebras and topological Markov chains*, J. Operator Th. 6 (1981), 55-74.

[H2] D. HANDELMAN, *Eventually positive matrices with rational eigenvectors*, Erg.Th.&Dyn.Syst. 7 (1987), 193-196.

[H3] D. HANDELMAN, *Positive polynomials, convex integral polytopes and a random walk problem*, Springer Lec. Notes 1282, Springer Verlag (1987).

[H4] D. HANDELMAN, *Eventual positivity and finite equivalence for matrices of polynomials*, Preprint.

[H5] D. HANDELMAN, *Polynomials with a positive power*, Symbolic dynamics and its applications, Contemporary Math. 135, Amer. Math. Soc. (1992), 229-230.

[H6] D. HANDELMAN, *Spectral radii of primitive integral companion matrices and log concave polynomials*, Symbolic dynamics and its applications, Contemporary Math. 135, Amer. Math. Soc. (1992), 231-238.

[HMS] C. HEEGARD, B. MARCUS & P. SIEGEL, *Variable length state splitting with applications to average run-length constrained (ARC) codes*, IEEE-Information Theory, v. 37 (1991), pp. 759-777.

[Hu] D. HUANG, *Flow equivalence of reducible shifts of finite type*, In preparation.

[Ke] R.B.KELLOGG, *Matrices similar to a positive or essentially positive matrix*, Lin.Alg.Applic. 4 (1971), 191-204.

[KR1] K.H.KIM & F.ROUSH, *Some results on decidability of shift equivalence*, J. Combinatorics, Info.Sys.Sci. 4 (1979), 123-146.

[KR2] K.H.KIM & F.W.ROUSH, *On strong shift equivalence over a Boolean semiring*, Erg.Th.& Dyn.Syst. 6 (1986), 81-97.

[KR3] K.H.KIM & F.W.ROUSH, *Decidability of shift equivalence. Dynamical Systems (Proceedings, Univ. of Maryland 1986-87)*, Springer Lec. Notes 1342, Springer(1988), 374-424.

[KR4] K.H.KIM & F.W.ROUSH, *An algorithm for sofic shift equivalence*, Erg. Th. & Dyn. Syst. 10 (1990), 381-393.

[KR5] K.H.KIM & F.W.ROUSH, *Strong shift equivalence of boolean and positive rational matrices*, Linear Algebra Appl. 161: 153-164 (1992).

[KR6] K.H.KIM & F.W.ROUSH, *Full shifts over R_+ and invariant tetrahedra*, PU.M.A. Ser. B, Vol.1 (1990), No. 4, pp. 251-256.

[KR7] K.H.KIM & F.W.ROUSH, *Williams' conjecture is false for reducible subshifts*, Journal AMS 5 (1992), 213-215.

[KR8] K.H.KIM & F.W.ROUSH, *Path components of matrices and strong shift equivalence over Q_+*, Linear Algebra Appl. 145: 177-186 (1991).

[KR9] K.H.KIM & F.W.ROUSH, *Strong shift equivalence over subsemirings of Q_+*, PU.M.A. Ser.B, Vol.2 (1991), 33-42.

[KRW] K.H.KIM, F.W.ROUSH & J.B.WAGONER, *Automorphisms of the dimension group and gyration numbers of automorphisms of a shift*, Journal AMS 5 (1992), 191-211.

[KMT] B. KITCHENS, B. MARCUS & P. TROW, *Eventual factor maps and compositions of closing maps*, Erg.Th.Dyn.Syst. 11 (1991), 85-113.

[Kr1] W. KRIEGER, *On a dimension for a class of homeomorphism groups*, Math. Ann. 252 (1980), 87-95.

[Kr2] W. KRIEGER, *On dimension functions and topological Markov chains*, Invent. Math. 56 (1980), 239-250.

[Kr3] W. KRIEGER, *On the subsystems of topological Markov chains*, Erg.Th.Dyn. Syst. 2 (1982), 195-202.

[L] D. LIND, *The entropies of topological Markov shifts and a related class of algebraic integers*, Erg. Th. Dyn. Syst. 6(1986),571-582.

[LL] R. LOEWY & D. LONDON, *A note on an inverse problem for nonnegative matrices*, Linear and Multilinear Algebra 6 (1978),83-90.

[M] B. MARCUS, *Factors and extensions of full shifts*, Monats. fur Mathematik 88 (1979), 239-247.

[MSW] B.H.MARCUS, P.H.SIEGEL & J.K.WOLF, *Finite-state modulation codes for data storage*, IEEE Journal of Selected Areas in Communications, Vol. 10 (1992), 5-37.

[MT1] B.MARCUS & S.TUNCEL, *The weight-per-symbol polytope and embeddings of Markov chains*, Ergod.Th. & Dyn.Syst. 11 (1991), 129-180.

[MT2] B.MARCUS & S.TUNCEL, *Entropy at a weight-per-symbol and embeddings of Markov chains*, Invent. Math. 102 (1990), 235-266.

[Mi] H. MINC, *Nonnegative Matrices*, Wiley Inter-Science (1988).

[N] M. NASU, *Topological conjugacy for sofic systems and extensions of automorphisms of finite subsystems of topological Markov shifts*, Proceedings of Maryland Special Year in Dynamics 1986-87. Springer Lecture Notes 1342 (1988), Springer-Verlag.

[Ne] M. NEWMAN, *Integral Matrices*, Academic Press, New York, 1972.

[P1] W. PARRY, *Intrinsic Markov chains*, Trans. AMS 112 (1964), 55-66.

[P2] W. PARRY, *Notes on coding problems for finite state processes*, Bull. London Math. Soc. 23 (1991) 1-33.

[PSc] W. PARRY & K. SCHMIDT, *Natural coefficients and invariants for Markov shifts*, Inventiones Math. 76 (1984), 15-32.

[PS] W. PARRY & D. SULLIVAN, *A topological invariant for flows on one-dimensional spaces*, Topology 14(1975), 297-299.

[PT1]   W. PARRY & S. TUNCEL, *Classification Problems in Ergodic Theory*, LMS Lecture Note Series Vol. 67. Cambridge Press, Cambridge, 1982.

[PT2]   W. PARRY & S. TUNCEL, *On the stochastic and topological structure of Markov chains*, Bull. London Math. Soc. 14 (1982) 16-27.

[PW]   W.PARRY & R.F.WILLIAMS, *Block coding and a zeta function for Markov chains*, Proc. LMS 35 (1977), 483-495.

[Pe]   D. PERRIN, *On positive matrices*, Theoretical Computer Science 94 (1992) 357-366.

[R]   H.J.RYSER, *Combinatorial Mathematics*, Carus Monographs No. 14, Math. Assoc. of America, 1963.

[Sh]   C. SHANNON, *A Mathematical Theory of Communication*, Bell Sys. Tech. J. 27 (1948) 379-423; 623-656.

[ShWe] C. SHANNON & W. WEAVER, *The Mathematical Theory of Communication*, University of Illinois Press (1949) (many reprintings).

[Su]   H.R. SULEIMANOVA, *Stochastic matrices with real eigenvalues*, Dokl. Akad. Nauk. SSSR 66 (1949), 343-345 (Russian).

[Tu]   S. TUNCEL, *A dimension, dimension modules and Markov chains*, Proc. LMS 3, Vol. 46 (1983), 100-116.

[Wa1]   J.WAGONER, *Markov partitions and K2*, Pub. Math. IHES No. 65 (1987), 91-129.

[Wa2]   J.B.WAGONER, *Topological Markov chains, C*-algebras and K_2.*, Advances in Math. Vol. 71, No. 2 (1988), 133-185.

[Wa3]   J.B.WAGONER, *Eventual finite generation for the kernel of the dimension group representation*, Trans. AMS 317 (1990), 331-350.

[Wa4]   J.B.WAGONER, *Triangle identities and symmetries of a subshift of finite type*, Pacific J. Math. Vol.44, No.1, 1990, 181-205.

[Wa5]   J.B.WAGONER, *Higher-dimensional shift equivalence and strong shift equivalence are the same over the integers*, Proc. AMS Vol 109, No.2, (1990), 527-536.

[W1]   R.F.WILLIAMS, *Classification of subshifts of finite type*, Annals of Math. 98 (1973), 120-153; Errata, Annals of Math. 99 (1974), 380-381.

[W2]   R.F.WILLIAMS, *Strong shift equivalence of matrices in GL(2,$\mathbf{Z}$)*, Symbolic dynamics and its applications, Contemporary Math. 135, Amer. Math. Soc. (1992), 445-451.

[W3]   R.F.WILLIAMS, *A new zeta function, natural for links*, To appear in the Proceedings of the Conference in honor of Smale's 60th birthday.

MIXED MATRICES:
IRREDUCIBILITY AND DECOMPOSITION

KAZUO MUROTA*

Abstract. This paper surveys mathematical properties of (layered-) mixed matrices with emphasis on irreducibility and block-triangular decomposition. A matrix A is a mixed matrix if $A = Q + T$, where Q is a "constant" matrix and T is a "generic" matrix (or formal incidence matrix) in the sense that the nonzero entries of T are algebraically independent parameters. A layered mixed (or LM-) matrix is a mixed matrix such that Q and T have disjoint nonzero rows, i.e., no row of $A = Q + T$ has both a nonzero entry from Q and a nonzero entry from T. The irreducibility for an LM-matrix is defined with respect to a natural admissible transformation as an extension of the well-known concept of full indecomposability for a generic matrix. Major results for fully indecomposable generic matrices such as Frobenius' characterization in terms of the irreducibility of determinant are generalized. As for block-triangularization, the Dulmage-Mendelsohn decomposition is generalized to the combinatorial canonical form (CCF) of an LM-matrix along with the uniqueness and the algorithm. Matroid-theoretic methods are useful for investigating a mixed matrix.

1. Introduction. The notion of a mixed matrix was introduced by Murota-Iri [40] as a mathematical tool for systems analysis by means of matroid-theoretic combinatorial methods. A matrix A is called a mixed matrix if $A = Q + T$, where Q is a "constant" matrix and T is a "generic" matrix (or formal incidence matrix) in the sense that the nonzero entries of T are algebraically independent [52] parameters (see below for the precise definition). A layered mixed (or LM-) matrix is defined (see below) as a mixed matrix such that Q and T have disjoint nonzero rows, i.e., no row of $A = Q + T$ has both a nonzero entry from Q and a nonzero entry from T.

The notion of a mixed matrix is motivated by the following physical observation. When we describe a physical system (such as an electrical network, a chemical plant) in terms of elementary variables, we can often distinguish following two kinds of numbers, together characterizing the physical system.

Inaccurate Numbers: Numbers representing independent physical parameters such as masses in mechanical systems and resistances in electrical networks. Such numbers are contaminated with noise and other errors and take values independent of one another; therefore they can be modeled as algebraically independent numbers, and

Accurate Numbers: Numbers accounting for various sorts of conservation laws such as Kirchhoff's laws. Such numbers stem from topological incidence relations and are precise in value (often ± 1); therefore they cause no serious numerical difficulty in arithmetic operations on them.

The "inaccurate numbers" constitute the matrix T whereas the "accurate numbers" the matrix Q. We may also refer to the numbers of the first kind as "system parameters" and to those of the second kind as "fixed constants". In this paper we do

* Research Institute of Mathematical Sciences, Kyoto University, Kyoto 606, Japan
email: murota@kurims.kyoto.u.ac.jp

not discuss physical/engineering significance of a mixed matrix, but concentrate on its mathematical properties. See [25], [27], [28], [30], [31], [37], [38], [39], [40], [42] for engineering applications of mixed matrices; and Chen [6], Iri [18], Recski [46], Yamada-Foulds [56] for graph/matroid theoretic methods for systems analysis.

Here is a preview of some nice properties enjoyed by a mixed matrix or an LM-matrix.

- The rank is expressed as the minimum of a submodular function (Theorem 2.5) and can be computed efficiently by a matroid-theoretic algorithm.
- A notion of irreducibility is defined with respect to a natural transformation of physical significance. The irreducibility for an LM-matrix is an extension of the well-known concept of full indecomposability for a generic matrix.
- An irreducible component thus defined satisfies a number of nice properties that justify the name of irreducibility (Theorems 4.3, 4.4, 4.5, 4.6). Many results for a fully indecomposable generic matrix are extended, including Frobenius' characterization in terms of the irreducibility of determinant.
- There exists a unique canonical block-triangular decomposition, called the combinatorial canonical form (CCF for short), into irreducible components (Theorem 3.1). This is a generalization of the Dulmage-Mendelsohn decomposition. The CCF can be computed by an efficient algorithm (see §6).

We now give the precise definitions of mixed matrix, layered mixed matrix and admissible transformation for a layered mixed matrix. For a matrix A, the row set and the column set of A are denoted by $\mathrm{Row}(A)$ and $\mathrm{Col}(A)$. For $I \subseteq \mathrm{Row}(A)$ and $J \subseteq \mathrm{Col}(A)$, $A[I, J] = (A_{ij} \mid i \in I, j \in J)$ means the submatrix of A with row set I and column set J. The rank of A is written as $\mathrm{rank}\, A$.

Let $\mathbf{K}$ be a subfield of a field $\mathbf{F}$. An $m \times n$ matrix A over $\mathbf{F}$ (i.e., $A_{ij} \in \mathbf{F}$) is called a *mixed matrix* with respect to $\mathbf{F}/\mathbf{K}$ if

$$(1) \qquad A = Q + T,$$

where

(M1) Q is an $m \times n$ matrix over $\mathbf{K}$ (i.e., $Q_{ij} \in \mathbf{K}$), and
(M2) T is an $m \times n$ matrix over $\mathbf{F}$ (i.e., $T_{ij} \in \mathbf{F}$) such that the set of its nonzero entries is algebraically independent [52] over $\mathbf{K}$.

The subfield $\mathbf{K}$ will be called the *base field*.

A mixed matrix A of (1) is called a *layered mixed matrix* (or an *LM-matrix*) with respect to $\mathbf{F}/\mathbf{K}$ if the nonzero rows of Q and T are disjoint. In other words, A is an LM-matrix, denoted as $A \in \mathrm{LM}(\mathbf{F}/\mathbf{K}) = \mathrm{LM}(\mathbf{F}/\mathbf{K}; m_Q, m_T, n)$, if it can be put into the following form with a permutation of rows:

$$(2) \qquad A = \begin{pmatrix} Q \\ T \end{pmatrix} = \begin{pmatrix} Q \\ O \end{pmatrix} + \begin{pmatrix} O \\ T \end{pmatrix},$$

where

(L1) Q is an $m_Q \times n$ matrix over $\mathbf{K}$ (i.e., $Q_{ij} \in \mathbf{K}$), and
(L2) T is an $m_T \times n$ matrix over $\mathbf{F}$ (i.e., $T_{ij} \in \mathbf{F}$) such that the set $\mathcal{T}$ of its nonzero entries is algebraically independent over $\mathbf{K}$.

Though an LM-matrix is, by definition, a special case of mixed matrix, the following argument would indicate that the class of LM-matrices is as general as the class of mixed matrices both in theory and in application. Consider a system of equations $Ax = b$ described with an $m \times n$ mixed matrix $A = Q + T$. By introducing an auxiliary variable $w \in F^m$ we can rewrite the equation as

$$\tilde{A} \begin{pmatrix} w \\ x \end{pmatrix} = \begin{pmatrix} b \\ 0 \end{pmatrix}$$

with a $(2m) \times (m + n)$ LM-matrix

(3)
$$\tilde{A} = \begin{pmatrix} \tilde{Q} \\ \tilde{T} \end{pmatrix} = \begin{pmatrix} I_m & Q \\ -\operatorname{diag}[t_1, \dots, t_m] & T' \end{pmatrix},$$

where $\operatorname{diag}[t_1, \dots, t_m]$ is a diagonal matrix with "new" variables $t_1, \dots, t_m (\in F)$, and $T'_{ij} = t_i T_{ij}$. Note that rank $\tilde{A} = \operatorname{rank} A + m$.

EXAMPLE 1. An equation described with a mixed matrix:

$$\begin{pmatrix} 2 + \alpha & 3 \\ \beta & 4 + \gamma \end{pmatrix} \begin{pmatrix} x_1 \\ x_2 \end{pmatrix} = \begin{pmatrix} b_1 \\ b_2 \end{pmatrix},$$

where $T = \{\alpha, \beta, \gamma\}$ is algebraically independent, can be rewritten as

$$\begin{pmatrix} 1 & 0 & 2 & 3 \\ 0 & 1 & 0 & 4 \\ -t_1 & 0 & t_1\alpha & 0 \\ 0 & -t_2 & t_2\beta & t_2\gamma \end{pmatrix} \begin{pmatrix} w_1 \\ w_2 \\ x_1 \\ x_2 \end{pmatrix} = \begin{pmatrix} b_1 \\ b_2 \\ 0 \\ 0 \end{pmatrix}$$

by means of an LM-matrix. □

For an LM-matrix $A \in \mathrm{LM}(F/K; m_Q, m_T, n)$ of (2) we define an *admissible transformation* to be a transformation of the form:

(4)
$$P_r \begin{pmatrix} S & O \\ O & I \end{pmatrix} \begin{pmatrix} Q \\ T \end{pmatrix} P_c,$$

where P_r and P_c are permutation matrices, and S is a nonsingular matrix over the base field K (i.e., $S \in \mathrm{GL}(m_Q, K)$).

An admissible transformation brings an LM-matrix into another LM-matrix and two LM-matrices are said to be *LM-equivalent* if they are connected by an admissible transformation. If A' is LM-equivalent to A, then $\mathrm{Col}(A')$ may be identified with $\mathrm{Col}(A)$ through the permutation P_c. Examples 2, 3 below illustrate the admissible transformation.

With respect to the admissible transformation (4) we can define the notion of irreducibility for LM-matrices, which is an extension of the well-studied concept of full indecomposability [5], [49]. First recall that a matrix A' is said to be *partially decomposable* if it contains a zero submatrix $A'[I, J] = O$ with $|I| + |J| = \max(|\mathrm{Row}(A')|, |\mathrm{Col}(A')|)$; otherwise, it is called *fully indecomposable*. An LM-matrix $A \in \mathrm{LM}(F/K; m_Q, m_T, n)$ is defined to be *LM-reducible* if it can be decomposed into smaller submatrices by means of the admissible transformation, or more precisely, if there exists a partially decomposable matrix A' which is LM-equivalent to A. On the other hand, A will be called *LM-irreducible* if it is not LM-reducible, that is, if any LM-matrix A' equivalent to A is fully indecomposable. Hence, if A is LM-irreducible, then it is fully indecomposable; but not conversely. By convention A is regarded as LM-irreducible if $\mathrm{Row}(A) = \emptyset$ or $\mathrm{Col}(A) = \emptyset$.

Let us consider the special case where $m_Q = 0$. Then $A = T$ and hence all the nonzero entries are algebraically independent. Such a matrix is called a *generic matrix* in Brualdi-Ryser [5]. The admissible transformation (4) reduces to $\bar{A} = P_r A P_c$, involving permutations only, and the LM-irreducibility is nothing but the full indecomposability. It is known that a fully indecomposable generic matrix enjoys a number of interesting properties. On the other hand, if a matrix is not fully indecomposable, it can be decomposed uniquely into fully indecomposable components. This is called the *Dulmage-Mendelsohn decomposition*, or the *DM-decomposition* for short. See [4], [5], [8], [21], [28], [33], [37] for more about the DM-decomposition.

In this paper we are mainly interested in whether these results for a generic matrix can be extended to a general LM-matrix. It will be shown that many major results for a fully indecomposable generic matrix are extended for an LM-irreducible matrix, and the DM-decomposition is extended to a canonical block-triangular decomposition under the admissible transformation (4). The canonical form is called the combinatorial canonical form (CCF) of an LM-matrix, which is illustrated in the following examples, whereas a precise description of the CCF will be given as Theorem 3.1 in §3.

EXAMPLE 2. Consider a 3×3 LM-matrix

$$A = \begin{pmatrix} Q \\ T \end{pmatrix} = \begin{pmatrix} 1 & 1 & 0 \\ 1 & 2 & 3 \\ 0 & t_1 & t_2 \end{pmatrix}$$

with

$$Q = \begin{pmatrix} 1 & 1 & 0 \\ 1 & 2 & 3 \end{pmatrix}, \quad T = \begin{pmatrix} 0 & t_1 & t_2 \end{pmatrix},$$

where $T = \{t_1, t_2\}$ is the set of algebraically independent parameters. This matrix is fully indecomposable (DM-irreducible) and cannot be decomposed into smaller blocks by means of permutations of rows and columns. By choosing $S = \begin{pmatrix} 1 & 0 \\ -1 & 1 \end{pmatrix}$ and $P_r = P_c = I$ in the admissible transformation (4), we can obtain a block-triangular decomposition:

$$\bar{A} = \begin{pmatrix} SQ \\ T \end{pmatrix} = \begin{pmatrix} 1 & 1 & 0 \\ & 1 & 3 \\ & t_1 & t_2 \end{pmatrix}.$$

Thus the admissible transformation is more powerful than mere permutations. □

EXAMPLE 3. Consider an LM-matrix $A = \begin{pmatrix} Q \\ T \end{pmatrix}$ of (2) defined by

$$
Q = \begin{array}{c}
\begin{array}{cccccccccc} \xi_1 & \xi_2 & \xi_3 & \xi_4 & \xi_5 & \eta_1 & \eta_2 & \eta_3 & \eta_4 & \eta_5 \end{array} \\
\begin{pmatrix}
0 & 0 & 1 & 1 & 1 & 0 & 0 & 0 & 0 & 0 \\
1 & 0 & 0 & 0 & -1 & 0 & 0 & 0 & 0 & 0 \\
0 & 1 & -1 & 0 & 0 & 0 & 0 & 0 & 0 & 0 \\
0 & 0 & 0 & 0 & 0 & 1 & 0 & 0 & -1 & 1 \\
0 & 0 & 0 & 0 & 0 & 0 & 1 & 1 & -1 & 0
\end{pmatrix}
\end{array},
$$

$$
T = \begin{array}{c}
\begin{array}{cccccccccc} \xi_1 & \xi_2 & \xi_3 & \xi_4 & \xi_5 & \eta_1 & \eta_2 & \eta_3 & \eta_4 & \eta_5 \end{array} \\
\begin{pmatrix}
r_1 & 0 & 0 & 0 & 0 & t_1 & 0 & 0 & 0 & 0 \\
0 & r_2 & 0 & 0 & 0 & 0 & t_2 & 0 & 0 & 0 \\
0 & 0 & 0 & 0 & 0 & \alpha & 0 & t_3 & 0 & 0 \\
0 & \beta & 0 & t_4 & 0 & 0 & 0 & 0 & 0 & 0 \\
0 & 0 & 0 & 0 & 0 & 0 & 0 & 0 & 0 & t_5
\end{pmatrix}
\end{array},
$$

where $T = \{r_1, r_2, \alpha, \beta; t_1, \cdots, t_5\}$ is the set of algebraically independent parameters. (See Example 16.2 in [28] for the physical meaning of this example.)

The combinatorial canonical form (CCF), i.e., the finest block-triangular form under the admissible transformation (4) is obtained as follows. Choosing

$$
S = \begin{pmatrix}
0 & -1 & 0 & 0 & 0 \\
0 & 0 & -1 & 0 & 0 \\
1 & 1 & 1 & 0 & 0 \\
0 & 0 & 0 & -1 & 0 \\
0 & 0 & 0 & -1 & 1
\end{pmatrix}
$$

in (4) we first transform Q to

$$
Q' = SQ = \begin{array}{c}
\begin{array}{cccccccccc} \xi_1 & \xi_2 & \xi_3 & \xi_4 & \xi_5 & \eta_1 & \eta_2 & \eta_3 & \eta_4 & \eta_5 \end{array} \\
\begin{pmatrix}
-1 & 0 & 0 & 0 & 1 & 0 & 0 & 0 & 0 & 0 \\
0 & -1 & 1 & 0 & 0 & 0 & 0 & 0 & 0 & 0 \\
1 & 1 & 0 & 1 & 0 & 0 & 0 & 0 & 0 & 0 \\
0 & 0 & 0 & 0 & 0 & -1 & 0 & 0 & 1 & -1 \\
0 & 0 & 0 & 0 & 0 & -1 & 1 & 1 & 0 & -1
\end{pmatrix}
\end{array},
$$

and then permute the rows and the columns of $\begin{pmatrix} Q' \\ T \end{pmatrix}$ with permutation matrices

$$
P_r = \begin{pmatrix}
0 & 1 & 0 & 0 & 0 & 0 & 0 & 0 & 0 & 0 \\
1 & 0 & 0 & 0 & 0 & 0 & 0 & 0 & 0 & 0 \\
0 & 0 & 0 & 1 & 0 & 0 & 0 & 0 & 0 & 0 \\
0 & 0 & 1 & 0 & 0 & 0 & 0 & 0 & 0 & 0 \\
0 & 0 & 0 & 0 & 1 & 0 & 0 & 0 & 0 & 0 \\
0 & 0 & 0 & 0 & 0 & 1 & 0 & 0 & 0 & 0 \\
0 & 0 & 0 & 0 & 0 & 0 & 1 & 0 & 0 & 0 \\
0 & 0 & 0 & 0 & 0 & 0 & 0 & 1 & 0 & 0 \\
0 & 0 & 0 & 0 & 0 & 0 & 0 & 0 & 1 & 0 \\
0 & 0 & 0 & 0 & 0 & 0 & 0 & 0 & 0 & 1
\end{pmatrix},
$$

$$P_{\mathrm{c}} = \begin{pmatrix} 0 & 0 & 0 & 1 & 0 & 0 & 0 & 0 & 0 & 0 \\ 0 & 0 & 0 & 0 & 1 & 0 & 0 & 0 & 0 & 0 \\ 1 & 0 & 0 & 0 & 0 & 0 & 0 & 0 & 0 & 0 \\ 0 & 0 & 0 & 0 & 0 & 1 & 0 & 0 & 0 & 0 \\ 0 & 1 & 0 & 0 & 0 & 0 & 0 & 0 & 0 & 0 \\ 0 & 0 & 0 & 0 & 0 & 0 & 1 & 0 & 0 & 0 \\ 0 & 0 & 0 & 0 & 0 & 0 & 0 & 1 & 0 & 0 \\ 0 & 0 & 0 & 0 & 0 & 0 & 0 & 0 & 1 & 0 \\ 0 & 0 & 1 & 0 & 0 & 0 & 0 & 0 & 0 & 0 \\ 0 & 0 & 0 & 0 & 0 & 0 & 0 & 0 & 0 & 1 \end{pmatrix}$$

to obtain an explicit block-triangular LM-matrix

$$\bar{A} = P_{\mathrm{r}} \begin{pmatrix} Q' \\ T \end{pmatrix} P_{\mathrm{c}} = \begin{array}{c} \begin{array}{cccccccccc} \xi_3 & \xi_5 & \eta_4 & \xi_1 & \xi_2 & \xi_4 & \eta_1 & \eta_2 & \eta_3 & \eta_5 \end{array} \\ \begin{pmatrix} 1 & & & & -1 & & & & & \\ & 1 & & -1 & & & & & & \\ & & 1 & & & & -1 & & & -1 \\ & & & 1 & 1 & 1 & 0 & 0 & 0 & \\ & & & 0 & 0 & 0 & -1 & 1 & 1 & -1 \\ & & & r_1 & 0 & 0 & t_1 & 0 & 0 & \\ & & & 0 & r_2 & 0 & 0 & t_2 & 0 & \\ & & & 0 & 0 & 0 & \alpha & 0 & t_3 & \\ & & & 0 & \beta & t_4 & 0 & 0 & 0 & \\ & & & & & & & & & t_5 \end{pmatrix} \end{array}.$$

This is the CCF of A, namely, the finest block-triangular matrix which is LM-equivalent to A. Hence A is LM-reducible whereas each diagonal block of $\bar{A}$ is LM-irreducible. The columns of $\bar{A}$ are partitioned into five blocks:

$$C_1 = \{\xi_3\}, \ C_2 = \{\xi_5\}, \ C_3 = \{\eta_4\}, \ C_4 = \{\xi_1, \xi_2, \xi_4, \eta_1, \eta_2, \eta_3\}, \ C_5 = \{\eta_5\}.$$

The zero/nonzero structure of $\bar{A}$ determines the following partial order among the blocks:

$$\begin{array}{c} C_5 \\ \uparrow \\ C_4 \\ \nearrow \quad \uparrow \quad \nwarrow \\ C_1 \quad\quad C_2 \quad\quad C_3 \end{array}$$

This partial order indicates, for example, that the blocks C_1 and C_2, having no order relation, could be exchanged in position without destroying the block-triangular form provided the corresponding rows are exchanged in position accordingly. This corresponds to the fact that the entry in the first row of the column ξ_5 is equal to 0. A precise description of the CCF and a combinatorial characterization of the partial order will be given in Theorem 3.1 in §3. The transformation matrices S, P_{r} and P_{c} can be found by the algorithm described in §6. $\quad\square$

2. Rank. In this section we consider combinatorial characterizations of the rank of an LM-matrix $A = \left(\begin{smallmatrix} Q \\ T \end{smallmatrix}\right) \in \mathrm{LM}(\boldsymbol{F}/\boldsymbol{K})$. We put $C = \mathrm{Col}(A)$, $R = \mathrm{Row}(A)$, $R_Q = \mathrm{Row}(Q)$ and $R_T = \mathrm{Row}(T)$; then $\mathrm{Col}(Q) = \mathrm{Col}(T) = C$, and $R = R_Q \cup R_T$.

Before dealing with a general LM-matrix, let us consider the special case of a generic matrix, i.e., where $A = T$ (with $m_Q = 0$) and hence all the nonzero entries are algebraically independent over $\mathbf{K}$. The zero/nonzero structure of T can be conveniently represented by a bipartite graph $G(T) = (\mathrm{Row}(T), \mathrm{Col}(T), T)$, which has $\mathrm{Row}(T) \cup \mathrm{Col}(T)$ as the vertex set and T (=set of nonzero entries of T) as the arc set. The *term-rank* of T, denoted as term-rank T, is equal to the maximum size of a matching in $G(T)$. In other words, term-rank T is the maximum size of a square submatrix $T[I, J]$ such that there exists a one-to-one correspondence $\pi : I \to J$ with $T_{i\pi(i)} \neq 0$ ($\forall i \in I$):

$$\text{term-rank}\, T = \max\{|I| \mid \exists\, \pi(\text{one-to-one}) : I \to J,\ \forall i \in I : T_{i\pi(i)} \neq 0\}.$$

The following fact is well known [5], [10]. See Lemma 2.3 below for the proof.

LEMMA 2.1. *For a generic matrix T, which has algebraically independent nonzero entries, we have*

$$\text{rank}\, T = \text{term-rank}\, T.$$

$\square$

The zero/nonzero structure of T is represented by the functions τ, $\gamma : 2R_T \times 2C \to \mathbf{Z}$ defined as

(5) $\qquad \tau(I, J) = \text{term-rank}\, T[I, J], \quad I \subseteq R_T, J \subseteq C,$

(6) $\qquad \Gamma(I, J) = \bigcup_{j \in J} \{i \in I \mid T_{ij} \neq 0\}, \quad I \subseteq R_T, J \subseteq C,$

(7) $\qquad \gamma(I, J) = |\Gamma(I, J)|, \quad I \subseteq R_T, J \subseteq C.$

Lemma 2.1 shows that $\tau(I, J) = \text{rank}\, T[I, J]$, whereas $\Gamma(I, J)$ stands for the set of nonzero rows of the submatrix $T[I, J]$, and $\gamma(I, J)$ for the number of nonzero rows of $T[I, J]$.

These functions τ, γ enjoy *bisubmodularity*, that is, they each satisfy an inequality of the following type:

(8) $\qquad f(I_1 \cup I_2, J_1 \cap J_2) + f(I_1 \cap I_2, J_1 \cup J_2) \leq f(I_1, J_1) + f(I_2, J_2).$

For a bisubmodular function f in general, $f_I \equiv f(I, \cdot) : 2C \to \mathbf{Z}$, for each I, is a *submodular* function:

(9) $\qquad f_I(J_1 \cap J_2) + f_I(J_1 \cup J_2) \leq f_I(J_1) + f_I(J_2), \quad J_i \subseteq C\ (i = 1, 2).$

The following fact is a version of the fundamental minimax relation concerning the maximum matchings and the minimum covers of a bipartite graph, which is often associated with J. Egerváry, G. Frobenius, D. König, P. Hall, R. Rado, O. Ore, and others [5], [20], [21], [53]. Note also that the function $\gamma(I, J) - |J|$ (with I fixed) is called the surplus function in Lovász-Plummer [21].

LEMMA 2.2. *For τ and γ defined by (5) and (7),*

$$\tau(I,J) = \min\{\gamma(I,J') - |J'| \mid J' \subseteq J\} + |J|, \quad I \subseteq R_T, J \subseteq C.$$

$\square$

We are now in the position to consider the rank of a general LM-matrix. The following lemma is a fundamental identity for an LM-matrix, an extension of Lemma 2.1 for a generic matrix. It will be translated first into a matroid-theoretic expression in Lemma 2.4, and then, with the aid of the matroid partition theorem, turned into the important minimax formulas in Theorem 2.5.

LEMMA 2.3. *For $A \in \mathrm{LM}(\boldsymbol{F}/\boldsymbol{K})$,*

(10) $$\mathrm{rank}\, A = \max\{\mathrm{rank}\, Q[R_Q, J] + \text{term-rank}\, T[R_T, C - J] \mid J \subseteq C\}.$$

Proof. First assume that A is square and consider the (generalized) Laplace expansion [16]:

$$\det A = \sum_{J \subseteq C} \pm \det Q[R_Q, J] \cdot \det T[R_T, C - J].$$

If $\det A \neq 0$, then both $Q[R_Q, J]$ and $T[R_T, C - J]$ are nonsingular for some J. The algebraic independence of T ensures the converse. This shows (10) for a square A. For a nonsquare matrix A, the same argument applies to its square submatrices. $\square$

For the matrix A of Example 3, we may take $J = \{\xi_5, \xi_3, \xi_4, \eta_4, \eta_3\}$ for the subset that attains the maximum ($=10$) on the right-hand side of (10). Therefore A is nonsingular.

Let us introduce some matroid-theoretic concepts to recast the identity in Lemma 2.3. Put $\mathcal{F} = \{J \subseteq C \mid \mathrm{rank}\, A[R, J] = |J|\}$, which denotes the family of linearly independent columns of A. As is easily verified, the family $\mathcal{F}$ satisfies the following three conditions:

(i) $\emptyset \in \mathcal{F}$,
(ii) $J_1 \subseteq J_2 \in \mathcal{F} \Rightarrow J_1 \in \mathcal{F}$,
(iii) $J_1 \in \mathcal{F}, J_2 \in \mathcal{F}, |J_1| < |J_2| \Rightarrow J_1 \cup \{j\} \in \mathcal{F}$ for some $j \in J_2 - J_1$.

In general, a pair $\boldsymbol{M} = (C, \mathcal{F})$ of a finite set C and a family $\mathcal{F}$ of subsets of C is called a *matroid* if it satisfies the three conditions above. C is called the *ground set* and $\mathcal{F}$ the *family of independent sets*. A maximal member (with respect to set inclusion) in $\mathcal{F}$ is called a *base*, and, by condition (ii), $\mathcal{F}$ is determined by the family $\mathcal{B}$ of bases. The size of a base is uniquely determined, which is called the *rank* of $\boldsymbol{M}$, denoted as $\mathrm{rank}\, \boldsymbol{M}$; i.e., $\mathrm{rank}\, \boldsymbol{M} = |B| = \max\{|J| \mid J \in \mathcal{F}\}$ for $B \in \mathcal{B}$. Given two matroids $\boldsymbol{M}_1 = (C, \mathcal{F}_1)$ and $\boldsymbol{M}_2 = (C, \mathcal{F}_2)$ with the same ground set C, another matroid, denoted as $(C, \mathcal{F}_1 \vee \mathcal{F}_2)$, is defined by

$$\mathcal{F}_1 \vee \mathcal{F}_2 = \{J_1 \cup J_2 \mid J_1 \in \mathcal{F}_1, J_2 \in \mathcal{F}_2\}.$$

This is called the *union* of $\boldsymbol{M}_1$ and $\boldsymbol{M}_2$, and denoted as $\boldsymbol{M}_1 \vee \boldsymbol{M}_2$. See [15], [20], [53], [54], [55] for more about matroids.

For an LM-matrix $A = \binom{Q}{T} \in \mathrm{LM}(\boldsymbol{F}/\boldsymbol{K})$ we consider the matroids $\boldsymbol{M}(A)$, $\boldsymbol{M}(Q)$, $\boldsymbol{M}(T)$ on C defined respectively by matrices A, Q, T with respect to the linear independence among column vectors. Then Lemma 2.3 is rewritten as follows.

LEMMA 2.4. *For* $A = \binom{Q}{T} \in \mathrm{LM}(\boldsymbol{F}/\boldsymbol{K})$, *we have* $\boldsymbol{M}(A) = \boldsymbol{M}(Q) \vee \boldsymbol{M}(T)$. □

This theorem makes it possible to compute the rank of an LM-matrix with $O(n^3 \log n)$ arithmetic operations (assuming $m = O(n)$ for simplicity) in the base field $\boldsymbol{K}$ by utilizing an established algorithm for matroid partition/union problem ([7], [9], [11], [15], [20], [53], [54], [55]). See the algorithm in §6.

As an extension of the surplus function for a generic matrix (cf. Lemma 2.2) we introduce a set function $p : 2^R \times 2^C \to \boldsymbol{Z}$ as follows. For $A \in \mathrm{LM}(\boldsymbol{F}/\boldsymbol{K})$ we define p by

(11) $$p(I, J) = \rho(I \cap R_Q, J) + \gamma(I \cap R_T, J) - |J|, \quad I \subseteq R, J \subseteq C,$$

where

(12) $$\rho(I, J) = \mathrm{rank}\, Q[I, J], \quad I \subseteq R_Q, J \subseteq C,$$

stands for the "constant" matrix Q, whereas γ (see (7) for the definition) represents the combinatorial structure of T. Note that, in the special case where $A = T$ (i.e., $m_Q = 0$), we have $p(I, J) = \gamma(I, J) - |J|$, which is the surplus function used in Lemma 2.2.

The function p is bisubmodular (cf. (8)) and therefore $p_I \equiv p(I, \cdot) : 2^C \to \boldsymbol{Z}$ is submodular for each $I \subseteq R$, namely,

(13) $$p_I(J_1 \cap J_2) + p_I(J_1 \cup J_2) \le p_I(J_1) + p_I(J_2), \quad J_i \subseteq C \ (i = 1, 2).$$

The submodular function p_R (i.e., p_I with $I = R$) is invariant under the LM-equivalence in the sense that, if A' is LM-equivalent to A, then $\mathrm{Col}(A')$ may be identified with $C = \mathrm{Col}(A)$ and the functions p and p' associated respectively with A and A' satisfy $p(\mathrm{Row}(A), J) = p'(\mathrm{Row}(A'), J)$ for $J \subseteq C$.

The following theorem gives two minimax expressions (14) and (15), similar but different, for the rank of an LM-matrix. The second expression (15) (or equivalently (16)), due to Murota [26] [28], Murota-Iri-Nakamura [41], is an extension of the min-imax relation between matchings and covers given in Lemma 2.2. In fact, the expression (15) with $\rho = 0$ reduces to Lemma 2.2 since then $\mathrm{rank}\, A[I, J] = \mathrm{rank}\, T[I, J] = \tau(I, J)$ by Lemma 2.1.

THEOREM 2.5. *Let* $A \in \mathrm{LM}(\boldsymbol{F}/\boldsymbol{K})$ *and* $I \subseteq R$, $J \subseteq C$. *Then*

(14) $$\mathrm{rank}\, A[I, J] = \min\{\rho(I \cap R_Q, J') + \tau(I \cap R_T, J') - |J'| \mid J' \subseteq J\} + |J|,$$

(15) $$\mathrm{rank}\, A[I, J] = \min\{\rho(I \cap R_Q, J') + \gamma(I \cap R_T, J') - |J'| \mid J' \subseteq J\} + |J|.$$

Using the function p_R *the latter formula for* $I = R$, $J = C$ *can be written as*

(16) $$\mathrm{rank}\, A = \min\{p_R(J) \mid J \subseteq C\} + |C|.$$

Proof. Lemma 2.4 shows that rank $A = $ rank $M(A) = $ rank $(M(Q) \vee M(T))$. On the other hand, the matroid union/partition theorem of Edmonds [9] (see also [11], [15], [20], [53], [54], [55]) says that

$$\text{rank}\,(M(Q) \vee M(T)) = \min\{\text{rank}\, Q(R_Q, J) + \text{rank}\, T(R_T, J) + |C - J| \mid J \subseteq C\},$$

which establishes (14) for $I = R, J = C$. The same argument applied to the submatrix $A[I, J]$ shows (14).

The right-hand sides of (14) and (15) are equal, since with the notations $I \cap R_Q = I_Q$, $I \cap R_T = I_T$, we have

$$\min_{J' \subseteq J}\{\rho(I_Q, J') + \tau(I_T, J') - |J'|\}$$
$$= \min_{J' \subseteq J}\{\rho(I_Q, J') + \min_{J'' \subseteq J'}\{\gamma(I_T, J'') - |J''|\}\}$$
$$= \min_{J'' \subseteq J}\{\min_{J' \supseteq J''}\{\rho(I_Q, J')\} + \gamma(I_T, J'') - |J''|\}$$
$$= \min_{J'' \subseteq J}\{\rho(I_Q, J'') + \gamma(I_T, J'') - |J''|\},$$

where the first equality is by Lemma 2.2 and the last equality is due to the monotonicity of $\rho(I_Q, J)$ with respect to J for a fixed I_Q. □

The two expressions in Theorem 2.5 look very similar, with τ in (14) replaced by γ in (15). Moreover, in both formulas, the functions to be minimized are submodular in J'. However, we will see in the next section that the second expression (15), not the first one, chimes in exact harmony with the admissible transformation (4), with respect to which we are to consider the block-triangular decomposition.

3. Decomposition (CCF)

3.1. Description of CCF. This section gives a precise description, Theorem 3.1 below, of the combinatorial canonical form (CCF), which has already been sketched informally in Examples 2, 3 in Introduction.

As stated in Theorem 2.5, the rank of $A[I, J]$ is expressed by the minimum of p_I. Then it would be natural to look at the family of minimizers:

$$(17) \qquad L(p_I) = \{J \subseteq C \mid p(I, J) \leq p(I, J'), \forall J' \subseteq C\}, \quad I \subseteq R,$$

which, for each $I \subseteq R$, forms a sublattice of $2C$ by virtue of the submodularity (13) of p_I. In fact, if both J_1 and J_2 attain the minimum value, say α, of p_I, then $2\alpha \leq p_I(J_1 \cap J_2) + p_I(J_1 \cup J_2) \leq p_I(J_1) + p_I(J_2) = 2\alpha$ shows that $J_1 \cap J_2 \in L(p_I)$ and $J_1 \cup J_2 \in L(p_I)$. The sublattice $L(p_R)$ plays a crucial role for the block-triangular decomposition, as explained below.

Here we make use of some fundamental results from lattice theory [2], [3]. Birkhoff's representation theorem implies that there exists a one-to-one correspondence between sublattices of $2C$ and pairs of a partition of C into blocks and a partial order among the blocks. This correspondence is given as follows.

Let L be a sublattice of $2C$. Take any maximal ascending chain:

$$X_0 \ (= \min L) \subset X_1 \subset \cdots \subset X_b \ (= \max L),$$

where $X_k \in L$, and put

(18)
$$
\begin{aligned}
C_0 &= X_0, \\
C_k &= X_k - X_{k-1} \quad (k = 1, \ldots, b), \\
C_\infty &= C - X_b.
\end{aligned}
$$

Then the family of the subsets $\{C_k \mid k = 1, \ldots, b\}$ is uniquely determined, being independent of the choice of the chain. A partial order $\preceq$ is introduced on $\{C_k \mid k = 1, \ldots, b\}$ by

$$C_k \preceq C_l \quad \Longleftrightarrow \quad [X \in L, \ C_l \subseteq X \ \Rightarrow \ C_k \subseteq X].$$

For convenience, we extend the partial order onto

$$\{C_0, C_\infty\} \cup \{C_k \mid k = 1, \ldots, b\}$$

by defining

$$C_0 \preceq C_k \quad (k = 1, \ldots, b) \quad \text{if} \quad C_0 \neq \emptyset,$$

$$C_k \preceq C_\infty \quad (k = 1, \ldots, b) \quad \text{if} \quad C_\infty \neq \emptyset.$$

We also introduce the following notation:

$$
\begin{aligned}
C_k \prec C_l &\quad \Longleftrightarrow \quad C_k \preceq C_l \text{ and } C_k \neq C_l; \\
C_k \prec \cdot \ C_l &\quad \Longleftrightarrow \quad \begin{cases} \text{(i) } C_k \prec C_l \text{ and} \\ \text{(ii) } \not\exists \, C_j \text{ such that } C_k \prec C_j \prec C_l. \end{cases}
\end{aligned}
$$

In this way, a sublattice L of $2C$ determines a pair of a partition $\{C_0; C_1, \ldots, C_b; C_\infty\}$ and a partial order $\preceq$, which we denote by

(19)
$$\mathcal{P}(L) = (\{C_0; C_1, \ldots, C_b; C_\infty\}, \preceq).$$

Note that $C_k \neq \emptyset$ for $k = 1, \ldots, b$, whereas C_0 and C_∞ are distinguished blocks that can be empty. It may also be mentioned that a pair of a partition of C and a partial order among the blocks is nothing but a quasi-order (=reflexive and transitive binary relation [2]) on C.

Conversely, given $\mathcal{P} = (\{C_0; C_1, \ldots, C_b; C_\infty\}, \preceq)$, a sublattice L is determined as follows: $X \in L$ if and only if $C_0 \subseteq X \subseteq C - C_\infty$ and, for $1 \leq l \leq b$,

$$X \cap C_l \neq \emptyset \quad \Rightarrow \quad \bigcup_{C_k \preceq C_l} C_k \subseteq X.$$

Namely, L is the family of (order-) ideals containing C_0 and contained in $C - C_\infty$. Note that $\min L = C_0$ and $\max L = C - C_\infty$. This correspondence between L and $\mathcal{P}$ is known to be a one-to-one correspondence.

According to this general principle, the sublattice $L(p_R)$ associated with an LM-matrix A determines $\mathcal{P}(L(p_R))$, a pair of a partition of C and a partial order $\preceq$. Note that by (18) the blocks are indexed consistently with the partial order in the sense that

$$(20) \qquad\qquad C_k \preceq C_l \quad\Rightarrow\quad k \leq l.$$

The following theorem, established in an unpublished report by Murota [26] in 1985 and published as Murota [28], Murota-Iri-Nakamura [41], claims the existence of the CCF of an LM-matrix. The construction of CCF is described in the next subsection along with an outline of the proof. A complete proof can be found in [26], [28], [41].

THEOREM 3.1. *For an LM-matrix $A \in \mathrm{LM}(\boldsymbol{F}/\boldsymbol{K})$ there exists another LM-matrix $\bar{A}$ which is LM-equivalent to A and satisfies the following properties.*

(B1) [Nonzero structure and partial order $\preceq$] *$\bar{A}$ is block-triangularized, i.e.,*

$$\bar{A}[R_k, C_l] = O \quad \text{if} \quad 0 \leq l < k \leq \infty,$$

where $\{R_0; R_1, \ldots, R_b; R_\infty\}$ and $\{C_0; C_1, \ldots, C_b; C_\infty\}$ are partitions of $\mathrm{Row}(\bar{A})$ and $\mathrm{Col}(\bar{A})$ respectively such that $R_k \neq \emptyset$, $C_k \neq \emptyset$ for $k = 1, \ldots, b$, whereas R_0, R_∞, C_0 and C_∞ can be empty.

Moreover, when $\mathrm{Col}(\bar{A})$ is identified with $\mathrm{Col}(A)$, the partition $\{C_0; C_1, \ldots, C_b; C_\infty\}$ agrees with that defined by the lattice $L(p_R)$ and the partial order on $\{C_1, \ldots, C_b\}$ induced by the zero/nonzero structure of $\bar{A}$ agrees with the partial order $\preceq$ defined by $L(p_R)$; i.e.,

$$\begin{aligned} \bar{A}[R_k, C_l] &= O \quad \text{unless} \quad C_k \preceq C_l \quad (1 \leq k, l \leq b); \\ \bar{A}[R_k, C_l] &\neq O \quad \text{if} \quad C_k \prec\!\!\cdot\, C_l \quad (1 \leq k, l \leq b). \end{aligned}$$

(B2) [Size of the diagonal blocks]

$$\begin{aligned} |R_0| &< |C_0| \quad \text{if} \quad R_0 \neq \emptyset, \\ |R_k| &= |C_k| \quad (> 0) \quad \text{for} \quad k = 1, \ldots, b, \\ |R_\infty| &> |C_\infty| \quad \text{if} \quad C_\infty \neq \emptyset. \end{aligned}$$

(B3) [Rank of the diagonal blocks]

$$\begin{aligned} \mathrm{rank}\,\bar{A}[R_0, C_0] &= |R_0|, \\ \mathrm{rank}\,\bar{A}[R_k, C_k] &= |R_k| = |C_k| \quad \text{for} \quad k = 1, \ldots, b, \\ \mathrm{rank}\,\bar{A}[R_\infty, C_\infty] &= |C_\infty|. \end{aligned}$$

(B4) [Uniqueness] *$\bar{A}$ is the finest block-triangular matrix with properties (B2) and (B3) that is LM-equivalent to A. Namely, if $\hat{A}$ is LM-equivalent to A which is block-triangularized with respect to certain partitions*

$$(\hat{R}_0; \hat{R}_1, \ldots, \hat{R}_q; \hat{R}_\infty), \quad (\hat{C}_0; \hat{C}_1, \ldots, \hat{C}_q; \hat{C}_\infty)$$

of Row($\hat{A}$) *and* Col($\hat{A}$) (= Col(A)) *with the diagonal blocks satisfying the conditions (B2) and (B3), then* $\hat{C}_k$ *is a union of the blocks defined by* $L(p_R)$. □

The matrix $\bar{A}$ above is the CCF of A. The CCF is uniquely determined so far as the partitions of the row and column sets as well as the partial order among the blocks are concerned, whereas there remains some indeterminacy, or degree of freedom, in the numerical values of the entries in the Q-part (for example, elementary row transformations within a block change numerical values without affecting the block structure). See $\bar{A}_1$ and $\bar{A}_2$ in Example 5 below. When the numerical indeterminacy is to be emphasized, such $\bar{A}$ will be called a CCF, instead of *the* CCF. We make use of such indeterminacy in Theorem 3.2.

The submatrices $\bar{A}[R_0, C_0]$ and $\bar{A}[R_\infty, C_\infty]$ are called the *horizontal tail* and the *vertical tail*, respectively. The tails are nonsquare if they are not empty, and (B1) and (B3) imply that

$$\text{rank } A = \text{rank } \bar{A} = |C| - \delta_0 = |R| - \delta_\infty$$

with

$$\delta_0 = |C_0| - |R_0|, \quad \delta_\infty = |R_\infty| - |C_\infty|.$$

Hence A is nonsingular if and only if $C_0 = R_\infty = \emptyset$. In Example 3 we have $C_0 = R_\infty = \emptyset$, and the number of square blocks $b = 5$.

EXAMPLE 4. Consider a 4×5 LM-matrix $A = \binom{Q}{T}$ with

$$
\begin{array}{cccccc}
 & x_1 & x_2 & x_3 & x_4 & x_5 \\
\end{array}
\qquad
\begin{array}{cccccc}
 & x_1 & x_2 & x_3 & x_4 & x_5 \\
\end{array}
$$

$$
Q = \begin{pmatrix} 1 & 1 & 1 & 1 & 0 \\ 0 & 2 & 1 & 1 & 0 \end{pmatrix}, \quad
T = \begin{pmatrix} t_1 & 0 & 0 & 0 & t_2 \\ 0 & t_3 & 0 & 0 & t_4 \end{pmatrix}.
$$

By choosing $S = \begin{pmatrix} 1 & -1 \\ 0 & 1 \end{pmatrix}$ in the admissible transformation (4), we obtain the CCF:

$$
\begin{array}{ccccc}
x_3 & x_4 & x_1 & x_2 & x_5 \\
\end{array}
$$

$$
\bar{A} = \begin{pmatrix} 1 & 1 & & 2 & \\ & & 1 & -1 & 0 \\ & & t_1 & 0 & t_2 \\ & & 0 & t_3 & t_4 \end{pmatrix}
$$

with a nonempty horizontal tail $C_0 = \{x_3, x_4\}$, a single ($b = 1$) square block $C_1 = \{x_1, x_2, x_5\}$, and an empty vertical tail $C_\infty = \emptyset$. It is not difficult to verify that $L(p_R) = \{C_0, C_0 \cup C_1\}$. Example 7 in §6 will illustrate how the CCF, as well as the matrix S, can be found efficiently. □

Here we mention an extension of the notion of LM-matrix and its CCF when the base field is replaced by a ring. Let D be an integral domain [52], and K the field of quotients of D; it is still assumed that K is a subfield of F. We say that a matrix $A = \binom{Q}{T}$ is an LM-matrix with respect to F/D, denoted as $A \in \text{LM}(F/D)$, if $A \in \text{LM}(F/K)$ and furthermore, Q is a matrix over D. Accordingly the admissible transformation over D is defined to be a transformation of the form (4) with S being a matrix over D with det $S \neq 0$. Then the matrix resulting from this transformation is again an LM-matrix with respect to F/D. Note, however, that an admissible transformation over D is not always invertible since the inverse of S may not exist among the matrices over D. The matrix S has its inverse $S-1$ over D if and only if $\det S$ is an invertible element of D, in which case S is called *unimodular* over D.

It follows easily from Theorem 3.1 (see also Example 5 below) that for $A \in$ LM(F/D) there exists an admissible transformation over D, which is not necessarily invertible, such that the resulting matrix $\bar{A}$ agrees with a CCF of A as an LM-matrix with respect to F/K.

When the invertibility is imposed upon the admissible transformation, we can still claim a similar statement when D is a principal ideal domain (PID) [52]; the ring of integers Z and the ring of univariate polynomials over a field are typical examples of a PID. It should be clear that a linear extension of a partial order means a linear order (=total order) that is compatible with the partial order, also called a topological sorting in computer science. Our indexing convention (20) for the blocks $\{C_k\}$ in the CCF of A represents a linear extension of the partial order $\preceq$ in the CCF. The following fact was observed by Murota [36] (see also [39]) in the case where D is a ring of polynomials. The proof will be given later in the next subsection.

THEOREM 3.2. *Let A be an LM-matrix with respect to F/D, where D is a PID. Let $\{C_k\}_{k=0}^{\infty}$ denote the partition of C in the CCF of A and $\preceq$ the partial order among the blocks (using the notation of Theorem 3.1). For any linear extension of $\preceq$, which is represented by the linear order of the index k of the blocks, there exist permutation matrices P_r and P_c, a unimodular matrix S over D, and a CCF $\hat{A}$ of A (as an LM-matrix with respect to F/K) such that*

$$\hat{A} = P_r \begin{pmatrix} S & O \\ O & I \end{pmatrix} \begin{pmatrix} Q \\ T \end{pmatrix} P_c$$

is in the same block-triangular form as $\bar{A}$, having the same diagonal blocks, i.e., $\hat{A}[R_k, C_l] = \bar{A}[R_k, C_l] = O$ for $k > l$ and $\hat{A}[R_k, C_k] = \bar{A}[R_k, C_k]$ for $k = 0, 1, \ldots, b, \infty$. (It is not claimed that $\hat{A}[R_k, C_l]$ coincides with $\bar{A}[R_k, C_l]$ for $k < l$.) □

EXAMPLE 5. Let $D = Z$, $K = Q$ and $F = Q(t_1, t_2)$, where t_1 and t_2 are indeterminates. Consider a 3×3 LM-matrix with respect to F/Z:

$$A = \begin{array}{c} \\ \\ \end{array} \begin{array}{ccc} x_1 & x_2 & x_3 \\ \end{array} \\ \begin{pmatrix} 2 & -2 & -4 \\ 3 & 1 & 2 \\ 0 & t_1 & t_2 \end{pmatrix}.$$

First regard A as a member of LM(F/Q). By choosing $S = S_1 = \begin{pmatrix} 1/4 & 1/2 \\ -3/2 & 1 \end{pmatrix}$ (with $\det S_1 = 1$) in the admissible transformation (4) we obtain a CCF:

$$\bar{A}_1 = \begin{array}{ccc} x_1 & x_2 & x_3 \\ \end{array} \\ \begin{pmatrix} 2 & & \\ & 4 & 8 \\ & t_1 & t_2 \end{pmatrix},$$

which has two square blocks $C_1 = \{x_1\}$ and $C_2 = \{x_2, x_3\}$ with no order relation between them.

The transformation using $S = S_1$ is not admissible over $\mathbf{Z}$. However, an admissible transformation over $\mathbf{Z}$ can be constructed easily by putting $S = S_2 = 4 \cdot S_1$, which yields another CCF:

$$\bar{A}_2 = \begin{pmatrix} \begin{array}{ccc} x_1 & x_2 & x_3 \\ 8 & & \\ & 16 & 32 \\ & t_1 & t_2 \end{array} \end{pmatrix}.$$

It is noted however that the admissible transformation with $S = S_2$ is not invertible since S_2 is not unimodular with $\det S_2 = 16$.

Restricting S to a unimodular matrix over $\mathbf{Z}$, we may take $S = S_3 = \begin{pmatrix} -1 & 1 \\ -3 & 2 \end{pmatrix}$ (with $\det S_3 = 1$) to transform A to a block-triangular matrix

$$\hat{A} = \begin{pmatrix} \begin{array}{ccc} x_1 & x_2 & x_3 \\ 1 & 3 & 6 \\ & 8 & 16 \\ & t_1 & t_2 \end{array} \end{pmatrix}$$

with order relation $C_1 = \{x_1\} \preceq C_2 = \{x_2, x_3\}$. This matrix $\hat{A}$ has the same diagonal blocks with

$$\bar{A}_3 = \begin{pmatrix} \begin{array}{ccc} x_1 & x_2 & x_3 \\ 1 & & \\ & 8 & 16 \\ & t_1 & t_2 \end{array} \end{pmatrix},$$

which is another CCF of A obtained with $S = S_3 = \begin{pmatrix} 1/8 & 1/4 \\ -3 & 2 \end{pmatrix}$. $\qquad\square$

3.2. Construction of CCF.

This subsection gives a sketch of the constructive proof of Theorem 3.1. A complete proof can be found in [26], [28], [41]. It should be emphasized that the following mathematical construction of the CCF can be polished up to a practically efficient algorithm, which will be described in §6.

First note that the admissible transformation (4) for $A \in \mathrm{LM}(\boldsymbol{F}/\boldsymbol{K}; m_Q, m_T, n)$ is equivalent to

$$P_{\mathrm{r}} \begin{pmatrix} S & O \\ O & P_T \end{pmatrix} \begin{pmatrix} Q \\ T \end{pmatrix} P_{\mathrm{c}},$$

which contains another permutation matrix P_T. In what follows we will find these four matrices $P_{\mathrm{c}}, S, P_T, P_{\mathrm{r}}$ that bring about the CCF.

[Matrix P_{c}]: As has been explained in §3.1, the submodular function p_R determines a sublattice $L(p_R)$, which in turn yields a pair

$$(21) \qquad \mathcal{P}(L(p_R)) = (\{C_0; C_1, \ldots, C_b; C_\infty\}, \preceq)$$

of a partition of $C = \mathrm{Col}(A) = \mathrm{Col}(Q) = \mathrm{Col}(T)$ and a partial order (see (19)). Recall the relation (18): $X_k = \cup_{l=0}^k C_l$ $(0 \leq k \leq b)$ as well as (20). The permutation matrix P_c is such that the column set C is reordered as $C_0, C_1, \ldots, C_b, C_\infty$, where the ordering within each block is arbitrary.

[Matrix S]: We use a short-hand notation $\rho(J) = \rho(R_Q, J) = \operatorname{rank} Q[R_Q, J]$ for $J \subseteq C$. Put $Q_0 = Q P_c$, where $\operatorname{Col}(Q_0)$ is identified with C through permutation P_c. Since $Q_0[\operatorname{Row}(Q_0), C_0]$ contains $\rho(X_0)$ independent row vectors and the others are linearly dependent on them, we can find a nonsingular matrix $S_0 \in \operatorname{GL}(m_Q, \boldsymbol{K})$ such that $Q_1 = S_0 Q_0$ satisfies

$$Q_1[\operatorname{Row}(Q_1) - R_{Q0}, C_0] = O,$$

$$\operatorname{rank} Q_1[R_{Q0}, C_0] = |R_{Q0}| = \rho(X_0)$$

for some $R_{Q0} \subseteq \operatorname{Row}(Q_1)$; that is,

$$Q_1 = S_0 Q_0 = \begin{array}{c} \\ R_{Q0} \\ \overline{R_{Q0}} \end{array} \begin{pmatrix} \overset{C_0}{[\equiv]} & \overset{\overline{C_0}}{*} \\ O & * \end{pmatrix},$$

where $\overline{R_{Q0}} = \operatorname{Row}(Q_1) - R_{Q0}$, $\overline{C_0} = C - C_0$ and $[\equiv]$ indicates a submatrix with independent rows.

Next, since $Q_1[\operatorname{Row}(Q_1), C_0 \cup C_1]$ contains $\rho(X_1)$ independent row vectors, we can find $S_1 \in \operatorname{GL}(m_Q, \boldsymbol{K})$ such that $Q_2 = S_1 Q_1$ satisfies

$$Q_2[\operatorname{Row}(Q_2) - R_{Q0}, C_0] = O,$$

$$Q_2[\operatorname{Row}(Q_2) - (R_{Q0} \cup R_{Q1}), C_0 \cup C_1] = O;$$

$$\begin{aligned} \operatorname{rank} Q_2[R_{Q0}, C_0] &= |R_{Q0}| = \rho(X_0), \\ \operatorname{rank} Q_2[R_{Q1}, C_1] &= |R_{Q1}| = \rho(X_1) - \rho(X_0) \end{aligned}$$

for some $R_{Q0}, R_{Q1} \subseteq \operatorname{Row}(Q_2)$ with $R_{Q0} \cap R_{Q1} = \emptyset$. That is,

$$Q_2 = S_1 Q_1 = \begin{array}{c} \\ R_{Q0} \\ R_{Q1} \\ \overline{R_{Q0} \cup R_{Q1}} \end{array} \begin{pmatrix} \overset{C_0}{[\equiv]} & \overset{C_1}{\triangle} & \overset{\overline{C_0 \cup C_1}}{*} \\ O & [\equiv] & * \\ O & O & * \end{pmatrix}.$$

We may further impose that

(22)   The nonzero row vectors of $Q_2[R_{Q0}, C_1]$ (indicated by $\triangle$ above) are linearly independent of the row vectors of $Q_2[R_{Q1}, C_1]$,

for otherwise we could eliminate the former with the latter.

Continuing such sweep-out operations, we can find a nonsingular matrix $S \in \operatorname{GL}(m_Q, \boldsymbol{K})$ and a partition of $\operatorname{Row}(\bar{Q})$:

(23) $(R_{Q0}; R_{Q1}, \ldots, R_{Qb}; R_{Q\infty})$

such that $\bar{Q} = SQP_c$ satisfies

(24) $$\bar{Q}[R_{Ql}, C_k] = O \quad (0 \le k < l \le \infty);$$

(25)
$$
\begin{aligned}
\text{rank } \bar{Q}[R_{Q0}, C_0] &= |R_{Q0}| = \rho(X_0), \\
\text{rank } \bar{Q}[R_{Qk}, C_k] &= |R_{Qk}| = \rho(X_k) - \rho(X_{k-1}) \quad (k = 1, \ldots, b), \\
|R_{Q\infty}| &= m_Q - \rho(X_b).
\end{aligned}
$$

We may further impose that

(26) For $0 \le k < l \le \infty$, the nonzero row vectors of $\bar{Q}[R_{Qk}, C_l]$ are linearly independent of the row vectors of $\bar{Q}[R_{Ql}, C_l]$.

[Matrix P_T]: Define a partition of R_T:

(27) $$(R_{T0}; R_{T1}, \ldots, R_{Tb}; R_{T\infty})$$

by

(28)
$$
\begin{aligned}
R_{T0} &= \Gamma(R_T, X_0), \\
R_{Tk} &= \Gamma(R_T, X_k) - \Gamma(R_T, X_{k-1}) \quad (k = 1, \ldots, b), \\
R_{T\infty} &= \text{Row}(T) - \Gamma(R_T, X_b)
\end{aligned}
$$

using the Γ of (6). Let P_T be a permutation matrix which permutes $\text{Row}(T)$ compatibly with (27). Then $\bar{T} = P_T T P_c$ is in an explicit block-triangular form:

(29) $$T[R_{Tl}, C_k] = \bar{T}[R_{Tl}, C_k] = O \quad (0 \le k < l \le \infty),$$

where we identify $\text{Row}(\bar{T}) = \text{Row}(T)$ and $\text{Col}(\bar{T}) = \text{Col}(T) = C$.

[Matrix P_r]: So far we have constructed two block-triangular matrices $\bar{Q}$ and $\bar{T}$, the former being block-triangularized with respect to the partitions (21) and (23) and the latter with respect to (21) and (27). Put these two matrices together:

$$\bar{A} = \begin{pmatrix} \bar{Q} \\ \bar{T} \end{pmatrix},$$

and consider a partition of $\text{Row}(\bar{A})$:

(30) $$(R_0; R_1, \ldots, R_b; R_\infty),$$

where $R_k = R_{Qk} \cup R_{Tk}$ for $k = 0, 1, \ldots, b, \infty$. By (24) and (29), $\bar{A}$ is (essentially) block-triangularized with respect to the partitions (21) and (30), namely,

$$\bar{A}[R_l, C_k] = O \quad (0 \le k < l \le \infty).$$

The matrix P_r is to rearrange $\text{Row}(\bar{A})$ compatibly with (30).

The block-triangular matrix $\bar{A}$ constructed in this way is obviously LM-equivalent to A. Based on the rank formula of Theorem 2.5 we can show that this matrix enjoys the properties (B2) to (B4). We will indicate the essense here, referring the reader to [28], pp. 177–179, for the complete proof. In addition to $\rho(J)$ we use another short-hand notation $\gamma(J) = \gamma(R_T, J)$ for $J \subseteq C$.

Consider the horizontal tail $\bar{A}[R_0, C_0]$. Since $C_0 \in L(p_R)$, $\rho(C_0) = |R_{Q0}|$ and $\gamma(C_0) = |R_{T0}|$, we have

$$0 = p_R(\emptyset) \geq \min p_R = p_R(C_0) = \rho(C_0) + \gamma(C_0) - |C_0| = |R_0| - |C_0|,$$

which, combined with Theorem 2.5, implies

$$\text{rank } \bar{A}[R_0, C_0] = \text{rank } \bar{A}[\text{Row}(\bar{A}), C_0] = \text{rank } A[R, C_0] = \min p_R + |C_0| = |R_0|.$$

This shows in particular that $|R_0| \leq |C_0|$. If the equality holds here, then $p_R(\emptyset) = \min p_R$, i.e., $\emptyset \in L(p_R)$. Since $C_0 = \min L(p_R)$, this implies $C_0 = \emptyset$ and therefore $R_0 = \emptyset$. Hence follow (B2) and (B3) for the horizontal tail.

For the first square block $\bar{A}[R_1, C_1]$ we note that $p_R(C_0) = p_R(C_0 \cup C_1) = \min p_R$. This shows

$$
\begin{aligned}
|R_1| &= |R_0 \cup R_1| - |R_0| \\
&= (\rho(C_0 \cup C_1) - \rho(C_0)) + (\gamma(C_0 \cup C_1) - \gamma(C_0)) \\
&= p_R(C_0 \cup C_1) - p_R(C_0) + |C_1| = |C_1|.
\end{aligned}
$$

It also follows from Theorem 2.5, as well as the relation: $\min p_R = |R_0| - |C_0|$ shown above, that

$$
\begin{aligned}
\text{rank } \bar{A}[R_0 \cup R_1, C_0 \cup C_1] &= \text{rank } \bar{A}[\text{Row}(\bar{A}), C_0 \cup C_1] \\
&= \text{rank } A[R, C_0 \cup C_1] \\
&= \min p_R + |C_0| + |C_1| = |R_0| + |R_1|.
\end{aligned}
$$

This shows $\text{rank } \bar{A}[R_1, C_1] = |R_1|$ since $\text{rank } \bar{A}[R_0, C_0] = |R_0|$.

The conditions (B2) and (B3) for the remaining blocks can be shown similarly. The invariance of p_R explained after (13) is the key to prove the uniqueness (B4); see [28], p. 179. We may mention that the argument above conforms with the Jordan-Hölder type decomposition principle for submodular functions developed by Iri [19], Nakamura [43], Tomizawa [51].

A similar argument establishes Theorem 3.2, which is concerned with the block-triangularization with respect to a unimodular transformation over a PID. The Hermite normal form [44], [50] under a unimodular transformation guarantees the existence of a unimodular matrix S such that $\bar{Q} = SQP_c$ satisfies (24) and (25). However, we cannot impose the further condition (22) or (26), which fact causes the discrepancy in the upper off-diagonal blocks of $\hat{A}$ and $\bar{A}$.

4. Irreducibility. In this section we investigate into the notion of LM-irreducibility. Most of the results below are natural extensions of the results concerning the full indecomposability (or DM-irreducibility) of a generic matrix. See Schneider [49] for a historical account on the notion of full indecomposability.

First recall the definition (see §1) of the LM-irreducibility with respect to the admissible transformation. Namely, an LM-matrix A is LM-irreducible (or simply irreducible) if it does not split into more than one nonempty block under the admissible transformation, or more precisely, if any LM-matrix A' that is LM-equivalent to A is fully indecomposable.

With reference to the CCF, $\bar{A}$, of A, we see that each block $\bar{A}[R_k, C_k]$ of the CCF is irreducible ($k = 0, 1, \ldots, b, \infty$) using the notation of Theorem 3.1. Hence, A is irreducible if (a) $b = 1$ and $C_0 = R_\infty = \emptyset$, (b) $b = 0$ and $R_\infty = \emptyset$, or (c) $b = 0$ and $C_0 = \emptyset$.

Combining this observation with Theorem 3.1(B1) we obtain the following characterization of LM-irreducibility in terms of the lattice $L(p_R)$ of minimizers of p_R. This is a kind of "dual" characterization of the LM-irreducibility as opposed to the "primal" characterization (definition) in terms of the indecomposability with respect to the admissible transformation.

THEOREM 4.1. *Let* $A \in \mathrm{LM}(\boldsymbol{F}/\boldsymbol{K})$.

(a) *In case* $|R| = |C|$*:* A *is LM-irreducible* $\Longleftrightarrow$ $L(p_R) = \{\emptyset, C\}$*;*

(b) *In case* $|R| < |C|$*:* A *is LM-irreducible* $\Longleftrightarrow$ $L(p_R) = \{C\}$*;*

(c) *In case* $|R| > |C|$*:* A *is LM-irreducible* $\Longleftrightarrow$ $L(p_R) = \{\emptyset\}$*.* □

This characterization will be rephrased in a more algorithmic statement later in Theorem 6.1.

The following theorem refers to the rank of submatrices of an LM-irreducible matrix. This is an extension of the result due to Marcus-Minc [22] and to Brualdi [4] for a generic matrix (cf. p.112 of [5]); see also Theorem 4.2.2 of [5].

THEOREM 4.2. *Let* $A \in \mathrm{LM}(\boldsymbol{F}/\boldsymbol{K})$ *be LM-irreducible.*

(a) *In case* $|R| = |C|$*:* $\mathrm{rank}\, A[R - \{i\}, C - \{j\}] = |R| - 1$ $(\forall i \in R, \forall j \in C)$*;*

(b) *In case* $|R| < |C|$*:* $\mathrm{rank}\, A[R, C - \{j\}] = |R|$ $(\forall j \in C)$*;*

(c) *In case* $|R| > |C|$*:* $\mathrm{rank}\, A[R - \{i\}, C] = |C|$ $(\forall i \in R)$*.*

Proof. (a) Put $R' = R - \{i\}$, $C' = C - \{j\}$ and suppose that $A[R', C']$ were singular. Then, by Theorem 2.5, $p(R', J') \leq -1$ for some J' $(\emptyset \neq J' \subseteq C')$. On the other hand, it follows from

$$p(R, J') - p(R', J') = \begin{cases} \rho(R_Q, J') - \rho(R_Q - \{i\}, J') & (\text{if } i \in R_Q) \\ \gamma(R_T, J') - \gamma(R_T - \{i\}, J') & (\text{if } i \in R_T) \end{cases}$$

that $p(R, J') - p(R', J') \leq 1$. Hence $p(R, J') \leq p(R', J') + 1 \leq 0$, which would imply $J' \in L(p_R)$, a contradition to Theorem 4.1(a). The proofs for (b), (c) are similar; see [29]. □

As immediate corollaries we obtain the following properties of a nonsingular irreducible LM-matrix. We regard the determinant of $A \in \mathrm{LM}(\boldsymbol{F}/\boldsymbol{K})$ as a polynomial in $\mathcal{T}$ (=set of nonzero entries of T) with coefficients from the base field $\boldsymbol{K}$.

THEOREM 4.3. *Let* $A \in \mathrm{LM}(\boldsymbol{F}/\boldsymbol{K})$ *be nonsingular and LM-irreducible.*

(1) A^{-1} *is completely dense, i.e.,* $(A^{-1})_{ji} \neq 0$*,* $\forall (i, j)$*.*

(2) *Each element of T appears in* det A. □

The following theorem of Murota [29] states to the effect that the combinatorial irreducibility (namely LM-irreducibility) is essentially equivalent to the algebraic irreducibility of the determinant. This is an extension of the result of Frobenius [12] for a generic matrix (see also [5], [47], [48], [49]).

THEOREM 4.4. *Let $A \in \mathrm{LM}(F/K)$ be nonsingular. The determinant* det A *is an irreducible polynomial in the ring $K[T]$ if A is LM-irreducible. Conversely, if* det A *is an irreducible polynomial, then there exists in the CCF of A at most one diagonal block which contains elements of T and all the other diagonal blocks are 1×1 matrices over K.*

Proof. The proof for the first half is long; see [29]. The second half follows easily from Theorem 3.1 and Theorem 4.3(2). □

A minor (=subdeterminant) of $A \in \mathrm{LM}(F/K)$ is also a polynomial in T over K. Let $d_k(T) \in K[T]$ denote the k-th determinantal divisor of A, i.e., the greatest common divisor of all minors of order k in A as polynomials in T over K. Note that $d_k(T) \in K* = K - \{0\}$ means $d_k(T)$ is a "constant" free from any variables in T.

THEOREM 4.5. *Let $A \in \mathrm{LM}(F/K)$ be LM-irreducible.*

(a) *In case $|R| = |C|$: $d_k(T) \in K*$ for $k = 1, \ldots, |R| - 1$;*

(b) *In case $|R| < |C|$: $d_k(T) \in K*$ for $k = 1, \ldots, |R|$;*

(c) *In case $|R| > |C|$: $d_k(T) \in K*$ for $k = 1, \ldots, |C|$.*

Proof. (a) It suffices to show that $d_k(T)$ is free from any $t \in T$ for $k = |R| - 1$. Suppose t appears at position (i, j). It follows from Theorem 4.2(a) that $\delta \equiv \det A[R - \{i\}, C - \{j\}] \neq 0$. Obviously δ does not contain t, and, a fortiori, $d_k(T)$ does not contain t, since $d_k(T)$ is a divisor of δ. (b) and (c) can be proven similarly using Theorem 4.2. □

For a general (reducible) LM-matrix Theorems 3.1, 4.4 and 4.5 together imply the following.

THEOREM 4.6. *Let r be the rank of $A \in \mathrm{LM}(F/K)$. Then $d_k(T) \in K*$ for $k = 1, \ldots, r - 1$, and the decomposition of the r-th determinantal divisor $d_r(T)$ of A into irreducible factors in the ring $K[T]$ is given by*

$$d_r(T) = \alpha \cdot \prod_{l=1}^{b} \det \bar{A}[R_l, C_l],$$

where $\bar{A}[R_l, C_l]$ $(l = 1, \ldots, b)$ are the irreducible square blocks in the CCF of A, and $\alpha \in K*$. (Exactly speaking, those factors on the right-hand side which belong to K should not be counted as irreducible factors in $K[T]$ since they are invertible elements in $K[T]$.) □

5. Further Properties

5.1. Principal structure of LM-matrices. A submatrix $A[I, C]$ (with $I \subseteq R$) of an LM-matrix A is again an LM-matrix, for which the CCF is defined. The CCF of $A[I, C]$, in turn, defines a partition of C, which varies with I. Let us denote by $\mathcal{P}_{\text{CCF}}(I)$ (cf. (21)) the pair of the partition of C and the partial order among the blocks in the CCF of the submatrix $A[I, C]$. Here we are interested in the family $\{\mathcal{P}_{\text{CCF}}(I) \mid I \in \mathcal{B}\}$ of partitions, where

$$\mathcal{B} = \{I \subseteq R \mid \operatorname{rank} A = \operatorname{rank} A[I, C] = |I|\},$$

which denotes the family of row-bases of A. The theorem below gives a concise characterization of the coarsest common refinement of $\{\mathcal{P}_{\text{CCF}}(I) \mid I \in \mathcal{B}\}$ in terms of the submodular function p_R associated with the whole matrix A.

The characterization refers to the notion of "principal structure of a submodular system" introduced by Fujishige [14], [15]. For $j \in C$ consider the family L_j of the minimizers of p_R over $\{J \subseteq C \mid J \ni j\}$:

$$L_j = \{J \subseteq C \mid J \ni j; \quad p_R(J) \le p_R(J') \quad \forall J' \ni j\},$$

which forms a sublattice of $2C$ because of the submodularity (13) of p_R. Denote by $D(j)$ the (uniquely determined) smallest set of L_j. The binary relation $\preceq_{p_R}$ on C defined by $[i \preceq_{p_R} j \iff i \in D(j)]$, or equivalently by $[i \preceq_{p_R} j \iff D(i) \subseteq D(j)]$, is a quasi-order, being reflexive and transitive. Then the equivalence relation defined by $[i \preceq_{p_R} j$ and $j \preceq_{p_R} i]$ determines a partition of C into blocks, among which a partial order is induced from the original quasi-order. This is called the *principal structure*, to be denoted as $\mathcal{P}_{\text{PS}}$, of the submodular system of (C, p_R).

The following theorem of Murota [32] shows that the coarsest common refinement of $\{\mathcal{P}_{\text{CCF}}(I) \mid I \in \mathcal{B}\}$ agrees with the principal structure of the submodular system of (C, p_R).

THEOREM 5.1.

$$\mathcal{P}_{\text{PS}} = \bigwedge_{I \in \mathcal{B}} \mathcal{P}_{\text{CCF}}(I),$$

where the right-hand side designates the coarsest partition of C which is finer than all $\mathcal{P}_{\text{CCF}}(I)$ with $I \in \mathcal{B}$. □

In view of the correspondence (as explained in §3.1) between the family of partitions $\{\mathcal{P}_{\text{CCF}}(I) \mid I \in \mathcal{B}\}$ and the family of sublattices $\{L(p_I) \mid I \in \mathcal{B}\}$, we can think of this theorem as a characterization of the sublattice generated by $\{L(p_I) \mid I \in \mathcal{B}\}$.

The essential content of the above theorem for the special case of a generic matrix $A = T$ (with $m_Q = 0$) has been obtained by McCormick [23] (without reference to the notion of principal structure).

EXAMPLE 6. Consider a 5×3 LM-matrix (with base field $\mathbf{Q}$):

$$A = \begin{array}{c} \\ r_1 \\ r_2 \\ r_3 \\ r_4 \\ r_5 \end{array} \begin{pmatrix} \begin{array}{ccc} x_1 & x_2 & x_3 \\ 1 & 2 & 1 \\ 1 & 1 & -1 \\ 0 & t_1 & t_2 \\ 0 & t_3 & t_4 \\ t_5 & t_6 & 0 \end{array} \end{pmatrix},$$

where $C = \{x_1, x_2, x_3\}$, $R = \{r_1, r_2, r_3, r_4, r_5\}$, and t_i $(i = 1, \ldots, 6)$ are indeterminates. This matrix is LM-irreducible, the whole matrix being a vertical tail.

For a nonsingular submatrix $A[I, C]$ with $I = \{r_1, r_2, r_3\}$, we obtain its CCF

$$\begin{array}{c} \\ r_1 \\ r_2 \\ r_3 \end{array} \begin{pmatrix} \begin{array}{ccc} x_1 & x_2 & x_3 \\ 1 & 2 & 1 \\ & -1 & -2 \\ & t_1 & t_2 \end{array} \end{pmatrix}$$

by subtracting row r_1 from row r_2 in $A[I, C]$. Hence, $\mathcal{P}_{\mathrm{CCF}}(I)$ is given by $\{x_1\} \prec \{x_2, x_3\}$.

By inspection we see that $\mathcal{B} = \{I \subset R \mid |I| = 3\}$. $\mathcal{P}_{\mathrm{CCF}}(I)$ for all $I \in \mathcal{B}$ are given as follows.

$I \in \mathcal{B}$	$\mathcal{P}_{\mathrm{CCF}}(I)$
$\{r_1, r_2, r_5\}$	$\{x_3\} \prec \{x_1, x_2\}$
$\{r_i, r_j, r_5\}$ $(i = 1, 2; j = 3, 4)$	$\{x_1, x_2, x_3\}$
Otherwise	$\{x_1\} \prec \{x_2, x_3\}$

This shows that $\bigwedge_{I \in \mathcal{B}} \mathcal{P}_{\mathrm{CCF}}(I)$ is given by $\{x_1\} \prec \{x_2\}$, $\{x_3\} \prec \{x_2\}$. On the other hand, we have $D(x_1) = \{x_1\}$, $D(x_2) = C$, $D(x_3) = \{x_3\}$ since $p_R(\emptyset) = 0$, $p_R(\{x_1\}) = 1$, $p_R(\{x_2\}) = 3$, $p_R(\{x_3\}) = 2$, $p_R(\{x_1, x_2\}) = 3$, $p_R(\{x_1, x_3\}) = 3$, $p_R(\{x_2, x_3\}) = 3$, $p_R(C) = 2$. Hence $\mathcal{P}_{\mathrm{PS}}$ agrees with $\{x_1\} \prec \{x_2\}$, $\{x_3\} \prec \{x_2\}$. Note also that $\mathcal{P}_{\mathrm{PS}} \neq \mathcal{P}_{\mathrm{CCF}}(I)$ for each $I \in \mathcal{B}$. □

5.2. Properties of mixed matrices. In this subsection $A = Q + T$ denotes a mixed matrix with respect to $\mathbf{F}/\mathbf{K}$, with $\mathrm{Row}(A) = R$ and $\mathrm{Col}(A) = C$.

If A is nonsingular, it can be decomposed into LU-factors as $P_r A P_c = L U$ with suitable permutation matrices P_r and P_c. In general the entries of the matrices L and U are rational functions in $\mathcal{T}$ (=set of nonzero entries of T) over $\mathbf{K}$. If all the diagonal entries of L and U belong to $\mathbf{K}$, then obviously $\det A \in \mathbf{K}$. The following theorem of Murota [24] asserts that the converse is also true (see [24], [28] for the proof). Recall the notation $\mathbf{K}* = \mathbf{K} - \{0\}$.

THEOREM 5.2. *Let $A = Q + T$ be a mixed matrix with base field $\mathbf{K}$. Then $\det A \in \mathbf{K}*$ if and only if there exist permutation matrices P_r and P_c, and LU-factors L and U: $P_r A P_c = L U$ such that (i) $L_{ii} = 1$ and $L_{ij} = 0$ for $i < j$; (ii) U_{ij} is a polynomial (of degree at most one) in $\mathcal{T}$ over $\mathbf{K}$ for $i > j$, $U_{ii} \in \mathbf{K}*$, and $U_{ij} = 0$ for $i > j$.* □

The final theorem of this section is an extension of the "determinantal version of the Frobenius-König theorem" due to Hartfiel-Loewy [17], who established it in the case where A is a square mixed matrix. Their original proof (for square case) is quite involved based on factorizations of determinants. Here we provide an alternative proof using the rank formula of Theorem 2.5 for LM-matrices.

THEOREM 5.3. *Let* $A = Q + T$ *be a mixed matrix. Then* rank $A < |C|$ *if and only if* (i) $T[I, J] = O$ *and* (ii) rank $Q[I, J] < |I| + |J| - |R|$ *for some* $I \subseteq R$ *and* $J \subseteq C$. *Similarly,* rank $A < |R|$ *if and only if* (i) $T[I, J] = O$ *and* (ii) rank $Q[I, J] < |I| + |J| - |C|$ *for some* $I \subseteq R$ *and* $J \subseteq C$.

Proof. Consider the LM-matrix $\tilde{A}$ of (3) for A and let $p : 2^{\tilde{R}} \times 2^{R \cup C} \to \mathbf{Z}$ be the function defined for $\tilde{A}$ as in (11), where $\tilde{R} = \text{Row}(\tilde{A})$ and we identify $\text{Col}(\tilde{A})$ with $R \cup C$. Then

$$p(\tilde{R}, (R - I) \cup J) = \text{rank } Q[I, J] + |(R - I) \cup \Gamma(R, J)| - |J|$$

with $\Gamma(R, J) = \{i \in R \mid \exists j \in J : T_{ij} \neq 0\}$ (cf. (6)), where it should be clear that $I \subseteq \text{Col}(\tilde{A})$ on the left-hand side and $I \subseteq \text{Row}(A)$ on the right-hand side. Theorem 2.5 applied to $\tilde{A}$ implies that rank $\tilde{A} < |R| + |C|$ if and only if $p(\tilde{R}, (R - I) \cup J) < 0$ for some $I \subseteq R$ and $J \subseteq C$. In the latter condition we may assume that $I \cap \Gamma(R, J) = \emptyset$, i.e., (i) $T[I, J] = O$, and then $p(\tilde{R}, (R - I) \cup J) < 0$ reduces to (ii) rank $Q[I, J] < |I| + |J| - |R|$. The proof for the first claim is completed by the obvious relation: rank $\tilde{A} = \text{rank } A + |R|$. The second claim follows from the first applied to A transposed. □

Most of the results for an LM-matrix can be carried over to those for a mixed matrix by way of the correspondence (3). In particular, we define an admissible transformation for a mixed matrix A to be a transformation of the form: $S \, A \, P_c$, where S is a nonsingular matrix over $\mathbf{K}$ and P_c a permutation matrix. See [28] and [29].

6. Algorithm for CCF. An efficient algorithm is described here which computes the CCF of an LM-matrix $A \in \text{LM}(\mathbf{F}/\mathbf{K}; m_Q, m_T, n)$ in $O(n3 \log n)$ time with arithmetic operations in the subfield $\mathbf{K}$ only, where $m = m_Q + m_T = O(n)$ is assumed for simplicity in this complexity bound. This section is an improved presentation of §3.2 of Murota [38].

In order to illustrate a connection between the CCF and the Dulmage-Mendelsohn decomposition we first restrict ourselves to a nonsingular LM-matrix A. In this case the CCF can be found as follows.

[Algorithm (outline) for the CCF of a nonsingular A]

Step 1: Find $J \subseteq C$ such that both $Q[R_Q, J]$ and $T[R_T, C - J]$ are nonsingular (such J exists by Lemma 2.3).

Step 2: Let S denote the inverse of $Q[R_Q, J]$ and put

$$A' := \begin{pmatrix} S & O \\ O & I \end{pmatrix} A.$$

Step 3: Find the Dulmage-Mendelsohn decomposition $\bar{A}$ of A', namely, $\bar{A} := P_r A' P_c$ with suitable permutation matrices P_r and P_c. ($\bar{A}$ is the CCF of A.) □

The first step (Step 1) is nothing but the well-studied problem of matroid partition and a number of efficient algorithms are available for it; see Edmonds [9] and Lawler [20]. The DM-decomposition in the last step (Step 3) can be computed by first finding a maximum (perfect) matching in the bipartite graph associated with A', i.e., the graph denoted as $G(A')$ at the beginning of §3.1, and then decomposing an auxiliary digraph into strongly connected components. See, e.g., [5], [21], [28] for more detail on the DM-decomposition.

For the LM-matrix of Example 3, which is nonsingular, we can take $J = \{\xi_5, \xi_3, \xi_4, \eta_4, \eta_3\}$ in Step 1. The transformation matrix S given in Example 3 is equal to the inverse of

$$
Q[R_Q, J] = \begin{array}{c}
\begin{array}{ccccc} \xi_5 & \xi_3 & \xi_4 & \eta_4 & \eta_3 \end{array} \\
\begin{pmatrix}
1 & 1 & 1 & 0 & 0 \\
-1 & 0 & 0 & 0 & 0 \\
0 & -1 & 0 & 0 & 0 \\
0 & 0 & 0 & -1 & 0 \\
0 & 0 & 0 & -1 & 1
\end{pmatrix}
\end{array}.
$$

For a general (not necessarily nonsingular) LM-matrix it has been shown that the CCF can be constructed by identifying the minimum cuts in an independent-flow problem. See Prop. 20.1 of [28] as well as [41] for this reduction and Fujishige [13], [15] for independent-flow problems.

The detail of the algorithm for a general LM-matrix $A \in \mathrm{LM}(\boldsymbol{F}/\boldsymbol{K}; m_Q, m_T, n)$ is now described. As before let $R_T = \mathrm{Row}(T)$ and $C = \mathrm{Col}(A)$. Furthermore let C_Q be a disjoint copy of C, where the copy of $j \in C$ will be denoted as $j_Q \in C_Q$. The algorithm works with a directed graph $G = (V, B)$ with vertex set $V = R_T \cup C_Q \cup C$ and arc set $B = B_T \cup B_C \cup B+ \cup M$, where

$$
B_T = \{(i, j) \mid i \in R_T, j \in C, T_{ij} \neq 0\}, \quad B_C = \{(j_Q, j) \mid j \in C\},
$$

and $B+$ and M are sets of arcs which are defined and updated in the algorithm; $B+$ consists of arcs from C_Q to C_Q and M from C to $R_T \cup C_Q$. The set of end-vertices of M (vertices incident to an arc in M) will be designated as ∂M ($\subseteq V$). Besides the graph G we use two matrices (or two-dimensional arrays) P and S, as well as a vector (or one-dimensional array) $base$. The array P represents a matrix over $\boldsymbol{K}$, of size $m_Q \times n$, where $P = Q$ at the beginning of the algorithm (Step 1 below). The other array S is also a matrix over $\boldsymbol{K}$, of size $m_Q \times m_Q$, which is set to the unit matrix in Step 1 and finally gives the matrix S in the admissible transformation (4). Variable $base$ is a vector of size m_Q, which represents a mapping (correspondence): $R_Q \to C \cup \{0\}$.

[Algorithm for the CCF of a general A]

Step 1:
$\qquad M := \emptyset; \quad base[i] := 0 \ (i \in R_Q); \quad P[i, j] := Q_{ij} \ (i \in R_Q, j \in C);$
$\qquad S := \text{unit matrix of order } m_Q.$

Step 2:
$\qquad I := \{i \in C \mid i_Q \in \partial M \cap C_Q\};$

$J := \{j \in C - I \mid \text{For all } i,\ base[i] = 0 \text{ implies } P[i,j] = 0\}$;

$S+_T := R_T - \partial M$; $S+_Q := \{j_Q \in C_Q \mid j \in C - (I \cup J)\}$; $S+ := S+_T \cup S+_Q$;

$S- := C - \partial M$;

$B+ := \{(i_Q, j_Q) \mid h \in R_Q, j \in J, P[h,j] \neq 0, i = base[h]\}$;

If there exists in G a directed path from $S+$ to $S-$ then go to Step 3; otherwise (including the case where $S+ = \emptyset$ or $S- = \emptyset$) go to Step 4.

Step 3:

Let $L\ (\subseteq B)$ be (the set of arcs on) a shortest path from $S+$ to $S-$ ("shortest" in the number of arcs);

$M := (M - L) \cup \{(j,i) \mid (i,j) \in L \cap B_T\} \cup \{(j,j_Q) \mid (j_Q,j) \in L \cap B_C\}$;

If the initial vertex ($\in S+$) of the path L belongs to $S+_Q$, then do the following:

{Let $j_Q\ (\in S+_Q \subseteq C_Q)$ be the initial vertex;

Find h such that $base[h] = 0$ and $P[h,j] \neq 0$;

$\qquad\qquad\qquad\qquad\qquad$ [$j \in C$ corresponds to $j_Q \in C_Q$]

$base[h] := j$; $w := 1/P[h,j]$;

$P[k,l] := P[k,l] - w \times P[k,j] \times P[h,l]$ ($h \neq k \in R_Q, l \in C$);

$S[k,l] := S[k,l] - w \times P[k,j] \times S[h,l]$ ($h \neq k \in R_Q, l \in R_Q$) };

For all $(i_Q, j_Q) \in L \cap B+$ (in the order from $S+$ to $S-$ along L) do the following:

{Find h such that $i = base[h]$; $\qquad$ [$j \in C$ corresponds to $j_Q \in C_Q$]

$base[h] := j$; $w := 1/P[h,j]$;

$P[k,l] := P[k,l] - w \times P[k,j] \times P[h,l]$ ($h \neq k \in R_Q, l \in C$);

$S[k,l] := S[k,l] - w \times P[k,j] \times S[h,l]$ ($h \neq k \in R_Q, l \in R_Q$) };

Go to Step 2.

Step 4:

Let $V_\infty\ (\subseteq V)$ be the set of vertices reachable from $S+$ by a directed path in G;

Let $V_0\ (\subseteq V)$ be the set of vertices reachable to $S-$ by a directed path in G;

$C_0 := C \cap V_0$; $C_\infty := C \cap V_\infty$;

Let G' denote the graph obtained from G by deleting the vertices $V_0 \cup V_\infty$ (and arcs incident thereto);

Decompose G' into strongly connected components $\{V_\lambda \mid \lambda \in \Lambda\}\ (V_\lambda \subseteq V)$;

Let $\{C_k \mid k = 1, \ldots, b\}$ be the subcollection of $\{C \cap V_\lambda \mid \lambda \in \Lambda\}$ consisting of all the nonempty sets $C \cap V_\lambda$, where C_k's are indexed in such a way that for $l < k$ there does not exist a directed path in G' from C_k to C_l;

$R_0 := (R_T \cap V_0) \cup \{h \in R_Q \mid base[h] \in C_0\}$;

$R_\infty := (R_T \cap V_\infty) \cup \{h \in R_Q \mid base[h] \in C_\infty \cup \{0\}\}$;

$R_k := (R_T \cap V_k) \cup \{h \in R_Q \mid base[h] \in C_k\}\ (k = 1, \ldots, b)$;

$\bar{A} := P_r \begin{pmatrix} P \\ T \end{pmatrix} P_c$, where the permutation matrices P_r and P_c are determined so that the rows and the columns of $\bar{A}$ are ordered as $(R_0; R_1, \ldots, R_b; R_\infty)$ and $(C_0; C_1, \ldots, C_b; C_\infty)$, respectively. $\square$

The subsets $I \subseteq C$ and $J \subseteq C$ represent the structure of the matroid $\boldsymbol{M}(Q)$ defined by the matrix Q; I is an independent set in $\boldsymbol{M}(Q)$ (i.e., rank $Q[R_Q, I] = |I|$), whereas $J \cup I$ is the closure (cf. [53], [54]) of I (i.e., $J = \{j \in C - I \mid \text{rank } Q[R_Q, I \cup$

$\{j\}] = |I|$). On the other hand, $(\partial M \cap C) - I$ is an independent set in the other matroid $\boldsymbol{M}(T)$ defined by the matrix T. Hence $\partial M \cap C \ (= I \cup ((\partial M \cap C) - I) \,)$ is independent in $\boldsymbol{M}(Q) \vee \boldsymbol{M}(T)$. Since $\boldsymbol{M}(Q) \vee \boldsymbol{M}(T) = \boldsymbol{M}(A)$, by Lemma 2.4, we have rank $A[R, \partial M \cap C] = |M|$. At each execution of Step 3 the size of M increases by one, and at the termination of the algorithm we have the relation: rank $A = |M|$.

The matrix $\bar{A}$ is the CCF of the input matrix A, where $\{R_0; R_1, \ldots, R_b; R_\infty\}$ and $\{C_0; C_1, \ldots, C_b; C_\infty\}$ give the partitions of the row set and the column set, respectively. The partial order among the blocks is induced from the partial order among the strongly connected components $\{V_\lambda \mid \lambda \in \Lambda\}$.

The shortest path in Step 3 and the strongly connected components in Step 4 can be found in time linear in the size of the graph G, which is $\mathrm{O}((n + m)2)$, by means of the standard graph algorithms; see, e.g., [1].

The updates of P in Step 3 are the standard pivoting operations [16] on P, which is a matrix over the subfield $\boldsymbol{K}$. The sparsity of P should be taken into account in actual implementations of the algorithm; for example, $P[h, j] = 0$ if $base[h] = 0$ and $j \in I \cup J$. Computational techniques developed for solving sparse linear programs can be utilized here. As indicated in Step 3, pivoting operations are required for each arc $(i_Q, j_Q) \in L \cap B+$. It is important to traverse the path L from $S+$ to $S-$, not from $S-$ to $S+$, to avoid unnecessary fill-ins. When the transformation matrix S is not needed, it may simply be eliminated from the computation without any side effect.

The above algorithm will be efficient enough also for practical applications. It would be still more efficient if we first compute the DM-decomposition by purely graph-theoretic algorithm and then apply the above algorithm to each of the DM-irreducible components; such two-stage procedure works since the CCF is a refinement of the DM-decomposition.

Finally we mention a characterization of the LM-irreducibility in terms of the graph used in the algorithm.

THEOREM 6.1. *Let A be a square LM-matrix. A is LM-irreducible (and hence nonsingular) if and only if in Step 4 of the algorithm both V_0 and V_∞ are empty and graph $G' (= G)$ is strongly connected.* □

EXAMPLE 7. The algorithm above is illustrated here for a 4×5 LM-matrix $A = \binom{Q}{T}$ with

$$Q = \begin{array}{c} \\ \\ \end{array} \begin{pmatrix} x_1 & x_2 & x_3 & x_4 & x_5 \\ 1 & 1 & 1 & 1 & 0 \\ 0 & 2 & 1 & 1 & 0 \end{pmatrix}, \quad T = \begin{array}{c} \\ f_1 \\ f_2 \end{array} \begin{pmatrix} x_1 & x_2 & x_3 & x_4 & x_5 \\ t_1 & 0 & 0 & 0 & t_2 \\ 0 & t_3 & 0 & 0 & t_4 \end{pmatrix},$$

where $\mathrm{Col}(A) = C = \{x_1, x_2, x_3, x_4, x_5\}$ and $\mathrm{Row}(T) = R_T = \{f_1, f_2\}$. We work with a 2×5 matrix P, a 2×2 matrix S, and a vector $base$ of size 2. The copy of C is denoted as $C_Q = \{x_{1Q}, x_{2Q}, x_{3Q}, x_{4Q}, x_{5Q}\}$.

The flow of computation is traced below.

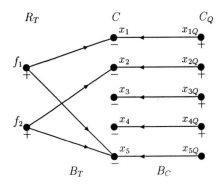

FIG. 1. *Graph* $G(0)$ *(+: vertices in $S+$; −: vertices in $S-$)*

Step 1: $M := \emptyset$;

$$base := \begin{matrix} r_1 \\ r_2 \end{matrix}\begin{pmatrix} 0 \\ 0 \end{pmatrix}, \quad P := \begin{matrix} r_1 \\ r_2 \end{matrix}\begin{matrix} \begin{matrix} x_1 & x_2 & x_3 & x_4 & x_5 \end{matrix} \\ \begin{pmatrix} 1 & 1 & 1 & 1 & 0 \\ 0 & 2 & 1 & 1 & 0 \end{pmatrix} \end{matrix}, \quad S := \begin{pmatrix} 1 & 0 \\ 0 & 1 \end{pmatrix}.$$

Step 2: $I := \emptyset$; $J := \{x_5\}$;
$S+_T := \{f_1, f_2\}$; $S+_Q := \{x_{1Q}, x_{2Q}, x_{3Q}, x_{4Q}\}$; $S+ := \{f_1, f_2, x_{1Q}, x_{2Q}, x_{3Q}, x_{4Q}\}$;
$S- := \{x_1, x_2, x_3, x_4, x_5\}$;
$B+ := \emptyset$;
There exists a path from $S+$ to $S-$. [See $G(0)$ in Fig.1]

Step 3: $L := \{(x_{1Q}, x_1)\}$; $M := \{(x_1, x_{1Q})\}$;
The initial vertex x_{1Q} of L is in $S+_Q$, and the matrices are updated (with $h = r_1$) to

$$base := \begin{matrix} r_1 \\ r_2 \end{matrix}\begin{pmatrix} x_1 \\ 0 \end{pmatrix}, \quad P := \begin{matrix} r_1 \\ r_2 \end{matrix}\begin{matrix} \begin{matrix} x_1 & x_2 & x_3 & x_4 & x_5 \end{matrix} \\ \begin{pmatrix} 1 & 1 & 1 & 1 & 0 \\ 0 & 2 & 1 & 1 & 0 \end{pmatrix} \end{matrix}, \quad S := \begin{pmatrix} 1 & 0 \\ 0 & 1 \end{pmatrix}.$$

Noting $L \cap B+ = \emptyset$ we return to Step 2.

Step 2: $I := \{x_1\}$; $J := \{x_5\}$;
$S+_T := \{f_1, f_2\}$; $S+_Q := \{x_{2Q}, x_{3Q}, x_{4Q}\}$; $S+ := \{f_1, f_2, x_{2Q}, x_{3Q}, x_{4Q}\}$;
$S- := \{x_2, x_3, x_4, x_5\}$;
$B+ := \emptyset$;
There exists a path from $S+$ to $S-$. [See $G(1)$ in Fig.2]

Step 3: $L := \{(x_{2Q}, x_2)\}$; $M := \{(x_1, x_{1Q}), (x_2, x_{2Q})\}$;
The initial vertex x_{2Q} of L is in $S+_Q$, and the matrices are updated (with $h = r_2$) to

$$base := \begin{matrix} r_1 \\ r_2 \end{matrix}\begin{pmatrix} x_1 \\ x_2 \end{pmatrix}, \quad P := \begin{matrix} r_1 \\ r_2 \end{matrix}\begin{matrix} \begin{matrix} x_1 & x_2 & x_3 & x_4 & x_5 \end{matrix} \\ \begin{pmatrix} 1 & 0 & 1/2 & 1/2 & 0 \\ 0 & 2 & 1 & 1 & 0 \end{pmatrix} \end{matrix}, \quad S := \begin{pmatrix} 1 & -1/2 \\ 0 & 1 \end{pmatrix}.$$

Noting $L \cap B+ = \emptyset$ we return to Step 2.

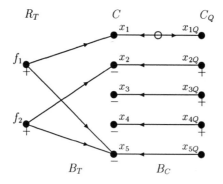

FIG. 2. *Graph G(1)*   (○: *arc in M; +: vertices in S+; −: vertices in S−*)

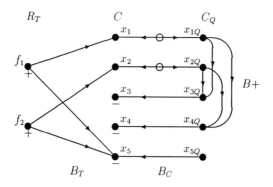

FIG. 3. *Graph G(2)*   (○: *arcs in M; +: vertices in S+; −: vertices in S−*)

Step 2: $I := \{x_1, x_2\}$; $J := \{x_3, x_4, x_5\}$;
  $S+_T := \{f_1, f_2\}$; $S+_Q := \emptyset$; $S+ := \{f_1, f_2\}$; $S- := \{x_3, x_4, x_5\}$;
  $B+ := \{(x_{1Q}, x_{3Q}), (x_{1Q}, x_{4Q}), (x_{2Q}, x_{3Q}), (x_{2Q}, x_{4Q})\}$;
  There exists a path from $S+$ to $S-$.   [See $G(2)$ in Fig.3]
Step 3: $L := \{(f_1, x_5)\}$; $M := \{(x_1, x_{1Q}), (x_2, x_{2Q}), (x_5, f_1)\}$;
  The initial vertex $f_1 \notin S+_Q$ and $L \cap B+ = \emptyset$, and therefore the matrices
  remain unchanged and we return to Step 2.
Step 2: $I := \{x_1, x_2\}$; $J := \{x_3, x_4, x_5\}$;
  $S+_T := \{f_2\}$; $S+_Q := \emptyset$; $S+ := \{f_2\}$; $S- := \{x_3, x_4\}$;
  $B+ := \{(x_{1Q}, x_{3Q}), (x_{1Q}, x_{4Q}), (x_{2Q}, x_{3Q}), (x_{2Q}, x_{4Q})\}$;
  There exists a path from $S+$ to $S-$.   [See $G(3)$ in Fig.4]
Step 3: $L := \{(f_2, x_2), (x_2, x_{2Q}), (x_{2Q}, x_{3Q}), (x_{3Q}, x_3)\}$;
  $M := \{(x_1, x_{1Q}), (x_3, x_{3Q}), (x_5, f_1), (x_2, f_2)\}$;
  The initial vertex $f_2 \notin S+_Q$ and $L \cap B+ = \{(x_{2Q}, x_{3Q})\}$, and the matrices

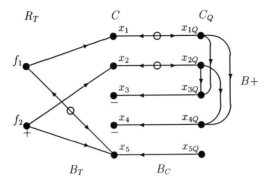

FIG. 4. *Graph* $G(3)$ (○: *arcs in* M; +: *vertex in* $S+$; −: *vertices in* $S-$)

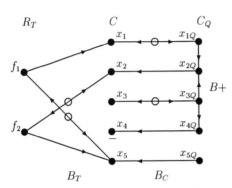

FIG. 5. *Graph* $G(4)$ (○: *arcs in* M; $S+ = \emptyset$, −: *vertex in* $S-$)

are updated (with $h = r_2$) to

$$base := \begin{array}{c} r_1 \\ r_2 \end{array}\!\begin{pmatrix} x_1 \\ x_3 \end{pmatrix}, \quad P := \begin{array}{c} \\ r_1 \\ r_2 \end{array}\!\begin{pmatrix} x_1 & x_2 & x_3 & x_4 & x_5 \\ 1 & -1 & 0 & 0 & 0 \\ 0 & 2 & 1 & 1 & 0 \end{pmatrix}, \quad S := \begin{pmatrix} 1 & -1 \\ 0 & 1 \end{pmatrix}.$$

Step 2: $I := \{x_1, x_3\}$; $J := \{x_2, x_4, x_5\}$;
$S+_T := \emptyset$; $S+_Q := \emptyset$; $S+ := \emptyset$; $S- := \{x_4\}$;
$B+ := \{(x_{1Q}, x_{2Q}), (x_{3Q}, x_{2Q}), (x_{3Q}, x_{4Q})\}$;
There exists no path from $S+ (= \emptyset)$ to $S-$.   [See $G(4)$ in Fig.5]

Step 4: $V_\infty := \emptyset$; $V_0 := \{x_3, x_4, x_{3Q}, x_{4Q}\}$;
$C_0 := \{x_3, x_4\}$; $C_\infty := \emptyset$;
Strongly connected components of G' (cf. Fig.6) are given by $\{V_{\lambda_1}, V_{\lambda_2}\}$,
where $V_{\lambda_1} = \{x_1, x_2, x_5, x_{1Q}, x_{2Q}, f_1, f_2\}$ and $V_{\lambda_2} = \{x_{5Q}\}$;
Since $C \cap V_{\lambda_2} = \emptyset$, we have $b := 1$ and $C_1 := C \cap V_{\lambda_1} = \{x_1, x_2, x_5\}$;
$R_0 := \{r_2\}$; $R_\infty := \emptyset$; $R_1 := \{r_1, f_1, f_2\}$;

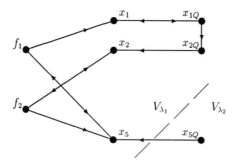

FIG. 6. *Graph G′*   *(V_{λ₁}, V_{λ₂}: strongly connected components)*

$$\bar{A} := P_r \begin{pmatrix} P \\ T \end{pmatrix} P_c = \begin{array}{c} r_2 \\ r_1 \\ f_1 \\ f_2 \end{array} \begin{array}{c} x_3 \quad x_4 \quad x_1 \quad x_2 \quad x_5 \\ \begin{pmatrix} 1 & 1 & 0 & 2 & 0 \\ & & 1 & -1 & 0 \\ & & t_1 & 0 & t_2 \\ & & 0 & t_3 & t_4 \end{pmatrix} \end{array}$$

is the CCF.   □

7. Conclusion. As a mathematical model for investigating the structure of linear dynamical systems, Murota [25], [28] proposed to consider a polynomial matrix $D(s)$ in indeterminate s over a field $F(\supset Q)$ which is represented as

$$D(s) = Q(s) + T(s),$$

where

(A1): The set of the nonzero coefficients of the entries of $T(s)$ is algebraically independent over Q, and

(A2): Every nonvanishing subdeterminant of $Q(s)$ is a monomial in s over Q.

Note that the assumption (A1) implies $D(s)$ is a mixed matrix with base field $K = Q(s)$. See [25], [28], [38] for physical backgrounds of the conditions (A1) and (A2); [27], [31], [42] for applications to control problems; and [30], [35], [39] for more recent results on such polynomial matrices.

As an extension of the CCF, Murota [34] considered the decomposition of an LM-matrix $A \in LM(F/K)$ with respect to a larger class of admissible transformations of the form: $S_r A S_c$ with S_r and S_c nonsingular matrices over K. This paper also considered the decomposition of A under this extended admissible transformation when A has certain symmetry expressed as an invariance with respect to a finite group.

Poljak [45] gave a combinatorial characterization to the rank of a power product, rank T^k, for a generic matrix T. It will be interesting to see whether his result can be extended to a mixed matrix.

Yamada-Luenberger [57] introduced the notion of "column-structured matrices" as a generalization of generic matrices.

Acknowledgements: Special thanks are due to Richard Brualdi, who invited me to the IMA Workshop on Combinatorial and Graph-Theoretical Problems in Linear Algebra. I owe Theorem 5.3 to the discussion with Raphael Loewy. I am grateful to John Gilbert, whose comment motivated me to formulate Theorem 3.2 in the present form. Comments from Clark Jeffries and Chris Lee were helpful to improve the presentation. Finally I appreciate the careful reading of the manuscript by Akihiro Sugimoto.

REFERENCES

[1] A. V. Aho, J. E. Hopcroft and J. D. Ullman: *The Design and Analysis of Computer Algorithms*, Addison-Wesley, 1974.

[2] M. Aigner: *Combinatorial Theory*, Springer-Verlag, 1979.

[3] G. Birkhoff: *Lattice Theory*, 3rd ed., American Math. Soc., 1979.

[4] R. A. Brualdi: "Term rank of the direct product of matrices," *Canadian J. Math.*, vol. 18, pp. 126–138, 1966.

[5] R. A. Brualdi and H. J. Ryser: *Combinatorial Matrix Theory*, Cambridge University Press, 1991.

[6] W.-K. Chen: *Applied Graph Theory — Graphs and Electrical Networks*, North-Holland, 1976.

[7] W. H. Cunningham: "Improved bounds for matroid partition and intersection algorithms," *SIAM J. Comput.*, vol. 15, pp. 948–957, 1986.

[8] A. L. Dulmage and N. S. Mendelsohn: "A structure theory of bipartite graphs of finite exterior dimension," *Trans. Roy. Soc. Canada*, Section III, vol. 53, pp. 1–13, 1959.

[9] J. Edmonds: "Minimum partition of a matroid into independent subsets," *Journal of National Bureau of Standards*, vol. 69B, pp. 67–72, 1965.

[10] J. Edmonds: "Systems of distinct representatives and linear algebra," *Journal of National Bureau of Standards*, vol. 71B, pp. 241–245, 1967.

[11] J. Edmonds: "Submodular functions, matroids and certain polyhedra," *Combinatorial Structures and Their Applications* (R. Guy et al. eds.), Gordon and Breach, pp. 69–87, 1970.

[12] G. Frobenius: "Über zerlegbare Determinanten," *Sitzungsber. Preuss. Akad. Wiss. Berlin*, pp. 274–277, 1917. (*Gesammelte Abhandlungen*, Vol. 3, Springer, No. 102, pp. 701–704, 1968.)

[13] S. Fujishige: "Algorithms for solving the independent-flow problems," *Journal of the Operations Research Society of Japan*, vol. 21, pp. 189–204, 1978.

[14] S. Fujishige: "Principal structure of submodular systems," *Discrete Appl. Math.*, vol. 2, pp. 77–79, 1980.

[15] S. Fujishige: *Submodular Functions and Optimization*, Annals of Discrete Math., vol. 47, North-Holland, 1991.

[16] F. R. Gantmacher: *The Theory of Matrices*, Chelsea, 1959.

[17] D. J. Hartfiel and R. Loewy: "A determinantal version of the Frobenius-König theorem," *Linear Multilinear Algebra*, vol. 16, pp. 155–165, 1984.

[18] M. Iri: "Applications of matroid theory," *Mathematical Programming - The State of the Art* (A. Bachem, M. Grötschel and B. Korte, eds.), Springer-Verlag, pp. 158–201, 1983.

[19] M. Iri: "Structural theory for the combinatorial systems characterized by submodular functions," *Progress in Combinatorial Optimization* (W. R. Pulleyblank, ed.), Academic Press, pp. 197–219, 1984.

[20] E. L. Lawler: *Combinatorial Optimization: Networks and Matroids*, Holt, Rinehart and Winston, 1976.

[21] L. Lovász and M. Plummer: *Matching Theory*, North-Holland, 1986.

[22] M. Marcus and H. Minc, "Disjoint pairs of sets and incidence matrices," *Illinois J. Math.*, vol. 7, pp. 137–147, 1963.

[23] S. T. McCormick, *A Combinatorial Approach to Some Sparse Matrix Problems*, Technical Report SOL 83-5, Stanford University, 1983.

[24] K. Murota, "LU-decomposition of a matrix with entries of different kinds," *Linear Algebra Appl.*, vol. 49, pp. 275–283, 1983.

[25] K. Murota, "Use of the concept of physical dimensions in the structural approach to systems analysis," *Japan J. Appl. Math.*, vol. 2, pp. 471–494, 1985.

[26] K. Murota, "Combinatorial canonical form of layered mixed matrices and block-triangularization of large-scale systems of linear/nonlinear equations," *Discussion Paper Series 257*, University of Tsukuba, 1985.

[27] K. Murota, "Refined study on structural controllability of descriptor systems by means of matroids," *SIAM J. Control Optim.*, vol. 25, pp. 967–989, 1987.

[28] K. Murota, *Systems Analysis by Graphs and Matroids — Structural Solvability and Controllability*, Algorithms and Combinatorics, vol. 3, Springer-Verlag, 1987.

[29] K. Murota, "On the irreducibility of layered mixed matrices," *Linear Multilinear Algebra*, vol. 24, pp. 273–288, 1989.

[30] K. Murota, "Some recent results in combinatorial approaches to dynamical systems," *Linear Algebra Appl.*, vol. 122/123/124, pp. 725–759, 1989.

[31] K. Murota, "A matroid-theoretic approach to structurally fixed modes of control systems," *SIAM J. Control Optim.*, vol. 27, pp. 1381–1402, 1989.

[32] K. Murota, "Principal structure of layered mixed matrices," *Discrete Appl. Math.*, vol. 27, pp. 221–234, 1990.

[33] K. Murota, "Combinatorial canonical form of mixed matrix and its application (in Japanese)," *Proceedings of Annual Symposium of Mathematical Society of Japan, Division of Applied Mathematics*, Okayama, pp. 222–241, 1990.

[34] K. Murota, "Hierarchical decomposition of symmetric discrete systems by matroid and group theories," *Mathematical Programming, Series A*, to appear.

[35] K. Murota, "On the Smith normal form of structured polynomial matrices," *SIAM J. Matrix Analysis Appl.*, vol. 12, no. 4, pp. 747–765, 1991.

[36] K. Murota, "On the Smith normal form of structured polynomial matrices II," *SIAM J. Matrix Analysis Appl.*, to appear.

[37] K. Murota, "Hierarchical decompositions of discrete systems — exploiting invariant structures by matroid (in Japanese)," *Bulletin Japan SIAM*, vol. 1, pp. 230–248, 1991.

[38] K. Murota, "A mathematical framework for combinatorial/structural analysis of linear dynamical systems by means of matroids," *Symbolic and Numerical Computation for Artificial Intelligence*, Edited by B. Donald, D. Kapur and J. Mundy, Academic Press, 1992. (To appear)

[39] K. Murota, "Matroids and systems analysis (in Japanese)," *Discrete Structures and Algorithms*, Edited by S. Fujishige, Kindai-Kagakusha, 1992. (To appear)

[40] K. Murota and M. Iri, "Structural solvability of systems of equations — A mathematical formulation for distinguishing accurate and inaccurate numbers in structural analysis of systems," *Japan J. Appl. Math.*, vol. 2, pp. 247–271, 1985.

[41] K. Murota, M. Iri and M. Nakamura, "Combinatorial canonical form of layered mixed matrices and its application to block-triangularization of systems of equations," *SIAM J. Algebraic Discrete Methods*, vol. 8, pp. 123–149, 1987.

[42] K. Murota and J. van der Woude, "Structure at infinity of structured descriptor systems and its applications," *SIAM J. Control Optim.*, vol. 29, pp. 878–894, 1991.

[43] M. Nakamura, "Structural theorems for submodular functions, polymatroids and polymatroid intersections," *Graphs and Combinatorics*, vol. 4, pp. 257–284, 1988.

[44] M. Newman, *Integral Matrices*, Academic Press, 1972.

[45] S. Poljak, "Maximum rank of powers of a matrix of a given pattern," *Proc. American Math. Soc.*, vol. 106, pp. 1137–1144, 1989.

[46] A. Recski, *Matroid Theory and Its Applications in Electric Network Theory and in Statics,*

Springer-Verlag, 1989.

[47] H. J. Ryser: "Indeterminates and incidence matrices," *Linear Multilinear Algebra*, vol. 1, pp. 149–157, 1973.

[48] H. J. Ryser: "The formal incidence matrix," *Linear Multilinear Algebra*, vol. 3, pp. 99–104, 1975.

[49] H. Schneider: "The concepts of irreducibility and full indecomposability of a matrix in the works of Frobenius, König and Markov," *Linear Algebra Appl.*, vol. 18, pp. 139–162, 1977.

[50] A. Schrijver: *Theory of Linear and Integer Programming*, Wiley, 1986.

[51] N. Tomizawa: "Strongly irreducible matroids and principal partition of a matroid into strongly irreducible minors (in Japanese)," *Trans. Inst. Electric Commun. Engineers*, vol. J59A, pp. 83–91, 1976.

[52] B. L. van der Waerden: *Algebra*, Springer-Verlag, 1955.

[53] D. J. A. Welsh: *Matroid Theory*, Academic Press, 1976.

[54] N. White: *Theory of Matroids*, Cambridge University Press, 1986.

[55] N. White: *Combinatorial Geometries*, Cambridge University Press, 1987.

[56] T. Yamada and L. R. Foulds: "A graph-theoretic approach to investigate structural and qualitative properties of systems: a survey," *Networks*, vol. 20, pp. 427–452, 1990.

[57] T. Yamada and D. G. Luenberger: "Generic properties of column-structured matrices," *Linear Algebra Appl.*, vol. 65, pp. 189–206, 1985.

A GEOMETRIC APPROACH
TO THE LAPLACIAN MATRIX OF A GRAPH

MIROSLAV FIEDLER*

Abstract. Let G be a finite undirected connected graph with n vertices. We assign to G an $(n-1)$-simplex $\Sigma(G)$ in the point Euclidean $(n-1)$-space in such a way that the Laplacian $L(G)$ of G is the Gram matrix of the outward normals of $\Sigma(G)$. It is shown that the spectral properties of $L(G)$ are reflected by the geometric shape of the Steiner circumscribed ellipsoid S of $\Sigma(G)$ in a simple manner. In particular, the squares of the half-axes of S are proportional to the reciprocals of the eigenvalues of $L(G)$. Also, a previously discovered relationship to resistive electrical circuits is mentioned.

1. Introduction. In the recent years, Laplacians of graphs have attracted much attention among both graph theorists and numerical analysts (cf. [9],[11]). The second smallest eigenvalue of the Laplacian has shown intimate relationship to the minimum cut [4] and the largest eigenvalue to the maximum cut [10]. Years ago, the author observed [6] that there are four parallel approaches to studying problems in this area: graph-theoretical, algebraic, geometric and via resistive electrical networks.

In the present paper, the geometric approach will be explained and applied to the case of the Laplacian matrix of a graph. An - up to congruence - uniquely defined simplex in the Euclidean point space will be assigned to the graph. Consequently, every geometric invariant of the simplex is at the same time an invariant of the graph. The main result shows that metric properties of the circumscribed Steiner ellipsoid of the simplex correspond in a simple manner to the spectral properties of the Laplacian.

2. Simplex geometry. Let us begin with some well known facts. We denote by $\mathcal{E}_n$ a fixed Euclidean n-dimensional vector space, by $(.,.)$ the inner product. [Recall that all Euclidean n-dimensional vector spaces are isomorphic.]

Given m vectors $a_1, ..., a_m$ in $\mathcal{E}_n$, we denote by $G(a_1, ..., a_m)$, or simply $G(a)$, its $m \times m$ Gram matrix $((a_i, a_k))$.

Fact 1. $G(a)$ is positive semidefinite. It is nonsingular if and only if $a_1, ..., a_m$ are linearly independent.

Fact 2. Every linear dependence relation among the vectors $a_1, ..., a_m$ is reflected by the same linear dependence relation among the columns (and rows) of $G(a)$, and conversely.

Fact 3. Every real symmetric positive semidefinite matrix of rank at most n is the Gram matrix of some system of vectors in $\mathcal{E}_n$.

Fact 4. If $a_1, ..., a_n$ are linearly independent vectors in $\mathcal{E}_n$, then there exists an

* Math. Inst. Acad., Prague

n-tuple $b_1, ..., b_n$ such that

$$(a_i, b_k) = \delta_{ik} = \left\{ \begin{array}{ll} 1 & \text{if } i = k, \\ 0 & \text{if } i \neq k. \end{array} \right.$$

The Gram matrices $G(a)$ and $G(b)$ are inverse to each other. The (ordered) sets $a_1, ..., a_n$ and $b_1, ..., b_n$ are also called biorthogonal bases.

We shall be considering now the *point* Euclidean n-space $\hat{\mathcal{E}}_n$ which contains two kinds of objects, points and vectors. The usual operations, addition and multiplication by scalars, for vectors is here completed by analogous operations for points with the following restriction:

A linear combination of points and vectors is allowed only in two cases:

(i) the sum of the coefficients at the points is one and the result is a point;

(ii) the sum of the coefficients at the points is zero and the result is a vector.

Thus, if A and B are points, then $1.B + (-1).A$, or simply $B - A$, is a vector (which can be considered as starting in A and ending in B). The point $\frac{1}{2}A + \frac{1}{2}B$ is the midpoint of the segment AB, etc.

The points $A_0, ..., A_p$ are called linearly independent if $\alpha_0 = ... = \alpha_p = 0$ is the only way how to express the zero vector as $\sum_{i=0}^{p} \alpha_i A_i$ with $\sum_{i=0}^{p} \alpha_i = 0$.

The dimension of a point Euclidean space is, by definition, the dimension of the underlying Euclidean vector space. It is equal to n if there are in the space $n + 1$ linearly independent points whereas any $n + 2$ points in the space are linearly dependent.

In a usual way, one can then define linear (point) subspaces of the point Euclidean space, halfspaces, convexity etc. A ray is a halfline, i. e. for some distinct points A, B, the set of all points of the form $A + \lambda(B - A)$, $\lambda \geq 0$.

An n-simplex in $\hat{\mathcal{E}}_n$ is usually defined as a convex hull of $n + 1$ linearly independent points, so called vertices, of $\hat{\mathcal{E}}_n$. (Thus a 2-simplex is a triangle, a 3-simplex a tetrahedron, etc.)

Let $A_1, ..., A_{n+1}$ be the vertices of such an n-simplex Σ in $\hat{\mathcal{E}}_n$. Define the vectors

(1) $$a_i = A_i - A_{n+1}, \quad i = 1, ..., n.$$

They are linearly independent so that there exist the vectors $b_1, ..., b_n$ which together with $a_1, ..., a_n$ form biorthogonal bases by Fact 4:

(2) $$(a_i, b_k) = \delta_{ik}, \quad i, k = 1, ..., n.$$

Let $\alpha_1, \alpha_2, ..., \alpha_{n+1}$ be the hyperplanes of the $(n - 1)$-dimensional faces of Σ:

α_k is the linear hull of the points A_i, $i \neq k$, α_k^+ the halfspace with boundary α_k and containing A_k. It is easily seen:

Theorem 2.1. *The vectors $b_1, ..., b_n$ and the vector $b_{n+1} = -\sum_{j=1}^{n} b_j$ are the vectors of inward normals of Σ, i. e. b_k is orthogonal to α_k and the ray $C_k + \alpha b_k, \alpha \geq 0$, for $C_k \in \alpha_k$, is contained in α_k^+.*

It will be more convenient, however, to introduce the vectors

(3) $$v_i = -b_i, \ i = 1, ..., n+1$$

and call them *normalized outward normals* of Σ. Observe that they satisfy the relation

(4) $$\sum_{i=0}^{n} v_i = 0.$$

Given the simplex Σ, one can define barycentric coordinates of points in $\hat{\mathcal{E}}_n$ as follows:

If $X \in \hat{\mathcal{E}}_n$ is a point, it can be, even uniquely, written as

$$X = \sum_{i=1}^{n+1} x_i A_i, \sum_{i=1}^{n+1} x_i = 1.$$

The numbers x_i are called barycentric coordinates of X and we write $X = (x_1, ..., x_{n+1})$.

If $u \in \hat{\mathcal{E}}_n$ is a vector, it can also be expressed, uniquely, as

$$u = \sum_{i=1}^{n+1} \lambda_i A_i, \sum_{i=1}^{n+1} \lambda_i = 0.$$

The numbers λ_i are called barycentric coordinates of u and we write $u = (u_1, ..., u_{n+1})$.

To unify these notions and use the technique of analytical projective geometry we redefine $\hat{\mathcal{E}}_n$ into a projective space $\bar{\mathcal{E}}_n$.

As usual, we introduce homogeneous barycentric coordinates as classes of nonzero (ordered) $(n+1)$-tuples of (real) numbers with respect to the equivalence: two such $(n+1)$-tuples $(x_1, ..., x_{n+1})$ and $(y_1, ..., y_{n+1})$ are equivalent if $x_i = \sigma y_i, i = 1, ..., n+1, \sigma \neq 0$. A point $(y_1, ..., y_{n+1})$ is called *proper* if and only if $\sum_{1}^{n+1} y_i \neq 0$ and it then coincides with the previously defined point with $\frac{y_i}{\sum_j y_j}$ as usual barycentric coordinates. A point $(y_1, ..., y_{n+1})$ is called *improper* if $\sum_{i=1}^{n+1} y_i = 0$ and it corresponds to the direction of any line determined by the nonzero vector $(y_1, ..., y_{n+1})$, in previous notation.

Linear dependence of these "generalized" points $P = (p_1, ..., p_{n+1})$, $Q = (q_1, ..., q_{n+1})$, ..., $R = (r_1, ..., r_{n+1})$, is reflected by the fact that the matrix

$$\begin{pmatrix} p_1 & \cdots & p_{n+1} \\ q_1 & \cdots & q_{n+1} \\ \vdots & & \\ r_1 & \cdots & r_{n+1} \end{pmatrix}$$

has the maximum row-rank. This enables to express linear dependence and to define linear subspaces. Every such linear subspace can be described either as a linear hull of points, or as intersection of $(n-1)$-dimensional subspaces, called hyperplanes; every hyperplane can be described as the set of all (generalized) points x the coordinates $(x_1, ..., x_{n+1})$ of which satisfy a linear equality

(5) $$\alpha_1 x_1 + \alpha_2 x_2 + ... + \alpha_{n+1} x_{n+1} = 0$$

where not all coefficients $\alpha_1, \ldots, \alpha_{n+1}$ are zero. The coefficients $\alpha_1, \ldots, \alpha_{n+1}$ can be viewed as (dual) coordinates of the hyperplane and the relation (5) as the incidence relation for the point (x) and the hyperplane (α).

In accordance with (5),

$$(6) \qquad \sum_{i=1}^{n+1} x_i = 0,$$

i. e. the condition that $x = (x_i)$ is improper, represents the equation of a hyperplane, so called *improper* hyperplane.

Two (proper) hyperplanes $\sum \alpha_i x_i = 0, \sum \beta_i x_i = 0$ are different if and only if the matrix

$$\begin{pmatrix} \alpha_1 & \cdots & \alpha_{n+1} \\ \beta_1 & \cdots & \beta_{n+1} \end{pmatrix}$$

has rank 2. They are then parallel if and only if the rank of the matrix

$$\begin{pmatrix} \alpha_1 & \cdots & \alpha_{n+1} \\ \beta_1 & \cdots & \beta_{n+1} \\ 1 & \cdots & 1 \end{pmatrix}$$

is again 2.

As usual, we define the (Euclidean) distance $\rho(A, B)$ of the points A, B in $\hat{\mathcal{E}}_n$ as the length $\sqrt{(B - A, B - A)}$ of the corresponding vector $B - A$.

3. Matrices assigned to an n-simplex. To the n-simplex Σ with the (ordered) vertices $A_1, \ldots, A_{n+1}$, we can assign an $(n + 1) \times (n + 1)$ matrix M the entries of which are the squares of the (Euclidean) distances of the points $A_1, \ldots, A_{n+1}$

$$(7) \qquad M = (m_{ij}), \quad m_{ij} = \rho^2(A_i, A_j), i, j = 1, \ldots, n + 1.$$

We shall call this matrix the *Menger matrix* of Σ. On the other hand, denote by Q the Gram matrix of the normalized outward normals $v_1, \ldots, v_{n+1}$

$$(8) \qquad Q = ((v_i, v_j)), \quad i, j = 1, \ldots, n + 1;$$

the vectors are constructed as above.

In the following theorem, we shall formulate the basic relation betweeen the matrices M and Q.

Theorem 3.1. *Let e be the column vector of $n+1$ ones. Then there exists a column $(n + 1)$-vector $q_0 = (q_{01}, \ldots, q_{0,n+1})^T$ and a number q_{00} such that*

$$(9) \qquad \begin{pmatrix} 0 & e^T \\ e & M \end{pmatrix} \begin{pmatrix} q_{00} & q_0^T \\ q_0 & Q \end{pmatrix} = -2I_{n+2}.$$

PROOF. Partition the matrices M and Q as

$$M = \begin{pmatrix} \widetilde{M} & m \\ m^T & 0 \end{pmatrix}, \quad Q = \begin{pmatrix} \tilde{Q} & q \\ q^T & \gamma \end{pmatrix},$$

where $\widetilde{M}, \tilde{Q}$ are $n \times n$.

Observe that by (1) and (2),

(10) $$\tilde{Q} = ((b_i, b_j)), \quad i, j = 1, \ldots, n,$$

and

$$\widetilde{M} = ((a_i - a_j, a_i - a_j)), \quad i, j = 1, \ldots, n.$$

Since

$$(a_i - a_j, a_i - a_j) = (a_i, a_i) + (a_j, a_j) - 2(a_i, a_j),$$

we obtain

(11) $$((a_i, a_j)) = \frac{1}{2}(m\tilde{e}^T + \tilde{e}m^T - \widetilde{M})$$

where

$$\tilde{e} = (1, \ldots, 1)^T$$

with n ones.

By Fact 4, the matrices $\tilde{Q}$ and $((a_i, a_j))$ are inverse to each other so that (11) implies

(12) $$\widetilde{M}\tilde{Q} = -2I_n + m\tilde{e}^T\tilde{Q} + \tilde{e}m^T\tilde{Q}.$$

Set

$$q_0 = \begin{pmatrix} \gamma \\ q \end{pmatrix}$$

where

(13) $$q = -\tilde{Q}m,$$

(14) $$\gamma = -2 + \tilde{e}^T\tilde{Q}m$$

and

(15) $$q_{00} = m^T\tilde{Q}m.$$

The left-hand side of (9) is then (the row-sums of Q are zero)

$$\begin{pmatrix} 0 & \tilde{e}^T & 1 \\ \tilde{e} & \widetilde{M} & m \\ 1 & m^T & 0 \end{pmatrix} \begin{pmatrix} m^T\tilde{Q}m & -m^T\tilde{Q} & -2 + \tilde{e}^T\tilde{Q}m \\ -\tilde{Q}m & \tilde{Q} & -\tilde{Q}\tilde{e} \\ -2 + \tilde{e}^T\tilde{Q}m & -\tilde{e}^T\tilde{Q} & \tilde{e}^T\tilde{Q}\tilde{e} \end{pmatrix}$$

78

which is by (12) easily checked to multiply to $-2I_{n+2}$.

Corollary 3.2. *The matrix*

$$\widehat{M} = \begin{pmatrix} 0 & e^T \\ e & M \end{pmatrix}$$

is nonsingular.

Observe that we started with an n-simplex and assigned to it the Menger matrix M and the Gram matrix Q of the normalized outward normals. In the following theorem we shall show that one can completely reconstruct the previous situation given the matrix Q.

Theorem 3.3. *Let Q be an $(n+1) \times (n+1)$ symmetric matrix such that*

(16) $\qquad\qquad\qquad Q$ *is positive semidefinite of rank n*

and

(17) $\qquad\qquad\qquad\qquad Qe = 0.$

Then there exists a (uniquely determined up to congruence) n-simplex for which Q is the Gram matrix of the normalized outward normals.

PROOF. By Fact 3, Q is the Gram matrix of some vectors $v_1, \ldots, v_{n+1}$ in a fixed Euclidean n-space $\mathcal{E}_n$. By Fact 2,

(18) $\qquad\qquad\qquad\qquad \sum_i v_i = 0$

and some n (in fact, any n) of these vectors, say, $v_1, \ldots, v_n$ are linearly independent.

Let U be a fixed point in $\mathcal{E}_n$. Denote by T_i the points

(19) $\qquad\qquad T_i = U + r\dfrac{v_i}{\sqrt{(v_i, v_i)}}, \quad i = 1, \ldots, n+1,$

where

(20) $\qquad\qquad\qquad\qquad r = \dfrac{1}{\sum_{i=1}^{n+1} \sqrt{(v_i, v_i)}}.$

The points T_i are points of the hypersphere H with center U and radius r. Denote by α_i the hyperplane tangent to H at T_i, $i = 1, \ldots, n+1$, by α_i^+ the halfspace determined by α_i and containing U. Let us show that the intersection

$$\Sigma \equiv \cap_{i=1}^{n+1} \alpha_i^+$$

is bounded, thus being an n-simplex.

Suppose the contrary so that Σ contains a halfline h. Without loss of generality, the halfline h can be assumed to have the end-point U:

$$h = \{U + \lambda z \mid \lambda \geq 0\}$$

where z is some nonzero vector.

Since $h \subset \alpha_i^+$,

(21) $$(v_i, z) \leq 0, \ i = 1, \ldots, n+1.$$

Let

$$z = \sum_{i=1}^{n} \xi_i v_i \quad \text{and set} \quad \xi_{n+1} = 0.$$

Define

(22) $$\mu = \min_{i=1,\ldots,n+1} \xi_i.$$

Then

$$z = \sum_{i=1}^{n} \xi_i v_i - \mu \sum_{i=1}^{n+1} v_i$$

by (18).

Therefore,

$$
\begin{aligned}
(z, z) &= \left(\sum_{i=1}^{n+1} (\xi_i - \mu) v_i, z \right) \\
&= \sum_{i=1}^{n+1} (\xi_i - \mu)(v_i, z) \\
&\leq 0
\end{aligned}
$$

by (21) and (22), a contradiction.

The n-simplex Σ has thus vertices $A_i = \cap_{j=1, j \neq i}^{n+1} \alpha_j$, $i = 1, \ldots, n+1$. Define

$$
\begin{aligned}
a_i &= A_i - A_{n+1}, \\
b_i &= -v_i, \ i = 1, \ldots, n.
\end{aligned}
$$

To prove the theorem, it suffices to show that a_i, b_i form biorthogonal bases, i. e. that

(23) $$(a_i, b_j) = \delta_{ij}.$$

By orthogonality of α_i and v_i,

(24) $$(A_j - A_k, v_i) = 0 \quad \text{whenever} \quad j \neq i \neq k, \ i, j, k = 1, \ldots, n+1.$$

Choose first $k = n+1, i, j = 1, \ldots, n, i \neq k$. We obtain (23) for $i \neq j, i, j = 1, \ldots, n$. Choosing now $i = n+1$, we have

$$(a_j - a_k, v_{n+1}) = 0 \quad \text{for} \quad j \neq k, \ j, k = 1, \ldots,$$

Thus, for some constant K,

(25) $$(a_j, v_{n+1}) = K, \ j = 1, \ldots, n.$$

Since

$$(a_j, \sum_{k=1}^{n+1} v_k) = 0$$

and since $(a_j, v_k) = 0$ for $j \neq k, \ j, k = 1, \ldots, n,$

$$(a_j, v_j) = (a_j, v_{n+1}) = 0$$

or,

$$(a_j, b_j) = K, \ j = 1, \ldots, n.$$

To show that $K = 1$, let

$$U = \sum_{k=1}^{n+1} \beta_k A_k, \sum_{k=1}^{n+1} \beta_k = 1.$$

Since

$$(A_k - T_i, v_i) = 0 \quad \text{for} \quad k \neq i,$$

we obtain by (19)

$$(A_k - U, v_i) = r\sqrt{(v_i, v_i)} \quad \text{for} \quad k \neq i, \ i, k = 1, \ldots, n+1.$$

Thus

$$((\sum_{j=1}^{n+1} \beta_j) A_k - \sum_{j=1}^{n+1} \beta_j A_j, v_i) = r\sqrt{(v_i, v_i)},$$

i. e.

(26) $$\sum_{j=1}^{n+1} \beta_j (A_k - A_j, v_i) = r\sqrt{(v_i, v_i)} \text{ for } k \neq i.$$

Choose $k = n+1, \ i \in \{1, \ldots, n\}$. We obtain

$$\sum_{j=1}^{n} \beta_j (a_j, b_i) = r\sqrt{(v_i, v_i)}$$

so that

$$\beta_i (a_i, b_i) = r\sqrt{(v_i, v_i)}$$

i. e.

(27) $$K\beta_i = r\sqrt{(v_i, v_i)}, \ i = 1, \ldots, n.$$

Choose now $i = n + 1$ and $k = 1$ in (26). We have

$$\sum_{j=1}^{n+1} \beta_j (A_1 - A_j, v_{n+1}) = r\sqrt{(v_{n+1}, v_{n+1})}$$

and by (25)

$$\beta_{n+1}(a_1, v_{n+1}) = r\sqrt{(v_{n+1}, v_{n+1})}$$

which yields

$$K\beta_{n+1} = r\sqrt{(v_{n+1}, v_{n+1})}.$$

Summing this equality with the equalities (27) for $i = 1, \ldots, n$, we obtain, by (20), $K = 1$.

To prove the uniqueness, observe that for $i \neq k$, the principal minor

$$\left(-\frac{1}{2}\right)^3 \det \begin{pmatrix} 0 & 1 & 1 \\ 1 & 0 & m_{ik} \\ 1 & m_{ik} & 0 \end{pmatrix}$$

of

$$-\frac{1}{2}\begin{pmatrix} 0 & e^T \\ e & M \end{pmatrix}$$

which equals $-\frac{1}{4}m_{ik}$ is also equal to

(28)
$$\frac{\det Q[N\backslash i \mid N\backslash k]}{\det \begin{pmatrix} q_{00} & q_o^T \\ q_0 & Q \end{pmatrix}}$$

Assume that $\widehat{M}, \hat{q}_{00}, \hat{q}_0$ also satisfy

(9′)
$$\begin{pmatrix} 0 & e^T \\ e & \widehat{M} \end{pmatrix}\begin{pmatrix} \hat{q}_{00} & \hat{q}_0^T \\ \hat{q}_0 & Q \end{pmatrix} = -2I_{n+2}.$$

By the observation, $-\frac{1}{4}\hat{m}_{ik}$ is equal to the expression corresponding to (28), thus

$$\widehat{M} = \sigma M \quad \text{for} \quad \sigma \quad \text{a constant}.$$

Since by (9) and (9')

$$eq_0^T + MQ = -2I_{n+1},$$

$$e\hat{q}_0^T + \sigma MQ = -2I_{n+1},$$

it follows that

$$e(\hat{q}_0^T - \sigma q_0^T) = -2(1 - \sigma)I_{n+1}.$$

Assuming $n \geq 2$, we obtain $\sigma = 1$ and $\widehat{M} = M$. Also, $\hat{q}_0 = q_0$ as well as $\hat{q}_{00} = q_{00}$.

In the sequel, we shall need some formulae for the distances and angles.

Theorem 3.4. *Let $X = (x_i), Y = (y_i), Z = (z_i)$ be proper points in $\widehat{\mathcal{E}}_n$, x_i, y_i, z_i their homogeneous barycentric coordinates, respectively, with respect to the simplex Σ. Then the inner product of the vectors $Y - X, Z - X$ is*

$$(29) \qquad (Y - X, Z - X) = -\frac{1}{2} \sum_{i,k=1}^{n+1} m_{ik}\left(\frac{x_i}{\Sigma x_j} - \frac{y_i}{\Sigma y_j}\right)\left(\frac{x_k}{\Sigma x_j} - \frac{y_k}{\Sigma y_j}\right).$$

PROOF. We can assume that $\Sigma x_j = \Sigma y_j = \Sigma z_j = 1$. Then

$$(Y - X, Z - X) = \left(\sum_{i=1}^{n+1}(y_i - x_i)A_i, \sum_{i=1}^{n+1}(z_i - x_i)A_i\right)$$

$$= \left(\sum_{i=1}^{n}(y_i - x_i)(A_i - A_{n+1}), \sum_{k=1}^{n}(z_k - x_k)(A_k - A_{n+1})\right).$$

Since

$$(A_i - A_{n+1}, A_k - A_{n+1}) = -\frac{1}{2}((A_i - A_k, A_i - A_k)$$

$$-(A_i - A_{n+1}, A_i - A_{n+1}) - (A_k - A_{n+1}, A_k - A_{n+1})),$$

we obtain

$$(Y - Z, Z - X) = -\frac{1}{2}\left(\sum_{i,k=1}^{n} m_{ik}(y_i - x_i)(z_k - x_k)\right.$$

$$-\sum_{k=1}^{n}(z_k - x_k)\sum_{i=1}^{n} m_{i,n+1}(y_i - x_i) - \sum_{i=1}^{n}(y_i - x_i)\sum_{k=1}^{n} m_{k,n+1}(z_k - x_k)\right)$$

$$= -\frac{1}{2}\sum_{i,k=1}^{n+1} m_{ik}(y_i - x_i)(z_k - x_k).$$

For homogeneous coordinates, this yields (28).

Corollary 3.5. *The square of the distance of the points $X = (x_i)$ and $Y = (y_i)$ in barycentric coordinates is*

$$(30) \qquad \rho^2(X,Y) = -\frac{1}{2}\sum_{i,k=1}^{n+1} m_{ik}\left(\frac{x_i}{\Sigma x} - \frac{y_i}{\Sigma y}\right)\left(\frac{x_k}{\Sigma x} - \frac{y_k}{\Sigma y}\right).$$

Theorem 3.6. *If the points $P = p_i$, $Q = (q_i)$ are both improper (i.e. $\Sigma p_i = \Sigma q_i = 0$), thus corresponding to directions of lines then these are orthogonal if and only if*

$$(31) \qquad \sum_{i,k=1}^{n+1} m_{ik}p_i q_k = 0.$$

PROOF. Let

$$Y = X + \lambda P,$$

$$Z = X + \mu Q$$

be any lines with directions P and Q, respectively. By Thm. 3.4, these lines are orthogonal if and only if

$$(Y_0 - X, \ Z_0 - X) = 0$$

for some $Y_0 = X + \lambda_0 P$, $Z_0 = X + \mu_0 Q$, $\lambda_0 \neq 0$, $\mu_0 \neq 0$. By (29), we obtain (31).

Corollary 3.7. *The equation of the circumscribed sphere to the simplex Σ is*

$$\sum_{i,k=1}^{n+1} m_{ik} x_i x_k = 0.$$

Its center is the point (q_{0i}), the square of its radius is $\frac{1}{4} q_{00}$, where the q_{0j}'s satisfy (9).

PROOF. The equation of the sphere with center q_{0i} and square of the radius $\frac{1}{4} q_{00}$ is by (30)

(32) $$-\frac{1}{2} \sum_{i,k=1}^{n+1} m_{ik} \Big(\frac{x_i}{\Sigma x_j} - \frac{q_{0i}}{\Sigma q_{0j}} \Big) \Big(\frac{x_k}{\Sigma x_j} - \frac{q_{0k}}{\Sigma q_{0j}} \Big) = \frac{1}{4} q_{00}.$$

Since by (9)

(33) $$\sum_{k=1}^{n+1} m_{ik} q_{0k} = -q_{00},$$

(34) $$\sum_{j=1}^{n+1} q_{0j} = -2,$$

the left-hand side of (32) is

$$-\frac{1}{2} \Big(\frac{\Sigma_{i,k} m_{ik} x_i x_k}{(\Sigma x_j)^2} - \frac{1}{2} q_{00} \Big).$$

This proves the assertion since this sphere contains all the points A_i.

Corollary 3.8. *The matrix $M = (m_{ik})$ has the property that*
(i) $m_{ii} = 0$, $i = 1,\ldots,n+1$,
(ii) $m_{ik} = m_{ki}$, $i,k = 1,\ldots,n+1$,
(iii) $\sum_{i,k=1}^{n+1} m_{ik} x_i x_k \leq 0$ whenever $(x_1,\ldots,x_{n+1}) \neq 0$ and $\sum_{i=1}^{n+1} x_i = 0$.

[These conditions are, in fact, characteristic for M to be a Menger matrix of a set of vertices of an n-simplex (cf. [7], [1], p.107, [2]).]

We include here two formulas without proof.

Theorem 3.9. *Let $A = (a_i)$ be a proper point in $\widehat{\mathcal{E}}_n$, $\alpha \equiv \sum_{i=1}^{n+1} \alpha_i x_i = 0$ equation of a hyperplane. Then the distance of A from α is given by*

$$(35) \qquad \rho(A, \alpha) = \left| \frac{\sum_{i=1}^{n+1} a_i \alpha_i}{\sqrt{\sum_{i,k=1}^{n+1} q_{ik}\alpha_i\alpha_k} \sum_{i=1}^{n+1} a_i} \right|.$$

In particular, $\frac{1}{\sqrt{q_{ii}}}$ is the height (i. e. the length of the altitude of Σ) corresponding to A_i.

Theorem 3.10. *Let α, β be proper hyperplanes in $\widehat{\mathcal{E}}_n$ with the equations*

$$\alpha \equiv \sum_i \alpha_i x_i = 0, \ \beta \equiv \sum_i \beta_i x_i = 0$$

in barycentric coordinates with respect to Σ. Then the cosine of their angle is given by

$$(36) \qquad \cos\widehat{\alpha\beta} = \frac{|\sum_{i,k=1}^{n+1} q_{ik}\alpha_i\beta_k|}{\sqrt{\sum_{i,k=1}^{n+1} q_{ik}\alpha_i\alpha_k}\sqrt{\sum_{i,k=1}^{n+1} q_{ik}\beta_i\beta_k}}.$$

Corollary 3.11. *Two proper hyperplanes $\alpha \equiv \sum_i \alpha_i x_i = 0$ and $\beta \equiv \sum_i \beta_i x_i = 0$ are orthogonal if and only if*

$$(37) \qquad \sum_{i,k=1}^{n+1} q_{ik}\alpha_i\beta_k = 0.$$

Corollary 3.12. *The improper point (direction) perpendicular to the proper hyperplane $\alpha \equiv \sum_{i=1}^{n+1} \alpha_i x_i = 0$ is*

$$(38) \qquad (\sum_k q_{1k}\alpha_k, \ \sum_k q_{2k}\alpha_k, \ \ldots, \ \sum_k q_{n+1,k}\alpha_k)$$

in homogeneous barycentric coordinates.

Let us present now another geometric interpretation of the matrix Q.

Theorem 3.13. *Let $D = diag(\sqrt{q_{ii}})$, let φ_{ik} $(i \neq k)$ denote the interior angle of Σ between the faces ω_i and ω_k,*

$$Q_0 = \begin{pmatrix} 1 & -\cos\varphi_{12} & \ldots & -\cos\varphi_{1,n+1} \\ -\cos\varphi_{12} & 1 & \ldots & -\cos\varphi_{2,n+1} \\ \ldots & \ldots & \ldots & \ldots \\ -\cos\varphi_{1,n+1} & -\cos\varphi_{2,n+1} & \ldots & 1 \end{pmatrix}.$$

Then,

$$(39) \qquad Q = DQ_0D.$$

PROOF. We have to show that

$$(Q)_{ik} = \sqrt{q_{ii}}(Q_0)_{ik}\sqrt{q_{kk}}.$$

This is clear for $i = k$. For $i \neq k$, it follows from the fact that the angle of the outward normals $\widehat{v_i v_k}$ and φ_{ik} add to π so that

$$\cos \varphi_{ik} = -\frac{(v_i, v_k)}{\sqrt{(v_i, v_i)}\sqrt{(v_k, v_k)}}$$

$$= -\frac{q_{ik}}{\sqrt{q_{ii}}\sqrt{q_{kk}}}.$$

It follows that

(40) $$\det Q_0 = 0.$$

[This generalizes the condition that in the triangle the sum of the angles is π.]

4. Some properties of simplices. Application to quadrics. Using the properties of the matrix Q qualitative properties of simplices were studied [2]. In particular, the following problem was completely solved:

To determine all possible distributions of acute, right and obtuse interior angles (of $(n-1)$-dimensional faces) in an n-simplex. By Thm. 3.13, the angle φ_{ik} is:

acute if and only if $q_{ik} < 0,$

right if and only if $q_{ik} = 0,$

obtuse if and only if $q_{ik} > 0.$

Therefore, all possible distributions of acute, right and obtuse interior angles correspond to all possible sign-patterns in the set of matrices Q_0, and thus Q by (39), satisfying (16) and (17). The sign-pattern of a real symmetric matrix $A = (a_{ik})$ is well described by the signed graph of A. If A is $m \times m$, $M = \{1, \ldots, m\}$, then its *signed graph* is $\Gamma(A) = (M, E_+, E_-)$, where

$$E_+ = \{(i, k); \ i, k \in M, \ a_{ik} > 0\},$$

$$E_- = \{(i, k); \ i, k \in M, \ a_{ik} < 0\}.$$

Lemma 4.1. *A symmetric sign-pattern $\Gamma = (M, E_+, E_-)$ is sign-pattern of a matrix Q satisfying (16) and (17) if and only if $m = n + 1$ and the graph (M, E_-) is connected.*

PROOF. Let $Q = (q_{ik})$ satisfy (16) and (17) and suppose that (M, E_-) is not connected. Let the set of indices in M_1 correspond to one component of (M, E_-). By the assumption,

(41) $$q_{ij} \geq 0 \quad \text{if } i \in M_1 \quad \text{and } j \in M\backslash$$

By (17), $Qe = 0$, or

$$\sum_{j \in M} q_{ij} = 0 \text{ for all } i \in M.$$

Summing for $i \in M_1$, we obtain

(42)
$$\sum_{i \in M_1, j \in M_1} q_{ij} + \sum_{i \in M_1, j \in M \setminus M_1} q_{ij} = 0.$$

Here, the second term is nonnegative by (41); the first term is also nonnegative since it is the value of the quadratic form with the matrix Q for the vector $\hat{e}$ having coordinate one for indices from M_1 and zero for the remaining indices. By (42), both terms are equal to zero. By positive semidefiniteness of Q, $Q\hat{e} = 0$. Since $0 \neq M_1 \neq M$, $\hat{e}$ is linearly independent of e, a contradiction with rank$Q = n$.

To prove the converse, let $\Gamma = (M, E_+, E_-)$ be a signed graph for which (M, E_-) is connected. The matrix Q of the quadratic form

$$\sum_{i,k,(i,k) \in E_-} (x_i - x_k)^2 - \varepsilon \sum_{i,k,(i,k) \in E_+} (x_i - x_k)^2$$

satisfies for $\varepsilon > 0$ sufficiently small (16) and (17) since the matrix corresponding to the first term has all principal submatrices of order n positive definite whereas the matrix Q is singular for every ε.

Corollary 4.2. *Every n-simplex has at least n acute interior angles.*

Corollary 4.3. *There exist n-simplices with exactly n acute interior angles and all the remaining $\frac{1}{2}(n-1)n$ angles right. The corresponding signed graph Γ is a tree with negative edges only.*

Let us remark that the n-simplices from Cor. 4.3 were called *right n-simplices* in [3]. It was proved there that the n edges opposite to the n acute interior angles (called *legs*) are mutually perpendicular. Consequently, there exists an orthogonal n-dimensional parallelepiped Π such that the n legs are among the edges of Π. One can even prove that if each acute angle φ_{ij} in a right n-simplex satisfies

(43)
$$\cos \varphi_{ij} = \frac{1}{\sqrt{d_i}\sqrt{d_j}}$$

where d_i is the degree of the node i in the tree from Cor. 4.3 then the parallelepiped Π is a cube.

In the second part of this section, we shall investigate metric properties of quadrics in $\hat{\mathcal{E}}_n$, in particular the important circumscribed Steiner ellipsoid S of the n-simplex Σ. This ellipsoid S is the quadric with the equation

(44)
$$\sum_{i<k} x_i x_k = 0.$$

It is easily seen that its center is the barycenter $(1, \ldots, 1)$ [or $(\frac{1}{n+1}, \ldots, \frac{1}{n+1})$ in non-homogeneous barycentric coordinates]; S contains all vertices A_i of Σ, the tangent plane to S at A_i has equation

$$\sum_{j \neq i} x_j = 0$$

and is parallel to ω_i.

Let us present a few general facts about quadrics in barycentric coordinates.

A quadric with the equation

(45)
$$\sum_{i,k=1}^{n+1} b_{ik} x_i x_k = 0, \quad B = (b_{ik}) = B^T,$$

is called *nonsingular* if $\det B \neq 0$.

The *polar* of a point $y = (y_i)$ w. r. to such a nonsingular quadric (45) is the hyperplane

$$\sum_{i,k=1}^{n+1} b_{ik} y_i x_k = 0;$$

indeed, not all coefficients at x_k are equal to zero.

A nonsingular quadric (45) is called *central* if there is a proper point (the center) the polar of which is the improper hyperplane (6).

Lemma 4.4. *A nonsingular quadric (45) is central if and only if*

(46)
$$e^T B^{-1} e \neq 0.$$

PROOF. If (45) is central and $C = (c_i)$ is the center then

(47)
$$\sum_i c_i \neq 0$$

and

$$\sum_{i,k} b_{ik} c_i x_k = 0$$

is the equation of the improper hyperplane, i. e.

(48)
$$\sum_i b_{ik} c_i = K, \text{ a nonzero constant, for all } k.$$

Thus in matrix form for $\hat{c} = (c_1, \ldots, c_{n+1})^T$,

$$B\hat{c} = Ke$$

which implies

$$\sum_i c_i = e^T \hat{c} = K e^T B^{-1} e.$$

By (47), we obtain (46).

Conversely, if (46) holds then the unique solution c_i to

$$\sum_i b_{ik} c_i = 1$$

satisfies (47) and defines a center of (45).

To find the axes of a nonsingular central quadric, let us use the following characteristic property of their improper points:

An improper point y is the improper point of an axis of a nonsingular central quadric if and only if the polar π of y is orthogonal to y, i. e. if and only if the orthogonal point to π coincides with y.

Theorem 4.5. *An improper point* $y = (y_1, \ldots, y_{n+1})$ *is the improper point of an axis for a nonsingular central quadric (45) if and only if the column vector* $\hat{y} = (y_1, \ldots, y_{n+1})^T$ *is an eigenvector of the matrix* QB *corresponding to a nonzero eigenvalue* λ *of* QB.

The square l^2 *of the length of the corresponding half-axis is then*

(49)
$$l^2 = -\frac{1}{\lambda e^T B^{-1} e}.$$

Remark. The square l^2 can even be negative (if (45) is not an ellipsoid).

PROOF. By the mentioned characterization, y satisfying $\sum_i y_i = 0$ is the improper point of an axis of (45) if and only if the orthogonal point (by (38)) $z = (z_i)$,

$$z_i = \sum_{j,k} q_{ij} b_{jk} y_k$$

to the polar of y

$$\sum_{j,k} b_{jk} y_k x_j = 0$$

coincides with y, i. e. if

$$z_i = \lambda y_i, \quad i = 1, \ldots, n+1,$$

$$\lambda \neq 0.$$

This yields for the column vector $\hat{y} \neq 0$

(50)
$$QB\hat{y} = \lambda \hat{y}, \quad \lambda \neq 0.$$

[The converse is also true.]

The length l of the corresponding half-axis is the distance from the center c of the intersection point, say u, of the line cy with the quadric [strictly speaking, if it is real].

Let thus, in a clear notation,

(51)
$$u = c + \xi y \text{ for some } \xi,$$

$$\sum_{i,k} b_{ik} u_i u_k = 0.$$

This yields

$$\sum_{i,k} b_{ik} c_i c_k + 2\xi \sum_{i,k} b_{ik} c_i y_k + \xi^2 \sum b_{ik} y_i y_k = 0.$$

The middle term being zero by (48), we obtain

(52)
$$\xi^2 = -\frac{\hat{c}^T B \hat{c}}{\hat{y}^T B \hat{y}}.$$

[Indeed, $\hat{y}^T B \hat{y} < 0$ by (iii) of Cor. 3.8.]

Without loss of generality, we can assume that in

(53)
$$B\hat{c} = Ke, \quad K \text{ constant},$$

$\hat{c}$ and K are such that $e^T \hat{c} = 1$, i. e. also

(54)
$$e^T B^{-1} e = \frac{1}{K}.$$

Then

$$e^T \hat{u} = 1$$

in (51) and the distance l of u and c satisfies by (30)

$$
\begin{aligned}
l^2 &= -\frac{1}{2} \sum_{i,k} m_{ik}(u_i - c_i)(u_k - c_k) \\
&= -\frac{1}{2}\xi^2 \sum_{i,k} m_{ik} y_i y_k \\
&= -\frac{1}{2}\xi^2 \hat{y}^T M \hat{y} \\
&= -\frac{\xi^2}{2\lambda} \hat{y}^T M Q B \hat{y}
\end{aligned}
$$

by (50).

Since

$$MQ = -2I_{n+1} - eq_0^T$$

by (9) and $\hat{y}^T e = 0$,

$$\hat{y}^T M Q = -2\hat{y}^T$$

and

$$l^2 = \frac{\xi^2}{\lambda}\hat{y}^T B\hat{y}$$
$$= -\frac{1}{\lambda}\hat{c}^T B\hat{c}$$
$$= -\frac{K}{\lambda}\hat{c}^T e$$
$$= -\frac{1}{\lambda e^T B^{-1} e}$$

by (53) and (54).

Corollary 4.6. *The squares l_i^2 of half-axes of the Steiner circumscribed ellipsoid S from (44) are given by*

(55)
$$l_i^2 = \frac{n}{n+1}\frac{1}{\lambda_i},$$

where the $\lambda_i's$ are the nonzero eigenvalues of the matrix Q. The directions of the corresponding axes coincide with the eigenvectors $\hat{y} = (y_i)$ of Q. Also

$$\sum_{i=1}^{n+1} y_i x_i = 0$$

are equations of the hyperplanes through the center of S orthogonal to the corresponding axes (hyperplanes of symmetry).

Remark. Since the $\lambda_i's$ are positive, S is indeed an ellipsoid.

PROOF. Follows immediately from Thm. 4.5 applied to $B = J - I, J = ee^T$ since

$$B^{-1} = \frac{1}{n}J - I,$$

$$QB = -Q$$

and

$$e^T B^{-1} e = \frac{n+1}{n}.$$

5. Simplex of a graph. In this section, we shall investigate undirected graphs without loops and multiple edges with n nodes. We assume that the set of nodes is $N = \{1, 2, \ldots, n\}$ and we write $G = (N, E)$ where E denotes the set of edges.

Recall that the *Laplacian matrix* of $G = (N, E)$, shortly *Laplacian*, is the real symmetric $n \times n$ matrix $L(G)$ whose quadratic form is

(56)
$$(L(G)x, x) = \sum_{i,k,i<k,(i,k)\in E} (x_i - x_k)^2.$$

Let us list a few elementary properties of $L(G)$:

(L1)
$$L(G) = D(G) - A(G),$$

where $A(G)$ is the adjacency matrix of G and $D(G)$ the diagonal matrix the i-th diagonal entry of which is d_i, the degree of the node i in G;

(L2) $L(G)$ is positive semidefinite and singular,

(57) $$L(G)e = 0;$$

(L3)   if G is connected, then $L(G)$ has rank $n - 1$;

(L4) for the complement $\bar{G}$ of G,

$$L(G) + L(\bar{G}) = nI - J,$$

where $J = ee^T$.

More generally, one can define the Laplacian of a weighted graph $G_C = (N, E, C)$ with nonnegative weight $c_{ij} = c_{ji}$ assigned to each edge $(i, j) \in E$ as follows:

$L(G_C)$ is the symmetric matrix of the quadratic form

(58) $$\sum_{(i,j)\in E, i<j} c_{ij}(x_i - x_j)^2.$$

Clearly, $L(G_C)$ is again positive semidefinite and singular with $L(G_C)e = 0$. If the graph with the node set N and the set of in C positively weighted edges is connected, the rank of $L(G_C)$ is $(n - 1)$. In this last case, we shall call such weighted graph *connected weighted graph*.

Let now G (or G_C) be a (weighted) graph with n vertices. The eigenvalues

$$\lambda_1 \leq \lambda_2 \leq \ldots \leq \lambda_n$$

of $L(G)$ (or $L(G_C)$) will be called Laplacian eigenvalues of G (G_C) (λ_1 the first, λ_2 the second etc.).

The second smallest Laplacian eigenvalue λ_2 was denoted as $a(G)$ and called algebraic connectivity in [4] since it has similar properties as the edge-connectivity $e(G)$, the number of edges (or, the sum of edge-weights) in the minimum cut.

In addition, the following inequalities were proved in [4]:

$$2(1 - \cos\frac{\pi}{n})e(G) \leq a(G) \leq e(G).$$

An important property of the eigenvector corresponding to λ_2 was proved in [5]. We state the result without proof:

Theorem 5.1. *Let G be a connected graph, let u be a (real) eigenvector of $L(G)$ corresponding to the algebraic connectivity $a(G)$. Then the subgraph of G induced on the node set corresponding to the nodes with nonnegative coordinates of u is connected.*

Observe now that the Laplacian $L(G)$ of a connected graph G as well as the Laplacian $L(G_C)$ of a connected weighted graph G_C satisfy the conditions (16) and (17) (with n instead of $n+1$) of the matrix Q.

Therefore, we can assign to G [or G_C] in $\hat{\mathcal{E}}_{n-1}$ an - up to congruence uniquely defined - $(n-1)$-simplex $\Sigma(G)$ [or, $\Sigma(G_C)$] which we shall call *simplex of the graph* G [G_C, respectively]. The corresponding Menger matrix M will be called the *Menger matrix* of G.

Applying Thm. 3.3 we have immediately, formulated just for the more general case of a weighted graph:

Theorem 5.2. *Let e be the column vector of n ones. Then there exists a unique column vector q_0 and a unique number q_{00} such that the symmetric matrix $M = (m_{ik})$ with $m_{ii} = 0$ satisfies*

(59)
$$\begin{pmatrix} 0 & e^T \\ e & M \end{pmatrix} \begin{pmatrix} q_{00} & q_0^T \\ q_0 & L(G_C) \end{pmatrix} = -2I_{n+1}.$$

The (unique) matrix M is the Menger matrix of G_C.

Example 5.3. Let G be the path P_4 with four nodes 1, 2, 3, 4 and edges (1,2), (2,3), (3,4). Then (59) is as follows:

$$\begin{pmatrix} 0 & 1 & 1 & 1 & 1 \\ 1 & 0 & 1 & 2 & 3 \\ 1 & 1 & 0 & 1 & 2 \\ 1 & 2 & 1 & 0 & 1 \\ 1 & 3 & 2 & 1 & 0 \end{pmatrix} \begin{pmatrix} 3 & -1 & 0 & 0 & -1 \\ -1 & 1 & -1 & 0 & 0 \\ 0 & -1 & 2 & -1 & 0 \\ 0 & 0 & -1 & 2 & -1 \\ -1 & 0 & 0 & -1 & 1 \end{pmatrix} = -2I_5.$$

$$M \qquad\qquad\qquad L(P_4)$$

Example 5.4. For a star S_n with nodes $1, 2, \ldots, n$ and the set of edges (i, k), $k = 2, \ldots, n$, the equality (59) reads

$$\begin{pmatrix} 0 & 1 & 1 & 1 & \ldots & 1 \\ 1 & 0 & 1 & 1 & \ldots & 1 \\ 1 & 1 & 0 & 2 & \ldots & 2 \\ 1 & 1 & 2 & 0 & \ldots & 2 \\ \cdot & \cdot & \cdot & \cdot & & \cdot \\ 1 & 1 & 2 & 2 & \ldots & 0 \end{pmatrix} \begin{pmatrix} n-1 & n-3 & -1 & -1 & \ldots & -1 \\ n-3 & n-1 & -1 & -1 & \ldots & -1 \\ -1 & -1 & 1 & 0 & \ldots & 0 \\ -1 & -1 & 0 & 1 & \ldots & 0 \\ \cdot & \cdot & \cdot & \cdot & \cdot & \cdot \\ -1 & -1 & 0 & 0 & \ldots & 1 \end{pmatrix} = -2I_{n+1}.$$

Remark. Observe that in both cases, the Menger matrix M is at the same time the *distance* matrix of G, i. e. the matrix $D = (D_{ik})$ for which D_{ik} means the distance of the nodes i and k in G_C, in general the minimum length of the lengths of all possible paths between i and k, the length of a path being the sum of the lengths of edges contained in the path. We intend to prove that $M = D$ for all weighted trees the length of each edge of which is appropriately chosen.

Theorem 5.5. *Let $T_C = (N, E, C)$ be a connected weighted tree, $N = \{1, \ldots, n\}$, $c_{ik} = c_{ki}$ denoting the weight of the edge $(i, k) \in E$. Let $L(T_C) = (q_{ik})$ be the Laplacian* *of T_C so that*

$$q_{ii} = \sum_{k,(i,k)\in E} c_{ik},$$

$$q_{ik} = -c_{ik} \text{ for } (i,k) \in E,$$

$$q_{ik} = 0 \text{ otherwise }, \ i,k \in N.$$

Denote further for $i, j, k \in N$

$$R_{ij} = \frac{1}{c_{ij}} \text{ for } (i,j) \in E,$$

$$R_{ii} = 0,$$

$$R_{ik}(= R_{ki}) = R_{ij_1} + R_{j_1 j_2} + \cdots + R_{j_{s-1},j_s} + R_{j_s,k}$$

whenever $(i, j_1, j_2, \ldots, j_s, k)$ is the (unique) path from i to k in T_C.

If, moreover,

$$R_{00} = 0,$$

$$R_{0i} = 1, \ i \in N,$$

$$q_{00} = \sum_{(i,k)\in E, i<k} R_{ik},$$

$$q_{0i} = d_i - 2, \ d_i \text{ being the degree of } i \in N \text{ in } T_C,$$

then the symmetric matrices

$$\tilde{R} = (R_{rs}), \ \tilde{Q} = (q_{rs}), \ r,s = 0,1,\ldots,n$$

satisfy

$$\tilde{R}\tilde{Q} = -2I_{n+1}.$$

PROOF. We have to show that

(60) $$\Sigma R_{rs} q_{st} = -2\delta_{rt}, \ r,t = 0,1,\ldots,n.$$

For $r = t = 0$, the left-hand side is $\sum_{i=1}^{n}(d_i - 2)$ which is -2 since $\Sigma d_i = 2(n-1)$, the number of edges in T_C.

For $r = 0$, $t \in N$, the left-hand side is zero.

For $r \in N$, $t = 0$

$$\sum_{s=0}^{n} R_{rs} q_{s0} = \sum_{(j,l)\in E, j<l} R_{jl} - \sum_{m \neq r}(2 - d_m)(R_{rj_1} + R_{j_1,j_2} + \cdots + R_{j_k m})$$

where $(r, j_1, \ldots, j_k, m)$ is the path in T_C from r to m. To show that the right-hand side is zero, it suffices to prove that in second sum, R_{jl} appears for every edge $(j,l) \in E$

with the coefficient 1. Thus, let $(j,l) \in E$. Denote by N_j, N_l respectively, the sets of edges in the components T_j, T_l of the graph obtained from T_C by deleting the edge (j,l). Let $i \in N_j$, say. Then R_{jl} appears in the retms corresponding to $m \in N_l$ only, with the coefficient $\sum_{m \in V_l}(2 - d_m)$. Now,

$$\sum_{m \in V_l} d_m = 1 + 2(|V_l| - 1)$$

since $m \in V_l$ has in T_l the same degree as in T_C with the exception of l for which it is by one less. Thus

$$\sum_{m \in V_l}(2 - d_m) = 2|V_l| - 1 - 2(|V_l| - 1)$$
$$= 1.$$

For $r = t \in N$ we obtain

$$\sum_{s=0}^{n} R_{rs}q_{sr} = q_{0r} + \sum_{j \in N, j \neq r} R_{rj}q_{jr}$$
$$= d_r - 2 + \sum_{j,(r,j) \in E} \frac{1}{c_{rj}}(-c_{rj})$$
$$= d_r - 2 - d_r$$
$$= -2.$$

Finally, for $r \in N, t \in N, r \neq t$

$$\sum_{s=0}^{n} R_{rs}q_{st} = q_{0t} + R_{rt}q_{tt} + \sum_{k \in N, r \neq k \neq t} R_{rk}q_{kt}$$
$$= d_t - 2 - R_{rt} \sum_{k,(k,t) \in E} c_{kt} - \sum_{k,(k,t) \in E} R_{rk}c_{kt}.$$

Denote by w the neighboring node to t in the path from r to t in T_C. Then,

$$R_{rw} = R_{rt} - R_{wt}$$

whereas

$$R_{rk} = R_{rt} + R_{kt}$$

for every other neighbor $k \neq w$ of t. Therefore, in the last expression,

$$\sum_{k,(k,t) \in E} R_{rk}c_{kt} = R_{rt} \sum_{k,(k,t) \in E} c_{kt} - R_{wt}c_{wt} + \sum_{k,(k,t) \in E, k \neq w} R_{kt}c_{kt}$$
$$= R_{rt} \sum_{k,(k,t) \in E} c_{kt} + d_t - 2$$

and (60) is again true.

For $c_{ik} = 1$ for all i, k, we obtain: **Corollary 5.6.** *Let $T = (N, E)$ be a tree with n nodes. Denote by $z = (z_i)$ the column n-vector with $z_i = d_i - 2$, d_i being the degree of the node i. Then:*
(i) the Menger matrix M of T coincides with the (usual) distance matrix D of T;
(ii) the equality (55) reads

$$(61) \qquad \begin{pmatrix} 0 & e^T \\ e & D \end{pmatrix} \begin{pmatrix} n-1 & z^T \\ z & L(T) \end{pmatrix} = -2I_{n+1}.$$

Let us return now to Thm. 5.2 and formulate a result which will relate the Laplacian eigenvalues with the Menger matrix M :

Theorem 5.7. *Let M be the Menger matrix of a graph G. Then the $n-1$ roots of the equation*

$$(62) \qquad \det \begin{pmatrix} 0 & e^T \\ e & M - \mu I \end{pmatrix} = 0$$

are the numbers $\mu_i = -\frac{2}{\lambda_i}$, $i = 2, \ldots, n$ where $\lambda_2, \ldots, \lambda_n$ are the nonzero Laplacian eigenvalues of G. If $y^{(i)}$ is an eigenvector of $L(G)$ corresponding to $\lambda_i \neq 0$ then $\begin{pmatrix} \frac{1}{\lambda_0} q_0^T y^{(i)} \\ y^{(i)} \end{pmatrix}$ is the corresponding annihilating vector of the matrix in (62).

PROOF. The first part follows immediately from the identity

$$(63) \qquad \begin{pmatrix} 0 & e^T \\ e & M - \mu I \end{pmatrix} \begin{pmatrix} q_{00} & q_0^T \\ q_0 & L(G) \end{pmatrix} = \begin{pmatrix} -2 & 0 \\ * & -2I - \mu L(G) \end{pmatrix}$$

where $*$ is some column vector.

Postmultiplying (63) for $\mu = -\frac{2}{\lambda_i}$ by $\begin{pmatrix} 0 \\ y^{(i)} \end{pmatrix}$, we obtain the second assertion.

Corollary 5.8. *Let $L(G) = ZZ^T$ be any full-rank-factorization, i. e. Z an $n \times (n-1)$ matrix. Then*

$$(64) \qquad Z^T M Z = -2I_{n-1}.$$

PROOF. By (59),

$$M L(G) = -2I_n - e q_0^T.$$

Since $Z^T e = 0$, we obtain premultiplying by Z^T

$$Z^T M Z Z^T = -2Z^T,$$

i. e. (64).

Corollary 5.9. *The roots of (62), i. e. the numbers $-\frac{2}{\lambda_i}$ where λ_i are the nonzero Laplacian eigenvalues of G, $i = 2, \ldots, n$, interlace the eigenvalues of the Menger matrix M of G.*

PROOF. Follows immediately from the following Lemma which is easily proved by transforming A to diagonal form by orthogonal transformation:

Lemma. *Let A be a real symmetric matrix, u a nonzero real vector, t a real number. Then the zeros of the equation*

$$\det \begin{pmatrix} t & u^t \\ u & A - xI \end{pmatrix} = 0$$

interlace the eigenvalues of A.

Remark. Corollaries 5.8 and 5.9 generalize the results of R. Merris [8] for the distance matrix of a tree.

Using the previous Lemma, an analogous proof to that of Thm. 5.7 gives:

Theorem 5.10. *The eigenvalues m_i of the Menger matrix M of G and the roots x_i of the equation*

$$\det \begin{pmatrix} q_{00} & q_0^T \\ q_0 & L(G) - xI \end{pmatrix} = 0$$

satisfy

$$m_i x_i = -2, \quad i = 1, \ldots, n.$$

The numbers x_i and the Laplacian eigenvalues λ_i of G interlace each other.

Let us return now to the geometric considerations from Section 4.

Theorem 5.11. *Let λ_i be a nonzero eigenvalue of $L(G_C)$, y a corresponding eigenvector. Then,*

$$\frac{n-1}{n} \cdot \frac{1}{\lambda_i}$$

is the square of a half-axis of the Steiner circumscribed ellipsoid of the simplex $\Sigma(G_C)$ of G_C and $\Sigma y_i x_i = 0$ is the equation of the hyperplane orthogonal to the corresponding axis. Also, y is the direction of the axis.

Corollary 5.12. *The smallest positive eigenvalue of $L(G_C)$ (the algebraic connectivity of G_C) corresponds to the largest half-axis of the Steiner circumscribed ellipsoid of $\Sigma(G_C)$.*

It has been conjectured [11] that the eigenvector $y = (y_i)$ corresponding to the second smallest eigenvalue $a(G)$ of G is due to the result presented here as Thm. 5.1 a good separator in the set of nodes N of G in the sense that the ratio of the cardinalities of the two parts is neither very small nor very large. The geometric meaning of the subsets N^+ and N^- of N is as follows:

Theorem 5.13. *Let $y = (y_i)$ be an eigenvector of $L(G)$ corresponding to $\lambda_2 (= a(G))$. Then*

$$N^+ = \{i \in N \mid y_i > 0\},$$

$$N^- = \{i \in N \mid y_i < 0\},$$

$$Z = \{i \in N \mid y_i = 0\}$$

correspond to the number of vertices of the simplex $\Sigma(G)$ in the decomposition of $\widehat{\mathcal{E}}_n$ with respect to the hyperplane of symmetry H of the Steiner circumscribed ellipsoid orthogonal to the largest half-axis: $\mid N^+ \mid$ is the number of vertices of $\Sigma(G)$ in one halfspace H^+, $\mid N^- \mid$ the number of vertices in H^-, and $\mid Z \mid$ the number of vertices in H.

6. Concluding remarks. Let us recall [6] the relationship between the resistive electrical circuits and simplices.

Let a connected resistive circuit C be given with nodes $1, 2, \ldots, n$ and branches (i, k), $i \neq k$, with conductivities $c_{ik}(= c_{ki})$; c_{ik} is the reciprocal of the resistance of the branch (if i and k are not joined by a direct branch, $c_{ik} = 0$). To C, we assign an $(n-1)$-simplex Σ whose normalized Gram matrix $Q = (q_{ik})$ is defined by

$$q_{ik} = -c_{ik}, \; i \neq k, \; i, k = 1, \ldots, n,$$
$$q_{ii} = \sum_{k \neq i} c_{ik}, \; i = 1, \ldots, n.$$

One can show that the total resistance R_{ik} between the nodes i and k, $i \neq k$, is equal to the (i, k)-entry of the corresponding Menger matrix. It follows from Thm. 3.13 and the nonnegativity of the c_{ik}'s that the simplex Σ has no obtuse angle, and, in fact, this completely characterizes all such simplices arising from resistive circuits. Thus knowing that in a black box with n outputs there are only resistors, then measuring mutual total resistances between the nodes gives us the Menger matrix of an $(n-1)$-simplex the corresponding normalized Gram matrix Q of which satisfying (9) yields conductivities in the branches (i, k) of a circuit with the same outputs as the given black box.

It would be desirable to find the interpretation of the numbers q_{0i} and q_{00} in the electrical and of q_{00} in the graph-theoretical model.

REFERENCES

[1] L. BLUMENTHAL, Theory and Applications of Distance Geometry. Clarendon Press, Oxford 1953.

[2] M. FIEDLER, Geometry of a simplex in E_n. I. Čas. pro pěst. mat. 79(1954), 297-320.

[3] M. FIEDLER, Über qualitative Winkeleigenschaften der Simplexe, Czechoslovak Math. J. 7(1957), 463-478.

[4] M. FIEDLER, Algebraic connectivity of graphs. Czechoslovak Math. J. 23(1973), 298-305.

[5] M. FIEDLER, A property of eigenvectors of non-negative symmetric matrices and its application to graph theory, Czechoslovak Math. J. 25(1975), 619-633.

[6] M. FIEDLER, Aggregation in graphs. In: Combinatorica (A. Hajnal, V. T. Sós - Editors), Coll. Math. Soc. J. Bolyai, vol. 18, 1976, 315-330.

[7] K. MENGER, Untersuchungen über allgemeine Metrik. Math. Annalen 100(1928), 75-163.

[8] R. MERRIS, The distance spectrum of a tree. J. Graph Th. 14(1990), 365-369.

[9] B. MOHAR, The Laplacian spectrum of graphs. In: Graph Theory, Combinatorics and Applications (Y. Alavi, et al., eds.), Wiley, New York, 1991, 871-898.

[10] B. MOHAR, S. POLJAK, Eigenvalues and the max-cut-problem. Preprint, Charles University, Prague 1988.

[11] A. POTHEN, H. SIMON, AND K.-P. LIOU, Partitioning sparse matrices with eigenvectors of graphs. SIAM J. Matr. Anal. Appl. 11(1990), 430-452.

QUALITATIVE SEMIPOSITIVITY

CHARLES R. JOHNSON*† AND DAVID P. STANFORD*

1. Introduction and background. A real m-by-n matrix A is called *semipositive* (SP) provided there is a real n-vector $x \geq 0$ such that $Ax > 0$. Vector and matrix inequalities here and throughout the paper denote entrywise inequalities.

The above condition for semipositivity is equivalent to the existence of an n-vector $x > 0$ such that $Ax > 0$, and in the case $m = n$, this condition can be written

$$\mathbf{R}^n_+ \cap A(\mathbf{R}^n_+) \neq \phi, \quad \text{with } \mathbf{R}^n_+ = \{x \in \mathbf{R}^n : x > 0\}.$$

A generalization of this condition, in which the open cone $\mathbf{R}^n_+$ is replaced by a more general open cone, is discussed in [BNS] in the context of M-matrices. It is observed there that, in this more general setting, a Z-matrix is a nonsingular M-matrix if and only if it is semipositive.

Semipositivity is also mentioned in [BP], and a more extensive study of semipositivity appears in [JKS]. The latter includes the following material, which is relevant to the present discussion.

1. If A is SP, then any matrix obtained from A by one of the following operations is also SP:

   (i) adjoining a column

   (ii) deleting a column with no positive entries

   (iii) deleting a row

   (iv) adjoining a row with no negative entries and at least one positive entry

   (v) multiplying A on either the right or the left by a permutation matrix

   (vi) multiplying A on either the right or the left by a diagonal matrix with all diagonal entries positive.

2. Let A be a real m-by-n matrix with $m \neq n$.

   (i) If $m > n$, then A is SP if and only if each n-by-n submatrix of A is SP.

   (ii) If $m < n$, then A is SP if and only if there is an m-by-m submatrix of A that is SP.

*Department of Mathematics, The College of William and Mary, Williamsburg, Virginia 23187-8795.

†The work of this author was supported in part by National Science Foundation grant DMS 90-00839 and by Office of Naval Research contract N00014-90-J1739.

DEFINITION. A semipositive matrix A is called *minimally semipositive* (MSP) provided that no matrix obtained from A by deleting one or more columns is semi-positive. A semipositive matrix that is not MSP is called *redundantly semipositive* (RSP).

3.   Operations (ii), (iv), (v), and (vi) of Part 1 preserve minimal semipositivity. Operations (i) and (iii) do not do so in general. Operations (i), (iii), (v), and (vi) preserve redundant semipositivity, while (ii) and (iv) do not do so in general.

4.   Let A be a real n-by-n matrix. Then A is MSP if and only if A is inverse nonnegative; i.e., A is nonsingular and $A^{-1} \geq 0$. Note that by 2(ii) any MSP matrix must have at least as many rows as columns.

We also note that semipositivity is a *monotonic* property of a matrix. If A, $B \in M_{m,n}(\mathbf{R})$, A is SP and $B \geq A$, then B is SP. In fact, if $\mathcal{K}(A) = \{x \geq 0 : Ax > 0\}$ then $\mathcal{K}(A) \subseteq \mathcal{K}(B)$. The proof is immediate. This fact implies also that redundant semipositivity is monotone. Minimal semipositivity is clearly not monotone for $n > 1$, since for an m-by-n MSP matrix A with $n > 1$, a matrix $B \geq 0$ can be chosen so that $A + B$ has a positive column, and so is RSP.

An m-by-n sign *pattern* is an m-by-n array of symbols chosen from $\{+, -, 0\}$. A *realization* of a sign pattern S is a real matrix A such that

$$a_{ij} > 0 \quad \text{when} \quad s_{ij} = +;$$
$$a_{ij} < 0 \quad \text{when} \quad s_{ij} = -;$$
$$\text{and } a_{ij} = 0 \quad \text{when} \quad s_{ij} = 0.$$

If P is a property that a matrix may have, then the sign pattern S *allows* P if there is a realization of S that enjoys property P, and S *requires* P if each realization of S enjoys P.

In [JLR], the sign patterns that can occur among the real, nonsingular, entrywise nonzero, inverse positive matrices are identified as follows. A square sign pattern S is called *complementary* provided there are permutation matrices P and Q such that PSQ has the partitioned form

$$\begin{bmatrix} * & + \\ - & * \end{bmatrix}.$$

Here, the diagonal blocks need not be square, and one block row or column may be empty. It is shown that a nonsingular, entrywise nonzero matrix is inverse positive if and only if its sign pattern is *not* complementary. This has been generalized to less restrictive sign patterns in [FG] and [J].

Our purpose here is to study semipositivity, and the related notions of minimal and redundant semipositivity, from the point of view of qualitative matrix theory. These properties have not previously been considered from a qualitative point of view. Qualitative matrix theory is the study of the sign patterns that require or allow a given property. The remaining sections are devoted to the require and allow questions for each of the properties SP, RSP and MSP, in turn. An easily checked characterization is given in each case, except that a complete understanding of the patterns that allow MSP in the non-square case remains open.

2. Semipositivity. This allow question for semipositivity is easily addressed.

THEOREM 1. *An m-by-n sign pattern S allows semipositivity if and only if each row of S contains a +.*

Proof. If each row of S contains $a+$, then there is a realization A of S in which these positive entries are large enough, and other entries small enough, so that all row sums of A are positive. Then, for $e = [1\ 1 \cdots 1]^T$, $Ae > 0$, so that A is SP. Conversely, if the ith row of S had only $-$'s and 0's, then $x \geq 0$ would imply that, for any realization A of S, $(Ax)_i \leq 0$, so that A could not be SP. □

Clearly any sign pattern that contains a column of $+$'s requires semipositivity. In order to describe all sign patterns that require semipositivity, we define the notion of a "generalized positive column".

DEFINITION 1. A sign pattern is said to have form G provided each row has a nonzero entry, the rightmost one being a $+$, and, in each row after the first the rightmost nonzero entry occurs in a position not to the left of the rightmost nonzero entry in the preceding row.

Thus an example of a sign pattern having form G is:

$$\begin{bmatrix} * & + & 0 & 0 & 0 & 0 & 0 \\ * & + & 0 & 0 & 0 & 0 & 0 \\ * & * & + & 0 & 0 & 0 & 0 \\ * & * & * & * & + & 0 & 0 \\ * & * & * & * & * & + & 0 \\ * & * & * & * & * & + & 0 \end{bmatrix}$$

The rightmost nonzero entry in each row of a sign pattern having form G is called a *frontal plus* of the row, and of the sign pattern.

DEFINITION 2. A generalized positive column is a sign pattern S which is permutation equivalent to a sign pattern having form G; i.e., $S = PS'Q$ for some sign pattern S' having form G and some permutation matrices P and Q.

We note that the sign pattern S is a generalized positive column if and only if no row consists entirely of 0's and some permutation of its columns produces a sign pattern in which the rightmost nonzero entry in each row is a $+$.

We will need the following lemma, which may be viewed as a qualitative theorem of the alternative:

LEMMA 1. *If S is a sign pattern that requires semipositivity, then S has a column in which each entry is nonnegative.*

Proof. If each column of S has a $-$, then there is a realization A of S in which each column sum is negative; i.e., $e^T A < 0$. Then, for any $x > 0$, $e^T Ax < 0$, so the sum of the entries of Ax is negative and Ax is not a positive vector. Thus A is not SP, so that S does not require semipositivity. □

THEOREM 2. *An m-by-n sign pattern S requires semipositivity if and only if S is a generalized positive column.*

Proof. Assume first that S is a generalized positive column. Let S' be permutation equivalent to S and have form G. We will show that S' requires semipositivity, and it will follow from $1(v)$ that S does. Let A be any realization of S'. We construct a vector $x \geq 0$ such that $Ax > 0$ as follows. Let $1 \leq j_1 < j_2 < \cdots < j_c \leq n$ be such that the columns of A that contain frontal plusses of rows of A are the columns numbered $j_1, j_2, \ldots, j_c$. For $1 \leq j \leq n$ and $j \notin \{j_1, j_2, \ldots, j_c\}$, let $x_j = 0$. Choose $x_{j_1} = 1$, and, independently of how the other x_{j_i} are chosen (because of the 0's to the right of a frontal $+$), those rows in A whose frontal plusses are in column j_1 will have a positive product with x. We may now select x_{j_2} large enough so that all rows of A with a frontal plus in column j_2 will have a positive product with x, independently of the choice of x_{j_i} for $i > 2$. We can continue in this way to select $x_{j_3}, \ldots, x_{j_c}$ so that each row A has a positive product with x, and so $Ax > 0$. Thus any realization of S' is semipositive, and S requires semipositivity.

Now suppose S is a sign pattern that requires semipositivity. We wish to show that there are permutation matrices P and Q such that PSQ has form G. By the qualitative theorem of the alternative, S has a column with each entry nonnegative. Permute this column to the right to form sign pattern S'. If each entry in this column is positive, S' has form G, and we are finished. Otherwise, permute the rows so that the last column has r zeros above $(n - r)+$'s, and call the resulting sign pattern S''. Now let S_1 be the subpattern of S'' lying in the first r rows and the first $n - 1$ columns. S_1 also requires semipositivity, since each realization of S_1 is obtained from a realization of S'' by deleting the last $n - r$ rows and then deleting the last column, which now consists of zeros. Thus S_1 contains a column in which each entry is nonnegative. We may repeat in S_1 the operations we have already performed on S, permuting entire rows and columns of S'' as we do so. Since S_1 does not intersect the last column of S'', that column will remain in the last place. Furthermore, only rows of S'' intersecting S_1 will be permuted at this stage, and all such rows have zero in the last position, so that the last column of S'' is unchanged as we operate on S_1. If the operation on S_1 did not produce a last column in S_1 with all $+$'s, we continue to form S_2, S_3, and so on. Since the number of rows in S_i decreases as i increases, this process terminates with some S_i having a positive last column, and at that point we have a sign pattern having form G that can be written in the form PSQ for some permutation matrices P and Q. $\square$

We note that, by following the proof, it may be decided algorithmically if S is permutation equivalent to a generalized positive column (and S may be put in this form if possible) in no more effort than the order of $p^2 q$, in which p is the smaller and q the larger of the number of rows and columns of S.

3. Redundant semipositivity. In this section we answer both the allow and require questions for the property of redundant semipositivity.

THEOREM 3. *The sign pattern S allows RSP if and only if each row has a $+$ and there is a column that does not contain the only $+$ in any row.*

Proof. Suppose S allows RSP. Then S allows SP, so each row has a $+$ by theorem 1. Let A be an RSP realization of S. Some column deleted submatrix B of A is SP, so the corresponding column deleted subpattern of S allows SP, so has a $+$ in each row. Hence the column that was deleted from S did not contain the only $+$ in any row.

Now suppose each row of S has a $+$, and column j of S does not contain the only $+$ in any row of S. The pattern $\widehat{S}$ resulting from the deletion of column j of S has a $+$ in each row, and so allows semipositivity. Let $\widehat{A}$ be an SP realization of $\widehat{S}$. Adjoin a properly signed column to $\widehat{A}$ to get an RSP realization A of S. Thus, S allows RSP. □

THEOREM 4. *The sign pattern S requires RSP if and only if S is permutation equivalent to a sign pattern S', where S' has form G and $s'_{1k} \neq 0$ for some $k > 1$.*

Proof. The "if" part is clear. To prove the "only if" part, suppose that S is an m-by-n sign pattern that requires RSP. For $t > 0$, let $A(t)$ be the realization of S whose i, j entry is $\begin{cases} \frac{1}{t} & \text{if } s_{ij} = + \\ -t & \text{if } s_{ij} = - \\ 0 & \text{if } s_{ij} = 0 \end{cases}$; let $A_j(t)$ denote the matrix obtained from $A(t)$ by deleting the jth column, and let $\mathcal{C}(t) = \{j : A_j(t) \text{ is } SP\}$. Since S requires RSP, $\mathcal{C}(t) \neq \phi$ for all $t > 0$. Clearly, $s > t > 0 \Rightarrow A(s) \leq A(t)$. We claim that $s > t > 0 \Rightarrow \mathcal{C}(s) \subset \mathcal{C}(t)$. Let $s > t > 0$ and suppose $j \in \mathcal{C}(s)$. Then $A_j(s)$ is SP, and $A_j(s) \leq A_j(t)$. Therefore, $A_j(t)$ is SP, by the monotonicity of semipositivity. Thus $j \in \mathcal{C}(t)$. Hence $\mathcal{C}(s) \subset \mathcal{C}(t)$. Now we claim that $\bigcap_{t>0} \mathcal{C}(t) \neq \phi$. Suppose $\bigcap_{t>0} \mathcal{C}(t) = \phi$. Then, for each $j = 1, 2, \ldots, n$, there is a positive number t_j such that $j \notin \mathcal{C}(t_j)$. Let $t > \max\{t_j : j = 1, 2, \ldots, n\}$. Then, for $j = 1, 2, \ldots, n$, $\mathcal{C}(t) \subset \mathcal{C}(t_j)$, so $j \notin \mathcal{C}(t)$. Hence $\mathcal{C}(t) = \phi$, contradicting the redundant semipositivity of $A(t)$. Thus $\bigcap_{t>0} \mathcal{C}(t) \neq \phi$. Let $j \in \bigcap_{t>0} \mathcal{C}(t)$. Then for each $t > 0$, the matrix $A_j(t)$ is SP. Let $\widehat{S}$ denote the sign pattern resulting from the deletion of the jth column of S, and let B be any realization of $\widehat{S}$. Choose $t > 0$ so that $-t$ is less than each entry of B and $\frac{1}{t}$ is less than each positive entry of B. Then $A_j(t) \leq B$. Since $A_j(t)$ is SP, B is SP by monotonicity. Hence $\widehat{S}$ requires SP. By permuting the jth column of S to the first column position and then permuting $\widehat{S}$ to a pattern having form G, we see that S is permutation equivalent to a sign pattern S' with the required properties. □

4. Minimal semipositivity. We consider first those sign patterns that require minimal semipositivity.

LEMMA 2. *A sign pattern S requires minimal semipositivity if and only if S requires semipositivity, and no column deleted subpattern of S allows semipositivity.*

Proof. Suppose S requires minimal semipositivity. Then S requires semipositivity. Now let T be a sign pattern obtained from S by deleting a column of S, and suppose T allows semipositivity. Let A be a realization of T that is semipositive.

We may insert into A a column whose sign pattern matches the column deleted from S to obtain T, to produce a realization B of S that is not minimally semipositive, contradicting the fact that S requires minimal semipositivity. Hence T does not allow semipositivity.

The converse is clear; if S requires semipositivity and no column-deleted subpattern of S allows semipositivity, then each realization of S is minimally semipositive. □

THEOREM 5. *The sign pattern S requires minimal semipositivity if and only if there are permutation matrices P and Q such that PSQ has form G, and each column of PSQ contains a frontal plus that is the only $+$ in its row.*

Proof. Suppose S requires minimal semipositivity. Then S requires semipositivity, so there are permutation matrices P and Q such that $S' = PSQ$ has form G. Suppose some column of S' contains no frontal entry of S' which is the only $+$ in its row. Then that column can be deleted, and the resulting sign pattern contains a $+$ in each row, and hence allows semipositivity. Thus S' does not require minimal semipositivity, so S does not. This contradiction establishes that PSQ has the required property.

Now suppose that S is a sign pattern and that P and Q are permutation matrices such that $S' = PSQ$ has form G, and each of its columns contains a frontal plus that is the only $+$ in its row. Then S' requires semipositivity, and furthermore if any column of S' is deleted, then the row of S' whose frontal plus is in that column and is the only $+$ in that row becomes a row with no $+$. Hence the column-deleted subpattern of S' does not allow semipositivity, so S' requires minimal semipositivity, and S does also. □

Because of 4, an n-by-n sign pattern requires MSP if and only if it requires inverse nonnegativity. In the square case, the condition of theorem 5 means that there is exactly one frontal plus in each column (as well as row) and, thus, no other positive entries. We may thus conclude:

COROLLARY. *An n-by-n sign pattern requires inverse nonnegativity if and only if it is permutation equivalent to a triangular matrix with positive diagonal entries and nonpositive off-diagonal entries.*

We turn now to consideration of those sign patterns that allow minimal semipositivity. We present only a necessary condition on a sign pattern S that it allow minimal semipositivity.

DEFINITION 3. A sign pattern S is called MS-complementary provided there are permutation matrices P and Q such that the pattern PSQ has the partitioned form $\begin{bmatrix} S_1 & S_2 \\ S_3 & S_4 \end{bmatrix}$, in which

a) S_3 is p-by-q with $p \geq 1$ and $q \geq 1$, and each entry of S_3 is 0 or $-$,

b) S_2 is r-by-s with $r \geq 0$ and $s \geq 0$, and each entry of S_2 is 0 or $+$. Furthermore, each row of S_2 has a $+$.

In this partition S_1 and S_4 are not required to be square.

THEOREM 6. *If the sign pattern S allows minimal semipositivity, then S is* **not** *MS-complementary.*

Proof. Suppose that S allows minimal semipositivity, and is MS-complementary. Then $S' = PSQ = \begin{bmatrix} S_1 & S_2 \\ S_3 & S_4 \end{bmatrix}$, with S_3 and S_2 as described above, allows minimal semipositivity.

If $r = 0$, $S' = [\, S_3 \quad S_4 \,]$, which fails to allow MSP since S_3 has a column with no $+$. If $s = 0$, $S' = \begin{bmatrix} S_1 \\ S_3 \end{bmatrix}$, which fails to allow MSP since S_3 has a row with no $+$. Hence $r \geq 1$ and $s \geq 1$.

Let $M = \begin{bmatrix} A & B \\ C & D \end{bmatrix}$ be a realization of S' that is minimally semipositive and that is partitioned according to the partitioning of S'. Using 1(iii) the matrix $[\, C \quad D \,]$ is semipositive. Then, using 1(ii) D is semipositive, and then using 1(iv) the matrix $\begin{bmatrix} B \\ D \end{bmatrix}$ is semipositive. This contradicts the minimal semipositivity of M. $\square$

We note that theorem 6 generalizes and gives another proof of the necessary condition in [JLR] because of the connection between minimal semipositivity and inverse nonnegativity mentioned in 4. The condition of Theorem 6 is not also sufficient.

The sign pattern

$$S = \begin{bmatrix} + & 0 & 0 \\ + & + & 0 \\ - & + & + \end{bmatrix}$$

neither allows MSP nor is MS-complementary. Thus, the question of which patterns allow MSP is open. However, by 4, the square case is covered by the results of [JLR], [FG], and [J], and, as an MSP matrix cannot have more columns than rows, only the case of more rows than columns need be considered.

REFERENCES

[BNS]   A. BERMAN, M. NEUMANN, AND R.J. STERN, *Nonnegative Matrices in Dynamic Systems*, John Wiley & Sons, 1989.

[BP]   A. BERMAN AND R.J. PLEMMONS, *Nonnegative Matrices in the Mathematical Sciences*, Academic Press, New York, 1979.

[FG]   M. FIEDLER AND R. GRONE, *Characterizations of Sign Patterns of Inverse Positive Matrices*, Linear Algebra and its Applications 40 (1981), pp. 237–245.

[J]   C.R. JOHNSON, *Sign Patterns of Inverse Nonnegative Matrices*, Linear Algebra and its Applications 55 (1983), pp. 69–80.

[JKS]   C.R. JOHNSON, M.K. KERR, AND D.P. STANFORD, *Semipositivity of Matrices*, The College of William and Mary, Department of Mathematics, Technical Report 91.09, 1991, submitted for publication.

[JLR]   C.R. JOHNSON, F.T. LEIGHTON, AND H.A. ROBINSON, *Sign Patterns of Inverse-Positive Matrices*, Linear Algebra and its Applications 24 (1979), pp. 75–83.

EIGENVALUES IN COMBINATORIAL OPTIMIZATION

BOJAN MOHAR* AND SVATOPLUK POLJAK†

Abstract. In the last decade many important applications of eigenvalues and eigenvectors of graphs in combinatorial optimization were discovered. The number and importance of these results is so fascinating that it makes sense to present this survey.

1. Introduction. The application of eigenvalue methods in combinatorial optimization has already a long history. The first eigenvalue bounds on the chromatic number were formulated by H. S. Wilf and A. J. Hoffman already at the end of sixties. Historically, the next applications, due to M. Fiedler and Donath and Hoffman in 1973, concerned the area of graph partition. A very important use of eigenvalues is the Lovász's notion of the theta-function from 1979. Using it, he solved the long standing Shannon capacity problem of the 5-cycle. Moreover, the theta-function provides the only known way to compute the chromatic number of perfect graphs in polynomial time.

Next strong result was the use of eigenvalues in the construction of superconcentrators and expanders by Alon and Milman in 1985. Their work motivated the study of eigenvalues of random regular graphs. Eigenvalues of random 01-matrices were studied already earlier by F. Juhász, who also analysed the behaviour of the theta-function on random graphs, and introduced the eigenvalues in the clustering. Isoperimetric properties of graphs have also a crucial role in designing random polynomial time algorithm for approximating volume of convex bodies (cf., e.g., [87]).

Recently, there is an increasing interest in the application of eigenvalues in combinatorial optimization problems. To mention only some of them, Burkard, Finke, Rendl, and Wolkowicz used the eigenvalue approach in the study of the quadratic assignment problem and general graph partition problems, Delorme and Poljak in the max-cut problem, and Juvan and Mohar in the labelling problems.

There are several ways of using eigenvalues in the combinatorial optimization. The first possibility consists in formulating concrete bounds which involve eigenvalues of some related matrices. Examples of such bounds are given by the bounds on the edge-connectivity, separation properties, bandwidth and cutwidth, and bounds on the chromatic number and stable sets in Sections 4.1, 4.2, and 4.3. Another way is to use the eigenvalues as a tool of transformation of combinatorial optimization problems to continuous optimization problems. Examples of this kind are provided by the bisection problem, max-cut problem, generalized partition problem, and the

---

* Supported by The Institute for Mathematics and Its Applications, University of Minnesota, and by the Ministry of Science and Technology of Slovenia. Department of Mathematics, University of Ljubljana, Jadranska 19 , 61 111 Ljubljana, Slovenia.

† On leave at the Institute für Diskrete Mathematik, Universität Bonn, supported by the A. von Humboldt Foundation. Department of Applied Mathematics, Charles University, Malostranské náměstí 25, 118 00 Praha 1, Czechoslovakia

theta-function. It seems that the finest estimates can be obtained in this way, in particular for the partition problems.

Different kind of applications is based on the properties of the Perron-Frobenius eigenvector of a nonnegative matrix. This technique is suitable for the ranking problems.

The common point of the most important among the presented applications is the possibility of the change to a "continuous optimization". In such cases there is a possibility of introducing a parameter $u \in \mathbf{R}^n$ and optimizing when U is restricted to be an element of a convex set $K \subseteq \mathbf{R}^n$. This way we get improved bounds or methods for the problems in question. A classical example is the Lovász' ϑ-function. Its use gives rise to polynomial time algorithms for determining the stability number, or the chromatic number of perfect graphs. Similar approach appears in relation to the following problems: bipartition width (Theorems 2.1 and 2.3, Corollary 2.4), partition (Theorem 2.18, Corollaries 2.19 and 2.20), max-cut (Lemma 2.10), stable sets and coloring (Theorems 4.11 and 4.15), bandwidth (Theorem 3.1 and the inequality (44)), etc.

Our survey is organized according to the types of the combinatorial optimization problems: partition, ordering, coloring and stable sets, routing. We also include a short section on the isometric embedding; especially because the bounds there rely on another property of eigenvalues, the Sylvester's inertia law. Appendix A contains some information about the computational aspects, and Appendix B collects known results on eigenvalues of random matrices. Some basic properties of eigenvalues are recalled in the following subsection.

There are several existing books and survey papers concerning graph eigenvalues, e.g., [16,34,33,47,96,98]. We do not intend to overlap our presentation with their contents. Therefore we restrict ourselves to some problems which can be classified as applications in combinatorial optimization, and that are not treated in any of the works mentioned above. In particular, we do not include discussion on expander graphs (which are accessible only via eigenvalue methods) and their applications, although they are quite important tool in the design of algorithms and several other areas of theoretical computer science.

The present text is biased by the viewpoint and the interests of the authors and can not be complete. Therefore we apologize to all those who feel that their work is missing in the references or has not been emphasized sufficiently in the text.

We thank Ch. Delorme, F. Rendl and M. Laurent, with whom we discussed parts of the topic presented here, and who also provided us with several references.

1.1. Matrices and eigenvalues of graphs.

Graphs are assumed to be finite and undirected (unless stated otherwise), multiple edges and loops are permitted. It is in the nature of some problems that only simple graphs make sense, e.g., when speaking about stable sets or colorings (cf. Section 4). In such cases graphs will be assumed to be simple. In some other cases, we will allow more general objects – *weighted graphs*, i.e., each edge $uv \in E(G)$ has a real weight $w(uv) > 0$. (We also set $w(uv) := 0$ if u and v are not adjacent.) Unweighted graphs are special case of weighted ones if we define $w(uv)$ as the number of edges between v and u. If G is a graph of order n, the *adjacency matrix* $A(G) = [a_{uv}]$ of G is an $n \times n$ matrix with rows and columns indexed by $V(G)$ and entries a_{uv} $(u, v \in V(G))$ equal to the number of edges between vertices u and v. Consistent with this is the definition of the adjacency matrix of a weighted graph with weights $w(uv)$ where $a_{uv} = w(uv)$. The *degree $deg(v)$* of a vertex $v \in V(G)$ is equal to the number of edges adjacent to v. In the weighted case we define the degree of v as the sum of weights of edges adjacent to v. Denote by $D(G)$ the diagonal matrix indexed by $V(G)$ and with vertex degrees on the diagonal, i.e., $d_{vv} = deg(v)$, $v \in V(G)$, and $d_{uv} = 0$ if $u \neq v$. The difference

$$(1) \qquad L(G) = D(G) - A(G)$$

is called the *(difference) Laplacian matrix* of G. It is easy to see that for a vector $x \in \ell^2(V(G))$ (a vector of dimension $|V(G)|$ with coordinates x_v, $v \in V(G)$, corresponding to vertices of G) we have

$$(2) \qquad x^T L(G) x = \sum_{u,v} a_{uv} (x_u - x_v)^2$$

which in case of a simple graph G reduces to

$$(3) \qquad x^T L(G) x = \sum_{uv \in E(G)} (x_u - x_v)^2 .$$

The matrix $L(G)$ is real symmetric, so it has $n = |V(G)|$ real eigenvalues $\lambda_1 \leq \lambda_2 \leq \ldots \leq \lambda_n$ (repeated according to their multiplicities). By (2) it is immediate that $L(G)$ is positive semidefinite. It also follows easily from (2) that $\lambda_1 = 0$ with a corresponding eigenvector $\mathbf{1} = (1, 1, \ldots, 1)^T$, and that $\lambda_2 > 0$ if and only if the graph is connected (see Section 2.2). We will use the notation $\lambda_k(G)$ to denote the k-th smallest eigenvalue of $L(G)$ (respecting the multiplicities), and generally, if M is a matrix with real eigenvalues, we denote by $\lambda_k(M)$ the k-th smallest eigenvalue of M. To denote the maximal eigenvalue of M we use the symbol $\lambda_{\max}(M)$. Consistently with this notation we will sometimes use $\lambda_{\min}(M)$ instead of $\lambda_1(M)$.

There are several useful min-max formulas for the expression of eigenvalues of a symmetric matrix and their sums. If M is a real symmetric matrix of order $n \times n$ then

$$(4) \qquad \begin{aligned} \lambda_1(M) &= \min\left\{ \frac{x^T M x}{\|x\|^2} \mid 0 \neq x \in \mathbf{R}^n \right\} \\ &= \min\left\{ x^T M x \mid x \in \mathbf{R}^n, \|x\| = 1 \right\} \end{aligned}$$

and similarly

$$(5) \qquad \lambda_{\max}(M) = \max\left\{ x^T M x \mid x \in \mathbf{R}^n, \|x\| = 1 \right\} .$$

The Rayleigh's characterization (4) has a generalization, the min-max characterization of $\lambda_k(M)$, known also as the Courant-Fisher's expression:

$$(6) \qquad \lambda_k(M) = \min_U \max_x \left\{ x^T M x \mid \|x\| = 1, \ x \in U \right\}$$

where the first minimum is over all k-dimensional subspaces U of $\mathbf{R}^n$. Another way of expressing (6) is

$$(7) \qquad \lambda_k(M) = \min\left\{ x^T M x \mid \|x\| = 1, x \perp x_i, 1 \leq i < k \right\}$$

where $x_1, \ldots, x_{k-1}$ are pairwise orthogonal eigenvectors of $\lambda_1, \ldots, \lambda_{k-1}$, respectively. We will also need a result known as Fan's Theorem [44] on the sum of the k smallest eigenvalues of M:

$$(8) \qquad \sum_{i=1}^{k} \lambda_i(M) = \min_{x_1, \ldots, x_k} \left\{ \sum_{i=1}^{k} x_i^T M x_i \mid \|x_i\| = 1, \ x_i \perp x_j, 1 \le i, j \le k, i \ne j \right\}$$

where the minimum runs over all pairwise orthogonal k-tuples of unit vectors $x_1, x_2, \ldots, x_k$.

The proofs of the Rayleigh's principle and the Courant-Fisher theorem can be found in standard books on matrix theory, e.g. [83]. A short proof of the Fan's theorem is given in [106].

If G is a (weighted) graph and $L = L(G)$ its Laplacian matrix then by (7)

$$(9) \qquad \lambda_2(G) = \min \{ x^T L x \mid \|x\| = 1, \ x \perp \mathbf{1} \}$$

since $\mathbf{1}$ is an eigenvector of $\lambda_1(G)$. Fiedler [46] used (9) to get a more useful expression for $\lambda_2(G)$:

$$(10) \qquad \lambda_2(G) = 2n \cdot \min \left\{ \frac{\sum_{uv \in E} a_{uv} (x_u - x_v)^2}{\sum_{u \in V} \sum_{v \in V} (x_u - x_v)^2} \mid x \ne c \cdot \mathbf{1} \ \text{for} \ c \in \mathbf{R} \right\}$$

where $V = V(G)$, $E = E(G)$, and $n = |V|$. The expression (10) implies a similar expression for $\lambda_{\max}(G)$

$$(11) \qquad \lambda_{\max}(G) = 2n \cdot \max \left\{ \frac{\sum_{uv \in E} a_{uv} (x_u - x_v)^2}{\sum_{u \in V} \sum_{v \in V} (x_u - x_v)^2} \mid x \ne c \cdot \mathbf{1} \ \text{for} \ c \in \mathbf{R} \right\}.$$

2. Partition Problems. The common question in the partition problems surveyed in this section is to find a partition of the vertex set of a graph into two parts S and $V \setminus S$ such that the edge cut

$$\delta S = \{ e = uv \in E(G) \mid u \in S, v \notin S \}$$

satisfies some specific extremal property. This property differs for graph bisection, edge connectivity, isoperimetric property, max-cut problem and clustering. It ranges from maximizing or minimizing $|\delta S|$ to optimization of functions that may depend both on δS and the partition $(S, V \setminus S)$.

It makes sense to introduce another notation which can also be used instead of δS. If A, B are disjoint subsets of $V(G)$ then we denote by $E(A, B)$ the set of edges with one end in A and the other end in B. We also write

$$e(A, B) = w(E(A, B)) = \sum_{e \in E(A,B)} w(e)$$

where w is the edge-weight function of the graph. Clearly, $e(A, B) = |E(A, B)|$ if G is unweighted.

The common initial point for the considered partition problems is to represent a partition $(S, V \setminus S)$ by a ± 1-vector $x_S = (x_{S,i}$ where $x_{S,i} = 1$ for $i \in S$ and $x_{S,i} = -1$

for $i \notin S$. If $w = (w_{ij})$ is an edge-weight function, then the weight $w(\delta S)$ of the cut induced by S can be expressed via the Laplacian matrix (cf. (2)) as follows.

$$(12) \qquad w(\delta S) := \sum_{i \in S, j \notin S} w_{ij} = \frac{1}{4} \sum_{i,j \in V} w_{ij}(x_{S,i} - x_{S,j})^2 = \frac{1}{4} x_S^T L(G) x_S \,.$$

The last subsection deals with multi-partition problems, which generalize the bisection problem of Subsection 2.1.

2.1. Graph bisection. Let $G = (V, E)$ be a graph with an even number of vertices. The *bisection width* $bw(G)$ is defined as the minimum number of edges whose deletion disconnects G into two parts of the same size, i.e.

$$bw(G) := \min\{w(\delta S) \mid S \subset V, |S| = \frac{1}{2}|V|\} \,.$$

The most applications of the graph bisection problem are in the area of VLSI (see [84]). The problem of determining $bw(G)$ is NP-complete ([53]), and hence approximative algorithms and bounds on $bw(G)$ come into interest.

The first eigenvalue lower bound on $bw(G)$ was formulated by Donath and Hoffman in [42]. Their bound has been later improved by Boppana [18] as follows. Let n be the number of vertices of G, and $u = (u_1, \ldots, u_n)^T$ a vector of length n.

THEOREM 2.1. *([18]) Let G be a graph of order n. Then*

$$(13) \qquad bw(G) \geq \frac{n}{4} \max_u \min_x x^T (L(G) + \operatorname{diag}(u)) x$$

where the maximum is taken over all vectors $u \in \mathbf{R}^n$ satisfying $\sum_{i=1}^n u_i = 0$, and the minimum is over all vectors $x \in \mathbf{R}^n$ satisfying $\sum_{i=1}^n x_i = 0$ and $\|x\| = 1$.

Proof. Let $(S, V \setminus S), |S| = \frac{1}{2}n$, be a partition which realizes the minimum bisection width, i.e. $w(\delta S) = bw(G)$. Let us define a vector $y = (y_i)$ by $y_i = 1$ for $i \in S$, and $y_i = -1$ for $i \notin S$. Further, let u be a vector with $\sum_{i=1}^n u_i = 0$ which realizes the maximum on the right-hand side of (13). Using (12), we get

$$bw(G) = w(\delta S) = \frac{1}{4} y^T L(G) y \,.$$

We also have

$$y^T \operatorname{diag}(u) y = \sum_{i=1}^n u_i y_i^2 = \sum_{i=1}^n u_i = 0$$

since $y_i^2 = 1$ for every i. Hence

$$\frac{1}{4} y^T (L(G) + \operatorname{diag}(u)) y = bw(G).$$

Finally, since $\|y\|^2 = n$ and $\sum_{i=1}^n y_i = 0$, we have

$$\frac{1}{4} y^T (L(G) + \operatorname{diag}(u)) y \geq \min_x \frac{n}{4} x^T (L(G) + \operatorname{diag}(u)) x$$

where the minimum is taken over all $x \in \mathbf{R}^n$ with $\|x\| = 1$ and $\sum_{i=1}^{n} x_i = 0$. This proves the theorem. □

The main result of [18] is that the lower bound given in the above theorem provides the actual value of $bw(G)$ with a high probability in a certain probabilistic model. Let $\mathcal{G}(n, m, b)$ be the set of graphs with n vertices, m edges, and the bisection width $bw(G) = b$.

THEOREM 2.2. ([18]) Suppose that $b \leq \frac{1}{2}m - \frac{5}{2}\sqrt{mn \log n}$. Then the lower bound of Theorem 2.1 is exact for a graph G from $\mathcal{G}(n, m, b)$ with probability at least $1 - O(\frac{1}{n})$. □

An important fact is that the bound of Theorem 2.1 is efficiently computable – see the Appendix A. It relies on the concavity of the function $f(u)$ given as

$$f(u) = \min_x x^T (L(G) + \mathrm{diag}(u))x$$

where the minimum is over all x with $\|x\| = 1$ and $\sum_{i=1}^{n} x_i = 0$.

For computational purposes, it is convenient to express the bound of Theorem 2.1 as the minimum eigenvalue of a certain matrix. Let $Q = (q_1, \ldots q_{n-1})$ be an $n \times (n-1)$ matrix such that the columns q_i are mutually orthogonal unit vectors satisfying $\mathbf{1}^T q_i = 0$, $i = 1, \ldots, n-1$.

COROLLARY 2.3. ([18]) We have

(14) $$bw(G) \geq \max_u \frac{n}{4} \lambda_{\min}(Q^T(L(G) + \mathrm{diag}(u))Q)$$

where the maximum runs over all vectors $u \in \mathbf{R}^n$ with $\sum_{i=1}^{n} u_i = 0$. □

The bound of Theorem 2.1 improves a previous bound of Donath and Hoffman, which is formulated in the following corollary.

COROLLARY 2.4. ([42]) We have

$$bw(G) \geq \max_u \frac{n}{4} \left(\lambda_1(Q^T(L(G) + \mathrm{diag}(u))Q) + \lambda_2(Q^T(L(G) + \mathrm{diag}(u))Q) \right)$$

where the maximum is taken over all vectors $u \in \mathbf{R}^n$ such that $\sum u_i = 0$.

Proof. The proof easily follows by the application of the Rayleigh quotient (6) to λ_1 and λ_2, and using the Fan's theorem (8). □

Let us note that the original proof of Corollary 2.4 was based on the following Hoffman-Wielandt inequality proved in [68]. Let M and N be symmetric matrices of size $n \times n$. Then

(15) $$\sum_{i=1}^{n} \lambda_i(M)\lambda_{n-i+1}(N) \leq tr(MN) \leq \sum_{i=1}^{n} \lambda_i(M)\lambda_i(N) .$$

Let us remark that the bounds of Corollaries 2.3 and 2.4 coincide when G is a vertex transitive graph, but there exist instances, e.g. the path of length 3, for which the bound of Boppana is strictly greater than that of Donath-Hoffman. Computational experiments with the Boppana's bound are reported in [118]. It appears that the bound provides a very good estimate on the bisection width also in practice. Some computational experiments with the Donath-Hoffman bound were done earlier in [42] and [30].

A probabilistic algorithm for the bisection width using *randomized rounding* was developed in [126]. It has been proved that, for r-regular graphs and every $0 < \varepsilon < 1$, the algorithm constructs a bisection such the number of edges in the cut does not exceed $bw(G)$ by more than $O\left(\sqrt{n \ln \frac{1}{\varepsilon}}\right)$. Another approximation algorithm is presented in [57].

2.2. Connectivity and separation. It was quite early when Fiedler [45] observed that the second smallest Laplacian eigenvalue $\lambda_2(G)$ measures the graph connectivity. He calls $\lambda_2(G)$ the *algebraic connectivity* of G. The use of the name is justified by the following results.

THEOREM 2.5. *([45]) Let G be a simple graph of order n different from the complete graph K_n. Denote by $\nu(G)$ and $\mu(G)$ its vertex- and edge-connectivity, respectively. Then*

(a) $\lambda_2(G) \leq \nu(G) \leq \mu(G)$, and
(b) $\lambda_2(G) \geq 2\mu(G)(1 - \cos(\pi/n))$.

In particular, G is connected if and only if $\lambda_2(G) > 0$. □

The algebraic connectivity has many properties similar to other measures of connectivity: $\lambda_2(G) \leq \lambda_2(G+e)$, a vertex deletion can reduce λ_2 by at most 1, $\lambda_2(G-v) \geq \lambda_2(G) - 1$, if G is a simple graph, etc. The reader is referred to a survey [96] for more details. Many additional properties obtained by M. Fiedler are also surveyed in [47]. Some properties of la_2 with respect to connectivity were found independently also by Anderson and Morley (see [7]).

By the work of Tanner [129] and later by several other authors (cf. [96]) it became clear that λ_2 measures the connectivity in the following sense: How difficult is to split the graph into two large pieces? More precisely, if $\lambda_2(G)$ is large, then any partition of $V(G)$ into classes $X \cup Y$, where X and Y are both large, has many edges between X and Y. Unfortunately, the converse is not true. For example, a highly connected graph together with an isolated vertex will have λ_2 equal to 0. The basic "folklore" result justifying the above statements is the following inequality:

PROPOSITION 2.6. *Let G be a (weighted) graph of order n. For a subset $X \subset V(G)$ let $w(\delta X)$ be the total weight of edges in δX. Then*

$$(16) \qquad w(\delta X) \geq \lambda_2(G) \frac{|X|(n - |X|)}{n}.$$

Proof. Let $x \in \ell^2(V)$ be given by $x_v = 1$ if $v \in X$, and $x_v = 0$ otherwise. Then

$$\sum_{u \in V} \sum_{u \in V} (x_u - x_v)^2 = 2|X|(n - |X|)$$

and

$$\sum_{uv \in E} a_{uv}(x_u - x_v)^2 = \sum_{uv \in \delta X} a_{uv} = w(\delta X).$$

By (10) we get the inequality of the proposition. $\qquad\qquad\qquad\qquad$ □

In the same way as above we get from (11):

PROPOSITION 2.7. *If G is a (weighted) graph of order n and $X \subset V(G)$ a subset of vertices then*

$$(17) \qquad\qquad w(\delta X) \le \lambda_n(G)\frac{|X|(n - |X|)}{n} \ .$$

$\qquad\qquad\qquad\qquad$ □

It is an important consequence of (16) and (17) that in a graph which has all non-trivial Laplace eigenvalues in a small interval ($\lambda_2(G)$ and $\lambda_n(G)$ close to each other), all vertex sets X of the same cardinality have approximately the same number of out-going edges $w(\delta X)$. In particular, this is true for random graphs or for random regular graphs (cf. Appendix B). The consequence of this fact is that many algorithms for problems involving some kind of separation behave very well for random graphs. In fact, to get good vertex partition into sets of given size in a graph, one does not need to be very clever – any partition will do a good job.

Related to the connectivity, but less trivial to establish are *separation properties* of graphs. A set $C \subset V(G)$ is said to *separate* vertex sets $A, B \subset V(G)$ if

(a) A, B, and C partition $V(G)$ and

(b) no vertex of A is adjacent to a vertex of B.

In applications one is interested in small sets C separating relatively large sets A and B. Usually we want that $|C| = o(n)$ and $|A| = \Omega(n)$, $|B| = \Omega(n)$. The following results show that graphs G with large $\lambda_2(G)$ do not contain small separators.

THEOREM 2.8. *Let G be a graph and $w : E(G) \to \mathbf{R}^+$ an arbitrary non-negative edge-weighting of G. Denote by $\lambda_2 = \lambda_2(G_w)$ the first non-trivial Laplace eigenvalue of the corresponding weighted graph G_w, and by $\Delta = \Delta(G_w)$ the maximal (weighted) degree of G_w. If $C \subset V(G)$ separates vertex sets A, B then*

$$(18) \qquad\qquad |C| \ge \frac{4\lambda_2|A||B|}{\Delta n - \lambda_2|A \cup B|}$$

and

$$(19) \qquad |C| \ge \frac{1}{2\lambda_2}\left(-n(\Delta - \lambda_2) + \sqrt{n^2(\Delta - \lambda_2)^2 + 16\lambda_2^2|A||B|} \right) \ .$$

Proof. Let $x \in \ell^2(V(G))$ be defined by $x_v = -1$ if $v \in A$, $x_v = 1$ if $v \in B$, and $x_v = 0$ if $v \in C$. By (10) we have:

$$(20) \qquad \lambda_2 \le 2n\frac{e(A, C) + e(B, C)}{8|A||B| + 2|C||A \cup B|} \le \frac{n\Delta|C|}{4|A||B| + |C||A \cup B|} \ .$$

By rearranging (20) we get (18). If we use in (18) the relation $|A \cup B| = n - |C|$ we get a quadratic inequality for $|C|$ which yields (19).   □

A slightly weaker version of (18) was obtained by Alon and Milman [6, Lemma 2.1].

Of course, using (18) or (19) makes sense only in case when $|A \cup B| \geq n/2$. Otherwise we should use in (20) the inequality $e(A, C) + e(B, C) \leq \Delta|A \cup B|$ instead of $e(A, C) + e(B, C) \leq \Delta|C|$ which was used above.

Theorem 2.8 implies that graphs containing small separators separating two large sets (e.g. graphs of bounded genus) have small λ_2. In other words, when λ_2 is large we see that any separator C separating large sets contains many vertices. For example, random graphs (edge probability $\frac{1}{2}$, cf. Appendix B) have $\lambda_2 = \frac{n}{2} - O(\sqrt{n \log n})$ and $\Delta = \frac{n}{2} + O(\sqrt{n \log n})$. If we want a separator C separating sets of sizes $c_1 n + o(n)$ and $c_2 n + o(n)$, respectively, we get from (19)

$$|C| \geq 4n\sqrt{c_1 c_2} + o(n) .$$

Concerning separation properties related to eigenvalues we also refer to [114].

2.3. Isoperimetric numbers. In [6] some inequalities of the isoperimetric nature relating λ_2 and some other quantities in graphs are presented. These results have analytic analogues [61] in the theory of Riemannian manifolds where the role of λ_2 is played by the smallest positive eigenvalue of the Laplacian differential operator on the Riemannian manifold. Approximately at the same time Buser [24] and Dodziuk [41] also discovered isoperimetric inequalities involving the Laplace eigenvalues of graphs.

In [5] expanders and graphs with large λ_2 are related. Expanders can be constructed from graphs which are called c-magnifiers ($c \in \mathbf{R}^+$). These are graphs which are highly connected according to the following property. For every set X of vertices of G with $|X| \leq \frac{n}{2}$, the neighbourhood $N(X)$ of X contains at least $c|X|$ vertices. In [5] it is shown that a graph G is $\frac{2\lambda_2}{\Delta + 2\lambda_2}$-magnifier and, conversely, if G is a c-magnifier then $\lambda_2(G) \geq \frac{c^2}{4 + 2c^2}$. The first result is based on Proposition 2.6, while the second one is a discrete version of the Cheeger's inequality [25] from the theory of Riemannian manifolds.

A strong improvement over the Alon's discrete version of the Cheeger's inequality was obtained by Mohar [95] in connection with another problem. The *isoperimetric number* $i(G)$ of a graph G is equal to

$$i(G) = \min \left\{ \frac{|\delta X|}{|X|} \mid X \subset V, 0 < |X| \leq \frac{|V|}{2} \right\}.$$

This graph invariant is NP-hard to compute, and even obtaining any lower bounds on $i(G)$ seems to be a difficult problem. The following easy derived bound

(21)
$$i(G) \geq \frac{\lambda_2(G)}{2}$$

is so important that it initiated a great interest in the study of eigenvalues of graphs. All started with an application of (21) in the construction of expander graphs (cf. [5,

89,90,15,88]) and this motivated much of the research surveyed in this paper. (Slightly before the interest in the Laplacian of graphs was influenced by its use in the analysis of the Laplace differential operator on Riemannian manifolds, cf. [22,24,41]).

The inequality (21) holds also for weighted graphs by the obvious change in the definition of $i(G)$. It follows easily from Proposition 2.6. It is notably important that there is also an upper bound on $i(G)$ in terms of $\lambda_2(G)$. One of the strongest such inequalities is the following [95]:

THEOREM 2.9. ([95]) Let G be a simple graph on at least 3 vertices. Then

$$(22) \qquad i(G) \leq \sqrt{\lambda_2(2\Delta - \lambda_2)}$$

where Δ is the maximal vertex degree in G, and $\lambda_2 = \lambda_2(G)$. □

Theorem 2.9 is a discrete version of Cheeger's inequality [25] relating the first non-trivial eigenvalue of the Laplace differential operator on a compact Riemannian manifold to an isoperimetric constant of the manifold. Discrete versions of the Cheeger's bound were found by Alon [5] (vertex version as mentioned above), Dodziuk [41] (for infinite graphs), Varopoulos [130] (also for the infinite case), Mohar [94,95], Sinclair and Jerrum [124] (in terms of Markov chains), Friedland [49]. Cf. also Diaconis and Stroock [39].

There are other definitions of isoperimetric constants of graphs. For example, define for $X \subseteq V(G)$

$$S(X) = \sum_{v \in X} \deg(v) \ .$$

Then one can define the following version of the isoperimetric number:

$$i'(G) = \min \Big\{ \frac{|\delta X|}{S(X)} \mid X \subset V, 0 < S(X) \leq |E(G)| \Big\}.$$

(Note that $|E(G)| = \frac{1}{2}S(V(G))$.) Similar eigenvalue bounds as (21) and (22) can be derived for $i'(G)$:

$$(23) \qquad i'(G) \geq \frac{1 - \lambda_{\max}(P)}{2}$$

where $P = [p_{uv}]$ is the transition matrix of the random walk on G, i.e., $p_{uv} = a_{uv}/\deg(u)$, and a_{uv} is the element of the adjacency matrix of G. The reader is referred to [94] for details. There is also the corresponding upper bound of Cheeger's type, derived for infinite graphs by Mohar [94], and for finite graphs by Sinclair and Jerrum [124]. Cf. also [39].

A similar isoperimetric quantity as $i'(G)$ was introduced by Friedland [49] who defines for $U \subset V(G)$, $U \neq \emptyset$,

$$\epsilon(U) = \min_{\emptyset \neq V \subseteq U} \frac{|\delta V|}{S(V)}$$

and shows that the smallest eigenvalue $\lambda_1(L_U)$ of the principal submatrix of $L(G)$ whose rows and columns are indexed by U is bounded by

$$(24) \qquad \lambda_1(L_U) \geq \min_{v \in U} \frac{\deg(v)}{2} \epsilon(U)^2 \ .$$

Friedland [49] also provides several norm estimates for $\lambda_1(L_U)$.

There are vertex oriented isoperimetric inequalities for graphs using eigenvalues which are appropriate for some other purposes, e.g., [5,87]. Lovász and Simonovits [87], and Dyer, Frieze and Kannan [43] used the isoperimetric number in a random polynomial time algorithm for estimating the volume of a convex body.

2.4. The maximum cut problem. A weighted graph G with an edge-weight function w will be denoted as a pair (G, w) in this subsection (which differs from our standard notation), because we need to consider some operations with the weight functions.

The *maximum cut* problem, or shortly the *max-cut* problem, is to find a subset $S \subset V$ for which the weight $w(\delta S) := \sum_{e \in \delta S} w(e)$ is maximum. Let $mc(G, w)$ denote the value of the maximum cut. An eigenvalue upper bound

$$(25) \qquad mc(G, w) \le \frac{n}{4} \lambda_{\max} L(G, w)$$

was first studied by Mohar and Poljak in [99]. Later, an optimized eigenvalue bound has been introduced by Delorme and Poljak in [35]:

$$(26) \qquad mc(G, w) \le \min_u \frac{n}{4} \lambda_{\max}(L(G, w) + \mathrm{diag}(u)) =: \varphi(G, w)$$

where the minimum is taken over all $u \in \mathbf{R}^n$ such that $\sum u_i = 0$. The validity of (26) is based on the following lemma.

LEMMA 2.10. *We have*

$$(27) \qquad mc(G, w) \le \frac{n}{4} \lambda_{\max}(L(G, w) + \mathrm{diag}(u))$$

for every vector $u \in \mathbf{R}^n$ satisfying $\sum u_i = 0$.

Proof. Let S be a subset of V for which $w(\delta S)$ is maximum. Let us consider a vector $y \in \mathbf{R}^n$ defined by $y_i = 1$ for $i \in S$ and $y_i = -1$ for $i \notin S$. Observe that y satisfies $\|y\|^2 = n$, and $\sum_{i=1}^n u_i y_i^2 = 0$ for a vector $u \in \mathbf{R}^n$ with $\sum_{i=1}^n u_i = 0$. By (12) we have:

$$
\begin{aligned}
mc(G, w) &= w(\delta S) = \frac{1}{4} y^T L y \\
&= \frac{1}{4}\left(y^T L y + \sum_{i=1}^n u_i y_i^2\right) \\
&\le \max_{\|x\|^2 = n} \frac{1}{4}(x^T L x + x^T \mathrm{diag}(u)x) \\
&= \frac{n}{4} \lambda_{\max}(L + \mathrm{diag}(u)) \, .
\end{aligned}
$$

$\square$

We call u with $\sum_{i=1}^n u_i = 0$ a *correcting vector*. Observe that the bound of (25) corresponds to the choice of correcting vector $u = 0$. Let $\varphi(G, w)$ denote the minimum in (26). In fact, the minimum is achieved for a unique correcting vector u. The optimum correcting vector u has a dual characterization by Theorem 2.11.

For a linear subspace $\mathcal{E} \subset \mathbf{R}^n$, let $\mathcal{C}(\mathcal{E})$ denote the convex cone generated by vectors $(x_1^2, x_2^2, \ldots, x_n^2)^T$ for $x = (x_1, x_2, \ldots, x_n)^T \in \mathcal{E}$.

THEOREM 2.11. *([35]) Let $\mathcal{E}$ be the eigenspace of the maximum eigenvalue λ_{max} of $L(G) + \mathrm{diag}(u)$ for a correcting vector u. Then u is the optimum correcting vector if and only if $(1, 1, \ldots, 1)^T \in \mathcal{C}(\mathcal{E})$.* $\square$

The bound $\varphi(G, w)$ has some pleasant properties. It can be computed in polynomial time with an arbitrarily prescribed precision, and it seems to provide a good estimate on $mc(G, w)$. Let $mc(G)$ and $\varphi(G)$ denote the max-cut and the eigenvalue bound for an unweighted graph G. Asymptotically, the ratio $\varphi(G)/mc(G)$ tends to 1 for a random graph G, but the worst case ratio of $\varphi(G, w)/mc(G, w)$ for $w \geq 0$ is not known. (It does not make sense to investigate the ratio with general w, since $mc(G, w)$ may become zero.) So far the worst known case is the 5-cycle C_5, for which the ratio is $\varphi(G)/mc(G) = \frac{25+5\sqrt{5}}{32} = 1.1306\ldots$. It remains open whether $\varphi(G, w)/mc(G, w) \leq 1.131$ for all graphs with non-negative weights. The conjecture was confirmed for planar graphs in [35], and other classes of graphs in [36,37]. We recall here the result for planar graphs. First we need an auxiliary result about subadditivity of φ with respect to amalgamation which we present without the proof.

Let (G_1, w^1) and (G_2, w^2) be a pair of weighted graphs on vertex sets V_1 and V_2. We define the *amalgam* $(G_1 + G_2, w)$ as the weighted graph on $V_1 \cup V_2$ where

$$
w_{ij} = \begin{cases}
w_{ij}^1 & \text{for } ij \in V_1 \\
w_{ij}^2 & \text{for } ij \in V_2 \\
w_{ij}^1 + w_{ij}^2 & \text{for } ij \in V_1 \cap V_2 \\
0 & \text{elsewhere .}
\end{cases}
$$

LEMMA 2.12. *([35]) We have $\varphi(G_1 + G_2, w) \leq \varphi(G_1, w^1) + \varphi(G_2, w^2)$ for any pair of weighted graphs (G_1, w^1) and (G_2, w^2).* $\square$

THEOREM 2.13. *([35]) Let $G = (V, E)$ be a planar graph with nonnegative edge weight function w. Then*

$$
\varphi(G, w) \leq 1.131 mc(G, w) .
$$

Proof. Let $\mathcal{C}$ denote the set of all odd cycles of G. Barahona (see [9]) proved that the solution of the max-cut problem for a nonnegatively weighted planar graph is given by the optimum solution of the following linear program (with variables x_e corresponding to edges of G):

(28) $$\max \textstyle\sum_{e \in E} w_e x_e ,$$

(29) $$\textstyle\sum_{e \in C} x_e \leq |C| - 1, \quad \text{for } C \in \mathcal{C} ,$$

(30) $$0 \leq x_e \leq 1, \qquad \text{for } e \in E .$$

Let us consider the dual linear program which reads

(31) $$\min \textstyle\sum_{C \in \mathcal{C}} (|C| - 1)\alpha_C + \sum_{e \in E} \beta_e ,$$

(32) $\beta_e + \sum\{\alpha_C \mid e \in C \in \mathcal{C}\} \geq w_e$ for $e \in E$,

(33) $\alpha \geq 0, \quad \beta \geq 0$

where α_C $(C \in \mathcal{C})$ and β_e $(e \in E)$ are dual variables corresponding to (29) and (30), respectively. Let α and β be the optimum dual solution. The dual constraint (32) can be interpreted as telling that the graph (G, w) is a subgraph of the amalgam of the collection of weighted odd cycles $\{(C, \alpha_C) \mid C \in \mathcal{C}\}$ and edges $\{(K_2, \beta_e) \mid e \in E\}$. Since the inequality $\varphi \leq 1.131mc$ holds for every member of the collection, it is true also for its amalgam (G, w) by the lemma. This proves the theorem.   □

More detailed discussion about the relations between the linear programming and eigenvalue approach is given in [112]. The max-cut problem is polynomially solvable for nonnegatively weighted planar graphs by an exact algorithm [66,104], and Theorem 2.13 shows that also eigenvalue approach provides a good estimate.

Another easy consequence of Lemma 2.12 is that

(34) $mc(G, w) = \varphi(G, w)$

for arbitrary non-negatively weighted graphs, because each such graph can be built by the amalgamation of single edges, considered as weighted graphs (e, w_e). (It is $\varphi(K_2) = mc(K_2) = 1$.) Let us call a graph *exact* when the equality in (34) holds. It has been proved in [36] that the recognition of exact weighted graphs is an NP-complete problem. The status of complexity is open for unweighted exact graphs.

The bound $\varphi(G, w)$ has several properties analogous to $mc(G, w)$. We present them in the following theorems.

Let S be a subset of vertices of a weighted graph (G, w). The *switching* w^S of the weight function w is defined as

$$w_{ij}^S = \begin{cases} -w_{ij} & \text{for } ij \in \delta S \\ w_{ij} & \text{otherwise.} \end{cases}$$

The operation of switching was studied in a connection with the cut polytope. In particular, it is well known (and easy to check) that $mc(G, w^S) = mc(G, w) - w(\delta S)$. We show that φ has an analogous property.

THEOREM 2.14. *([36]) We have $\varphi(G, w^S) = \varphi(G, w) - w(\delta S)$.*   □

The next operation that we consider is *vertex splitting*. Let $p_1, \ldots, p_n$ be integers. We define a weighted graph $(\tilde{G}, \tilde{w})$ by splitting each vertex i of G into p_i independent vertices $v_{i1}, \ldots, v_{ip_i}$, and the original weight w_{ij} is equally divided among the new edges $p_i p_j$ between the splitted vertices, i.e., $\tilde{w}_{is,jt} = \frac{w_{ij}}{p_i p_j}$ for $i, j = 1, \ldots, n$, $s = 1, \ldots, p_i$, $t = 1, \ldots, p_j$. It is not difficult to see that $mc(\tilde{G}, \tilde{w}) = mc(G, w)$ for any splitting.

THEOREM 2.15. *([36]) We have $\varphi(\tilde{G}, \tilde{w}) = \varphi(G, w)$ for any splitting $(\tilde{G}, \tilde{w})$ of (G, w). Moreover, there exists an eigenvector $\tilde{x}$ corresponding to the optimized maximum eigenvalue of $\varphi(\tilde{G}, \tilde{w})$ such that the entries of $\tilde{x}$ coincide on each splitted vertex.*   □

The splitting operation can be used to get an alternative definition of $\varphi(G, w)$, because for every nonnegatively weighted graph there (asymptotically) exists a splitting such that the optimum correcting vector for $(\tilde{G}, \tilde{w})$ is $u = 0$.

We define the *contraction* of a pair k and ℓ as follows. Let $G^{k \sim \ell}$ be the graph obtained from G by identifying vertices k and ℓ, and summing the weights on the identified edges. Formally, the vertex set of $G^{k \sim \ell}$ is $V(G) \backslash \{\ell\}$, and

$$
w_{ij}^{k \sim \ell} = \begin{cases} w_{i\ell} + w_{ik} & \text{for } j = k \text{ (the identified vertex)} \\ w_{\ell j} + w_{kj} & \text{for } i = k \text{ (the identified vertex)} \\ w_{ij} & \text{for } i, j \notin \{k, \ell\} \, . \end{cases}
$$

THEOREM 2.16. *([36]) We have $\varphi(G^{k \sim \ell}, w^{k \sim \ell}) \leq \varphi(G, w)$ for any contracted pair k and ℓ.* □

The operations of contraction and switching are useful in practical solving the max-cut problem by branch and bound technique. The contraction is used when a pair of nodes should be fixed in the same partition class. When two nodes i and j are fixed to belong to different classes, we first switch the weight function, and then consider the contraction of the pair i and j in the switched graphs. Theorems 2.14 and 2.16 ensure that the upper bound is nonincreasing in the branching process. Computational experiments with computing the upper bound $\varphi(G, w)$, and exactly solving the max-cut problem has been done by Poljak and Rendl [113]. A 'typical' gap between $mc(G, w)$ and $\varphi(G, w)$ is 4–5%.

2.5. Clustering. Stimulated by the work of Fiedler [45], Juhász and Mályusz [77] discovered an application of eigenvalue bounds to a clustering problem. The formulation of this problem is close to the bisection problem, but has a slightly different objective function. Let G be a (weighted) graph and A its (weighted) adjacency matrix. If $c = (S_1, S_2)$ is a partition of $V(G)$ into non-empty sets S_1, S_2 (called *clusters* of the partition c), denote by

$$
f(c) = \frac{|S_1||S_2|}{|V(G)|}(d_{11} - 2d_{12} + d_{22})
$$

where

$$
d_{ij} = \frac{1}{|S_i||S_j|} e(S_i, S_j), \quad i, j \in \{1, 2\} \, .
$$

The task is to find a partition c for which $f(c)$ is maximum.

THEOREM 2.17. *([77]) Let G be a (weighted) graph of order n. Let P be the orthogonal projection parallel to $(1, \ldots, 1)^T$, i.e. $P = (\delta_{ij} - \frac{1}{n})$ where δ_{ij} is the Kronecker δ. Then*

$$
f(c) \leq \lambda_{\max}(PAP)
$$

for any partition c of $V(G)$.

Proof. We reproduce the proof (following [77]) since in [77] the theorem is proved for unweighted graphs only.

Given a partition $c = (S_1, S_2)$ of $V = V(G)$, let $x \in \ell^2(V)$ be the vector with coordinates

$$
x_v = \begin{cases} \left(\frac{|S_2|}{|S_1||V|}\right)^{\frac{1}{2}} & \text{if } v \in S_1 \\[2ex] -\left(\frac{|S_1|}{|S_2||V|}\right)^{\frac{1}{2}} & \text{if } v \in S_2 . \end{cases}
$$

Then $Px = x$ and $\|x\| = 1$. Moreover,

$$
\begin{aligned}
(PAPx, x) &= (Ax, x) = \sum_{u,v \in V} a_{uv} x_u x_v \\
&= \frac{|S_2|}{|S_1||V|} \sum_{u,v \in S_1} a_{uv} - \frac{2}{|V|} \sum_{u \in S_1, v \in S_2} a_{uv} + \frac{|S_1|}{|S_2||V|} \sum_{u,v \in S_2} a_{uv} \\
&= \frac{|S_1||S_2|}{|V|}(d_{11} - 2d_{12} + d_{22}) = f(c) .
\end{aligned}
$$

By the Rayleigh's principle (5) we get the inequality of the theorem. $\qquad \square$

The functional f usually characterizes good clusterings. However, it is difficult to find an optimal partition. Therefore Juhász and Mályuzs [77,75] proposed relaxations $F_1(c)$, $F_2(c)$, and $F_3(c)$ of $f(c)$, which can be easier optimized. A function $\rho \in \ell^2(V)$ is a *weight-function* if $\rho(v) \geq 0$ for $v \in V$ and $\sum_{v \in V} \rho(v)^2 = 1$. Then $\rho(v)$ can be viewed as the weight of the vertex v, and $\rho(v)\rho(u)$ is the weight of the edge vu. For $U \subseteq V$ define $\rho(U) = \sum_{u \in U} \rho(u)$, the weight of U. Then we define, for $c = (S_1, S_2)$:

$$
F_1(c) = \max_{\rho(S_1) = \rho(S_2)} \{\rho(S_1)\rho(S_2)(d_{11}(\rho) - 2d_{12}(\rho) + d_{22}(\rho))\}
$$

where

$$
d_{ij}(\rho) = \frac{1}{\rho(S_i)\rho(S_j)} \sum_{u \in S_i} \sum_{v \in S_j} \rho(u)\rho(v) a_{uv} , \qquad i,j \in \{1,2\}.
$$

It is proved in [77] that $f(c) \leq F_1(c) \leq \lambda_{\max}(PAP)$. Further, it is shown that all optimal partitions for F_1 can be obtained by partitioning $V(G)$ according to the signs of coordinates of eigenvectors of PAP corresponding to its maximal eigenvalue.

The functional $F_2(c)$ is defined as an optimum over two weight-functions, δ and ρ:

$$
F_2(c) = \min_{\delta} \max_{\rho} \left\{ \rho(S_1)\rho(S_2) \left(\frac{\rho(S_1)}{\rho(S_2)} d_{11}(\rho) - 2d_{12}(\rho) + \frac{\rho(S_2)}{\rho(S_1)} d_{22}(\rho) \right) \right\}
$$

where the maximum is over all weight functions ρ such that $(\delta\rho)(S_1) = (\delta\rho)(S_2)$ with $(\delta\rho)(U) := \sum_{u \in U} \delta(u)\rho(u)$. Then ([77])

$$
F_1(c) \geq F_2(c) \geq \lambda_{|V|-1}(A).
$$

An optimal partition c with respect to F_2 can be obtained by the signs of the coordinates of the second largest eigenvalue $\lambda_{|V|-1}(A)$ of A.

The functional $F_3(c)$ is defined similarly:

$$F_3(c) = \min_{\rho} \left\{ \rho(S_1)\rho(S_2)\left(\frac{\rho(S_1)}{\rho(S_2)}d_{11}(\rho) - 2d_{12}(\rho) + \frac{\rho(S_2)}{\rho(S_1)}d_{22}(\rho)\right)\right\}.$$

One can prove ([76]) that

$$f(c) \geq F_3(c) \geq \lambda_{\min}(A)$$

and that an optimal partition c with respect to F_3 can be obtained by the signs of the coordinates of an eigenvector corresponding to $\lambda_{\min}(A)$. A behaviour of the clustering based on the above functionals for random graphs is studied in [76].

A clustering problem (with a fixed number $k \geq 2$ of clusters) was also studied by Bolla [17]. Let $v_1,\ldots,v_n$ be binary random variables taking values 0 and 1, and let $e'_1,\ldots,e'_m (m >> n)$ be a sample for the variables v_i. Then one can form a hypergraph $H = (V,E)$ with vertices $V = \{v_1,\ldots,v_n\}$ and hyperedges $E = \{e_1,\ldots,e_m\}$ where e_j contains all those vertices for which the value of the j-th object e'_j is equal 1. It is of practical interest [17] to partition E into k clusters, $E = E_1 \cup E_2 \cup \ldots \cup E_k$ such that the following criterion function is minimized:

$$Q = \sum_{i=1}^{k} Q(H_i)$$

where $H_i = (V_i, E_i)$ is the hypergraph corresponding to the i-th cluster E_i and

$$Q(H_i) = \min_{1 \leq d \leq n} \left\{ c \cdot 2^{n-d} + \sum_{j=1}^{d} \lambda_j(H_i)\right\}$$

with $c > 0$ a constant (its choice depends on the size of the problem), and $\lambda_j(H_i)$ the j-th eigenvalue of an appropriately defined Laplacian matrix $L(H)$ of the hypergraph H_i as described below. It is shown in [17] that the eigenvalue sum in $Q(H_i)$ is related to a combinatorial property of H_i. It measures how distinct the clusters H_i are from each other.

The Laplacian matrix $L(H)$ of a hypergraph $H = (V,E)$ on n vertices is defined by Bolla [17] as an $n \times n$ matrix with its ij-th entry ℓ_{ij} equal to

$$\ell_{ij} = \begin{cases} -\sum_{e \in E, v_i, v_j \in e} \frac{1}{|e|} & \text{if } i \neq j \\[2mm] \sum_{e \in E, v_i \in e} \frac{|e|-1}{|e|} & \text{if } i = j. \end{cases}$$

Clearly, $|e|$ denotes the number of vertices incident with the hyperedge e. Note that if H is a graph then $L(H)$ defined above is equal to $\frac{1}{2}$ of the usual Laplacian matrix of the graph.

2.6. Graph partition. The graph partition problem asks for a partition of the vertex set of a weighted graph into a fixed number of classes of given sizes, so that the number (resp. the weight) of edges between is minimum. Hence, the problem is a generalization of the bisection problem of section 2.1. The first eigenvalue bound on this problem is by Donath and Hoffman [42], which we recall in Theorem 2.21.

Further work on the problem include [10,11,12,13], where also several relaxation to the transportation problem were considered. We survey here only the recent results of Rendl and Wolkowicz [118] because they give the tightest eigenvalue bound.

The eigenvalue lower bound on the bisection width has been extended to a more general graph partition problem by Rendl and Wolkowicz [118]. The problem reads as follows.

GRAPH PARTITION PROBLEM

INSTANCE: A graph $G = (V, E)$, integers $m_1, \ldots, m_k$ such that $\sum_{i=1}^{k} m_i = n = |V|$.

TASK: Find a partition $(S_1, \ldots, S_k)$ of V such that $|S_i| = m_i$, $i = 1, \ldots, k$, and the number of edges whose end vertices are in distinct partition classes is minimum.

Assume that $m_1, \ldots, m_k$ are fixed, and let $(S_1, \ldots, S_k)$ be an optimal partition. Let us denote by E_{cut} (and E_{uncut}) the set of edges whose endvertices belong to distinct classes (to the same class) of the partition. Let A denote the adjacency matrix of G. Let $1_j = (1, \ldots, 1)^T$ be the all 1's vector of length j, and $m = (m_1, \ldots, m_k)^T$.

THEOREM 2.18. *([118]) We have*

$$|E_{uncut}| \leq \min_{u} \max_{x} \frac{1}{2} tr(X^T(A + \mathrm{diag}(u))X)$$

where the minimum is taken over all vectors $u \in R^n$ satisfying $\sum_{i=1}^{n} u_i = 0$, and the maximum is over all $n \times k$ matrices X satisfying

(i) $X1_k = 1_n$,
(ii) $X^T 1_n = m$,
(iii) $X^T X = \mathrm{diag}(m)$.

Proof. Consider the matrix $X = (x_{ij})$ defined by $x_{ij} = 1$ if $i \in S_j$, and $x_{ij} = 0$ otherwise. It is straightforward to check that $|E_{uncut}| = \frac{1}{2} tr(X^T AX)$, and X satisfies (i), (ii) and (iii). It is also easy to see that $tr(X^T \mathrm{diag}(u)X) = 0$ for any vector $u = (u_i)$ with $\sum_{i=1}^{n} u_i = 0$. $\qquad \square$

The optimization problem formulated in Theorem 2.18 can be efficiently solved in two special cases, but a general solution is not known. The special cases are given in Corollaries 2.19 and 2.20. Let $Q = (q_1, \ldots, q_{n-1})$ be an $n \times (n-1)$ matrix whose columns $q_1, \ldots, q_{n-1}$ satisfy $q_i^T q_j = 0$ for $i \neq j$, $\|q_i\| = 1$ and $1_n^T q_i = 0$ for $i = 1, \ldots, n-1$.

COROLLARY 2.19. *([118]) Let G be a graph with adjacency matrix A whose vertices are partitioned into sets of sizes $m_1, \ldots, m_k$, respectively, and let Q be a matrix as described above. Assume that $m_1 = \ldots = m_k = \frac{n}{k}$. Then $|E_{uncut}|$ is bounded above by*

$$\min_{u} \max_{X} \left\{ \frac{1}{2} tr(X^T(A + \mathrm{diag}(u))X) \mid X \text{ satisfies } (i), (ii), (iii) \right\}$$

(35)
$$= \min_{u} \frac{n}{2k} \sum_{j=1}^{k-1} \lambda_{n-j+1}(Q^T(A + \mathrm{diag}(u))Q) + \frac{1}{k}|E|$$

where the minimum runs over all vectors $u \in \mathbf{R}^n$ with $\sum u_i = 0$.

Observe that $Q^T A Q$ is of size $(n-1) \times (n-1)$, and hence $\lambda_{n-1}, \ldots, \lambda_{n-k+1}$ are the $k-1$ largest eigenvalues of $Q^T A Q$. We give the proof only for the special case of $k = 2$.

Proof. Assume $m_1 = m_2 = \frac{n}{2}$. Then the conditions on X are equivalent with $X = (x, \mathbf{1}_n - x)$ where $x = (x_i)$ is a vector satisfying $\sum_{i=1}^{n} x_i = \sum_{i=1}^{n} x_i^2 = \frac{n}{2}$. Set $M = A + \mathrm{diag}(u)$, and $\mathbf{1} = \mathbf{1}_n$. We have

$$\frac{1}{2} tr(X^T M X) = \frac{1}{2}(x^T M x + (\mathbf{1} - x)^T M (\mathbf{1} - x)) =$$

$$x^T M x - \mathbf{1}^T M x + \frac{1}{2} \mathbf{1}^T M \mathbf{1} = \frac{1}{4}(\mathbf{1} - 2x)^T M (\mathbf{1} - 2x) + \frac{1}{4} \mathbf{1}^T M \mathbf{1}.$$

Observe that

$$\frac{1}{4} \mathbf{1}^T M \mathbf{1} = \frac{1}{4} \mathbf{1}^T A \mathbf{1} + \frac{1}{4} \mathbf{1}^T \mathrm{diag}(u) \mathbf{1} = \frac{1}{2}|E| + \frac{1}{4} \sum_{i=1}^{n} u_i = \frac{1}{2}|E|.$$

Further, $y = \mathbf{1} - 2x$ satisfies $y^T y = n$ by the properties of x. Hence

$$\min_X \frac{1}{2} tr(X^T M X) = \min_y \left\{ \frac{1}{4} y^T M y + \frac{1}{2}|E| \mid y^T y = n \right\} = \frac{n}{4} \lambda_1(M) + \frac{|E|}{2}$$

by the Rayleigh's principle. $\qquad \square$

The next corollary deals with the case of a 2-partition in possibly nonequal parts. Let $m_1 + m_2 = n$, and denote

$$C(u) = \frac{1}{n} m_1 m_2 Q^T(A + \mathrm{diag}(u))Q,$$

$$c(u) = \frac{m_1 - m_2}{n} \sqrt{\frac{m_1 m_2}{n}} Q^T(A + \mathrm{diag}(u))Q,$$

$$c' = |E|(m_1^2 + m_2^2) \frac{1}{2n^2}.$$

COROLLARY 2.20. *([118]) Suppose we have a 2-partition of $V(G)$ into sets of sizes m_1 and m_2, respectively. Then $|E_{uncut}|$ is bounded above by*

$$\min_u \max_X \left\{ \frac{1}{2} tr(X^T(A + \mathrm{diag}(u))X) \mid X \text{ satisfies } (i),(ii),(iii) \right\} =$$

$$\min_u \max_z \left\{ z^T C(u) z + c(u)^T z + c' \mid z \in \mathbf{R}^{n-1}, \|z\| = 1 \right\}$$

where the minimum runs over all vectors $u \in \mathbf{R}^n$ with $\sum u_i = 0$. $\qquad \square$

Both Corollaries 2.19 and 2.20 can be viewed as generalizations of the Boppana's bound of Corollary 2.3. This can be shown as follows. By Corollary 2.19 (or 2.20) we have (with $U = \mathrm{diag}(u)$)

$$|E_{uncut}| \leq \min_u \frac{n}{4} \lambda_{max}(Q^T(A + U)Q) + \frac{|E|}{2},$$

and hence

$$|E_{cut}| = |E| - |E_{uncut}| \geq \frac{|E|}{2} - \frac{n}{4} \min_u \lambda_{\max}(Q^T(A+U)Q).$$

Let $\overline{d} = \sum d_i/n = 2|E|/n$ denote the average vertex degree of G. Then

$$\frac{|E|}{2} - \frac{n}{4}\lambda_{\max}(Q^T(A+U)Q) = \frac{n}{4}\lambda_{\min}(\overline{d}I_{n-1} - Q^T(A+U)Q) =$$
$$\frac{n}{4}\lambda_{\min}(Q^T(\overline{d}I_n - A - U)Q) = \frac{n}{4}\lambda_{\min}(Q^T(D - A + \operatorname{diag}(u'))Q)$$

where $D = \operatorname{diag}(d_1,\dots,d_n)$ is the diagonal matrix defined by the degrees of the graph G, and $u_i' = \overline{d} - d_i - u_i$, $i = 1,\dots,n$. Observe that $\sum u_i' = \sum u_i + n\overline{d} - \sum d_i = \sum u_i = 0$, and $L = D - A$ is the Laplacian matrix of G. This proves the equivalence of the bounds.

A bound on the general graph partition problem has already been formulated by Donath and Hoffman in [42].

THEOREM 2.21. *([42]) Assume that $m_1 \geq m_2 \geq \dots \geq m_k$. Then*

(36)
$$|E_{cut}| \geq \max_u \frac{1}{2}\sum_{i=1}^{k} m_i \lambda_i(L(G) + \operatorname{diag}(u))$$

where the maximum runs over all vectors $u \in \mathbf{R}^n$ with $\sum u_i = 0$. $\qquad\square$

However, the bound given by this theorem is weaker than that of Theorem 2.18. It should be mentioned that Donath and Hoffman derived another partition result [42, Theorem 3].

3. Ordering. A number of recent papers use eigenvalues and eigenvectors of matrices associated to graphs to obtain orderings (labellings) of vertices of a graph which give rise to acceptable approximations for several optimization problems on graphs. Applications to cutwidth, bandwidth and the min-p-sum problem use the Laplacian matrix, its second smallest eigenvalue and the corresponding eigenvectors. Applications to ranking and scaling use the Perron-Frobenius eigenvector.

3.1. Bandwidth and min-p-sum problems. A *labelling* of a (weighted) graph $G = (V, E)$ is a 1-1 mapping $\psi : V \to \{1,\dots,|V|\}$. For a real number p, $0 < p < \infty$, we define its *p-discrepancy* $\sigma_p(G,\psi)$ as

$$\sigma_p(G,\psi) := \Big(\sum_{uv \in E} a_{uv}|\psi(u) - \psi(v)|^p\Big)^{1/p},$$

and for $p = \infty$,

$$\sigma_\infty(G,\psi) := \max_{uv \in E} |\psi(u) - \psi(v)|.$$

Notice that $\sigma_\infty(G,\psi)$ is independent of the edge weights a_{uv}. The minimal value

$$\sigma_p(G) := \min_\psi \sigma_p(G,\psi), \qquad 0 < p \leq \infty,$$

is called the *min-p-sum* of the graph G. For the case $p = \infty$, $\sigma_\infty(G)$ is also known as the *bandwidth* of the graph G. The reader can find more information about the min-1-sum and the bandwidth in [27,26]. Let us just mention that the bandwidth and the min-1-sum problem are NP-complete in general [54,110].

As for the comments in the introduction, good lower bounds on $\sigma_p(G)$ are extremely important and nontrivial to obtain. Let us present a result of Juvan and Mohar [79].

THEOREM 3.1. *([79]) Let G be a graph of order n with at least one edge, and let*

$$\beta(G) := \sup_w \left\lceil \frac{n}{\lambda_2(G_w)} \left(-\Delta(G_w) + \sqrt{\Delta^2(G_w) + \lambda_2^2(G_w)} \right) \right\rceil$$

where the supremum is taken over all non-negative weightings w of the edges of G, and where $\lambda_2(G_w)$ is the second smallest eigenvalue of the corresponding weighted Laplacian matrix and $\Delta(G_w)$ denotes the maximal w-weighted degree of G. Then

$$\sigma_\infty(G) \geq \begin{cases} \beta(G) - 1 & \text{if } \beta(G) \equiv n \ (mod\,2) \\ \beta(G) & \text{otherwise.} \end{cases}$$

Proof. We will use the separation result of Section 2.2. Fix an arbitrary weighting function w, and let $\lambda_i = \lambda_i(G_w)$. Also, choose an optimal labelling ψ for the bandwidth of G. Define $A, B, C \subset V(G)$ as follows:

$$A := \{v \in V(G) \mid \psi(v) \leq (n-k)/2\}$$
$$B := \{v \in V(G) \mid \frac{n-k}{2} < \psi(v) \leq (n+k)/2\}$$
$$C := \{v \in V(G) \mid \psi(v) > (n+k)/2\}$$

where $1 \leq k < n$ is a number to be determined later. We will assume that $k \equiv n \ (mod\,2)$. Notice that $|A| = |C| = (n-k)/2$ and $|B| = k$. The sum of the weights of the edges between particular parts A, B, C can be estimated using Proposition 2.6 as follows:

$$(37) \qquad e(A, B \cup C) \geq \lambda_2 \frac{|A||B \cup C|}{n}$$

$$(38) \qquad e(A \cup B, C) \geq \lambda_2 \frac{|C||A \cup B|}{n}$$

$$(39) \qquad e(B, A \cup C) \leq \Delta|B| \ .$$

Since

$$(40) \qquad 2e(A, C) = e(A, B \cup C) + e(A \cup B, C) - e(B, A \cup C)$$

we get from (37)–(39) that

$$(41) \qquad 2e(A, C) \geq \frac{\lambda_2}{2n}(n^2 - k^2) - \Delta k \ .$$

It follows that $e(A, C) > 0$ if $k < \frac{n}{\lambda_2}(-\Delta + \sqrt{\Delta^2 + \lambda_2^2})$ which implies that there is at least one edge between A and C. Consequently, $\sigma_\infty(G) \geq k + 1$ where

$$k := \left\lceil \frac{n}{\lambda_2} \left(-\Delta + \sqrt{\Delta^2 + \lambda_2^2} \right) \right\rceil - \varepsilon$$

and $\varepsilon = 1$ or 2 (chosen so that $k \equiv n \ (mod \ 2)$). $\qquad\qquad\qquad\square$

Notice that if λ_2 is small compared to Δ, the lower bound of Theorem 3.1 behaves like $\lambda_2 n/(2\Delta)$ and for $\lambda_2 = \Delta(1 - o(1))$ it behaves like $(\sqrt{2} - 1)n(1 - o(1))$. If we replace the inequality (39) with

$$(42) \qquad\qquad e(B, A \cup C) \leq \lambda_n \frac{k(n - k)}{n}$$

(cf. Proposition 2.7) we get another lower bound [79] on the bandwidth:

$$(43) \qquad\qquad \sigma_\infty(G) \geq \begin{cases} \beta'(G) - 1 & \text{if } \beta'(G) \equiv n \ (mod \ 2) \\ \beta'(G) & \text{otherwise} \end{cases}$$

where $\beta'(G)$ is defined as

$$(44) \qquad\qquad \beta'(G) := \sup_w \left\lceil \frac{\lambda_2(G_w)}{2\lambda_n(G_w) - \lambda_2(G_w)} n \right\rceil .$$

The supremum is again over all non-negative edge weighting functions w. It is easy to see that the supremum in the definitions of $\beta(G)$ and $\beta'(G)$ is always attained. Unfortunately, unlike some other similar optimizations (cf. Sections 2 and 4) we do not know of an efficient algorithm for computing $\beta(G)$ or $\beta'(G)$.

Further improvements are possible. The first possibility is to use in (37) and (38) improved bounds on partitions as outlined in Sections 2.1 and 2.2. Another approach has the following background. It may happen that $\lambda_2(G_w) = 0$ for every weighting w and therefore also $\beta'(G) = 0$ (and $\beta(G) = 0$ with the proper interpretation). Of course, this will happen if and only if G is disconnected. Well, if one of the components of G is large, the others very small, the bandwidth will depend on the large component. One may assume that in such a case we reduce the bandwidth problem to the connected components. However, the same will happen if there is a small part of the graph which is "loosely" connected to the rest. Possible improvements will be to add (weighted) edges to the graph in order to get rid of such such anomalies. The proof of Theorem 3.1 will go through with the same arguments. The only difference will be to replace the condition $e(A, C) > 0$ with

$$(45) \qquad\qquad e(A, C) > w^+ := \sum_{uv \notin E(G)} w(uv) .$$

In this case we get the bound:

$$\sigma_\infty(G) \geq \left\lceil \frac{n}{\lambda_2} \left(-\Delta_w + \sqrt{\Delta_w^2 + \lambda_2(\lambda_2 - 2w^+/n)} \right) \right\rceil - 1$$

where $\lambda_2 = \lambda_2(G_w)$ and $\Delta_w = \Delta(G_w)$ is the maximal (weighted) degree of G_w. There is a similar bound along the lines of (43) and (44).

There is another possibility of optimization. One may try to find a subgraph G' of G for which any of the previous bounds will be better than for G. It is obvious by the Courant-Fisher principle that $\lambda_{\max}(G') \leq \lambda_{\max}(G)$ but it may happen that $\lambda_2(G') > \lambda_2(G)$ if G contains fewer vertices than G.

In [78] also bounds on $\sigma_1(G)$ and $\sigma_2(G)$, the most useful among the parameters $\sigma_p(G)$, $0 < p < \infty$, are obtained.

THEOREM 3.2. *([78]) Let G be a (weighted) graph of order n and let $\overline{d}(G)$ denote the average (weighted) degree of G. Then*

(a)
$$\lambda_2(G)\frac{n^2 - 1}{6} \leq \sigma_1(G) \leq \overline{d}(G)\frac{n(n + 1)}{6},$$

(b)
$$\lambda_2(G)\frac{n(n^2 - 1)}{12} \leq \sigma_2(G)^2 \leq \overline{d}(G)\frac{n^2(n + 1)}{12}.$$

$\square$

Juvan and Mohar [78] suggested to use eigenvectors corresponding to $\lambda_2(G)$ to determine an approximation to an optimal labelling for any of the problems of calculating $\sigma_p(G)$, $0 < p \leq \infty$. Their calculations on several classes of graphs show fairly successful behaviour. A labelling corresponding to a vector $x \in \ell^2(V(G))$ is determined by the increase of components $x_v, v \in V(G)$, of x. The vertex v with the smallest value x_v will be labelled 1, the second smallest 2, etc. The heuristic argument behind this algorithm is that eigenvectors of $\lambda_2(G)$ minimize the sum $\sum_{uv \in E}(x_u - x_v)^2$ (under the constraint $x^t \mathbf{1} = 0$, $\|x\| = 1$ which guarantee that x is not a "multiple of $\mathbf{1}$"), and $\sigma_2(G)$ minimizes $\sum_{uv \in E}(\psi(u) - \psi(v))^2$.

3.2. Cutwidth. The *cutwidth* $c(G)$ of a (weighted) graph G of order n is defined as the minimum of $c(G, \psi)$ over all labellings ψ of G where

$$c(G, \psi) = \max_{1 \leq i < n} e(\psi^{-1}(\{1, 2, \ldots, i\}), \psi^{-1}(\{i + 1, \ldots, n\})).$$

In other words, we want a labelling ψ where the maximal number (sum of the weights) of edges between vertices with $\psi(v) \leq i$ and vertices with $\psi(u) > i$ $(1 \leq i < n)$, is as small as possible. It is known that the cutwidth computation is NP-hard [53]. The following simple result is outlined in [78]:

THEOREM 3.3. *([78]) Let G be a graph of order n and ψ a labelling of G. Then*

$$\lambda_2(G)\frac{\lfloor \frac{n}{2} \rfloor \lceil \frac{n}{2} \rceil}{n} \leq c(G, \psi) \leq \lambda_n(G)\frac{\lfloor \frac{n}{2} \rfloor \lceil \frac{n}{2} \rceil}{n}.$$

Consequently,

$$\lambda_2(G)\frac{\lfloor \frac{n}{2} \rfloor \lceil \frac{n}{2} \rceil}{n} \leq c(G) \leq \lambda_n(G)\frac{\lfloor \frac{n}{2} \rfloor \lceil \frac{n}{2} \rceil}{n}.$$

$\square$

3.3. Ranking. An idea of using eigenvalues and eigenvectors for the tournament ranking problem appears in the book of Berge [14]. Let $G = (V, F)$ be a *partial tournament*, i.e., a directed graph such that $ij \in F$ implies that $ji \notin F$. The existence of an arc ij means that the player i has beaten the player j. One can rank the players in a partial tournament using the outdegrees (number of wins). But it may happen that "while one player has beaten a large number of very weak players, another player has beaten only a few very strong players." To discover such "anomalies" one can use quite successfully the following ranking procedure based on the *power indices* as defined below.

Let $p_i^j(k)$ denote the number of walks of length k from the vertex i to j, and let

$$p^j(k) = p_1^j(k) + \cdots + p_n^j(k)$$

where $n = |V|$ as usual. The *power index* of the j-th player is defined as

$$\pi^j = \lim_{k \to \infty} \frac{p^j(k)}{p^1(k) + \cdots + p^n(k)} \ .$$

THEOREM 3.4. *The power indices of players in a partial tournament are given by the positive eigenvector of the adjacency matrix of the tournament normalized in the ℓ^1-norm.*

Proof. It is well-known that the entries of the k-th power of the adjacency matrix A determine the number of walks of length k, i.e. $A^k = [p_i^j(k)]_{i,j \in V}$. Therefore $p^j(k) = e_j A^k \mathbf{1}$ where e_j is the vector with j-th coordinate equal to 1, all others equal to 0. By the Perron-Frobenius theorem, A has a positive eigenvector, and its eigenvalue has the largest modulus among the eigenvalues of A. Moreover, by the power method, $A^k \mathbf{1} / \|A^k \mathbf{1}\|$ converges to the Perron-Frobenius eigenvector. Therefore also a different normalization $A^k \mathbf{1} / 1 A^k \mathbf{1}$, which approaches the vector of power indices, converges to a multiple of the normalized Perron-Frobenius eigenvector. $\square$

Applications in geography. Similar idea appeared also in the geographic literature, where it was used to compare accessibility of nodes in a transportation network. The accessibility of a node i is expressed by its index (called *Gould index*) π^i defined by

$$\pi^i = \frac{x_i}{\|x\|}$$

where x is the eigenvector of $\lambda_{\max}(A(G))$ and G is an undirected graph representing the network. The references to applications can be found in the survey paper [127].

3.4. Scaling. A method of scaling based on the maximum eigenvector has been proposed by T. L. Saaty in [121]. A matrix $A = (a_{ij})$ is *positive and reciprocal* if $a_{ij} = a_{ji}^{-1} > 0$ for $i, j = 1, \ldots, n$. The goal is to find a positive vector $w = (w_1, \ldots, w_n)$, called a *scaling*, so that the entries a_{ij} of A are, in some sense, well approximated by the ratios $\frac{w_i}{w_j}$. This will be denoted as $a_{ij} \approx \frac{w_i}{w_j}$.

In practical applications, the matrix A may be obtained as a matrix of pairwise comparisons, and the entries a_{ij} are obtained by consulting a board of experts. As a heuristical method of scaling, T. L. Saaty proposed to take the eigenvectors corresponding to the maximum eigenvalue $\lambda_{\max}(A)$. By the Perron-Frobenius theorem, this eigenvector is unique, and all its entries are positive. T. L. Saaty has shown the following properties of w.

THEOREM 3.5. *([121]) Let A be a positive reciprocal matrix of size $n \times n$ and w its Perron-Frobenius eigenvector. Then the following holds:*

(i) $\lambda_{\max}(A) \geq n$,

(ii) $\lambda_{\max}(A) = n$ if and only if $a_{ij}a_{jk} = a_{ik}$ for all $i, j, k = 1, \ldots, n$ (i.e., A is already scaled).

(iii) If $a_{1j} \leq a_{2j} \leq \ldots \leq a_{nj}$ for all j, then $w_1 \leq w_2 \leq \ldots \leq w_n$.

Proof. (i) Since $\lambda_{\max}$ is the eigenvalue with the eigenvector w, we have

$$\sum_{j=1}^{n} a_{ij}w_j = \lambda_{\max} w_i, \quad i = 1, \ldots, n,$$

which gives

$$\sum_{i=1}^{n}\sum_{j=1}^{n} a_{ij}w_j w_i^{-1} = n\lambda_{\max} \ .$$

Applying the inequality $x + 1/x \geq 2$ with $x = a_{ij}w_j w_i^{-1}$, we get $x + 1/x = a_{ij}w_j w_i^{-1} + a_{ji}w_i w_j^{-1} \geq 2$, which together with $a_{ii} = 1$ gives

$$
\begin{aligned}
n\lambda_{\max} &= \sum_{i,j=1}^{n} a_{ij}w_j w_i^{-1} \\
&= \sum_{i=1}^{n} a_{ii} + \sum_{1 \leq i < j \leq n} (a_{ij}w_j w_i^{-1} + a_{ji}w_i w_j^{-1}) \geq n + 2\binom{n}{2} = n^2 \ .
\end{aligned}
$$

Hence $\lambda_{\max} \geq n$.

(ii) It follows from the above inequality that $\lambda_{\max} = n$ if and only if $x + 1/x = a_{ij}w_j w_i^{-1} + a_{ji}w_i w_j^{-1} = 2$, for all pairs i, j. This holds only if $a_{ij}w_j w_i^{-1} = 1$ for all $i, j = 1, \ldots, n$, and this is equivalent to the condition in (ii).

(iii) Fix $i < j$. Since $w_k > 0$, and $a_{ik} \leq a_{jk}$ for $k = 1, \ldots, n$, we have

$$\lambda_{\max} w_i = \sum_{k=1}^{n} a_{ik}w_k \leq \sum_{k=1}^{n} a_{jk}w_k = \lambda_{\max} w_j \ .$$

Hence $w_i \leq w_j$. $\qquad\qquad\qquad\qquad\qquad\qquad\qquad\qquad\qquad\qquad\qquad\quad$ $\square$

Parts (i) and (ii) of Theorem 3.5 indicate that the difference between $\lambda_{\max}$ and n depends on how far the matrix A is from a scaled matrix. In fact, Saaty [121] has shown that the value of $(\lambda_{\max} - n)/(n - 1)$ corresponds to the variance of $\varepsilon_{ij} = a_{ij}w_j w_i^{-1}$.

3.5. The quadratic assignment problem. The quadratic assignment problem is one of the most interesting combinatorial optimization problems, with many applications in the layout theory. Several other known hard combinatorial optimization problems, like the travelling salesman, the bisection width, or the min-1-sum problem are special cases of the quadratic assignment problem. (The bisection width and the min-1-sum problem are considered in Sections 2.1 and 3.1, respectively.) The problem reads as follows.

QUADRATIC ASSIGNMENT PROBLEM
INSTANCE: Matrices A, B and C of size $n \times n$.
TASK: Find a permutation π of $\{1, 2, \ldots, n\}$ which minimizes

$$(46) \qquad \sum_{i=1}^{n} c_{i\pi(i)} + \sum_{i=1}^{n} \sum_{j=1}^{n} a_{ij} b_{\pi(i)\pi(j)} \ .$$

The crucial application of the eigenvalue approach is estimating the quadratic term in (46). The following theorem, concerning the symmetric case, is due to Finke, Burkard, and Rendl [48].

THEOREM 3.6. ([48]) Let A and B be symmetric $n \times n$ matrices with eigenvalues $\lambda_1 \leq \ldots \leq \lambda_n$ and $\mu_1 \leq \ldots \leq \mu_n$, respectively. Then

$$(47) \qquad \min_{\pi} \sum_{i=1}^{n} \sum_{j=1}^{n} a_{ij} b_{\pi(i)\pi(j)} \geq \sum_{i=1}^{n} \lambda_i \mu_{n-i+1}$$

where the minimum is taken over all permutations π of $\{1, 2, \ldots, n\}$.

Proof. Let π be the permutation which realizes the minimum in (47), and let X be the permutation matrix corresponding to π. Since B and XBX^T are similar, the eigenvalues of XBX^T are also $\mu_1 \leq \ldots \leq \mu_n$. We have

$$\sum_{i=1}^{n} \sum_{j=1}^{n} a_{ij} b_{\pi(i)\pi(j)} = tr(AXBX^T)$$

and using the Hoffman-Wielandt inequality (15) we finally get:

$$tr(A(XBX^T)) \geq \sum_{i=1}^{n} \lambda_i \mu_{n-i+1} \ .$$

$\square$

Further results and generalizations to non-symmetric case appear in [63,64,117]. A survey of applications of the quadratic assignment problem is in [23].

4. Stable Sets and Coloring. In this section we survey eigenvalue bounds on the chromatic number $\chi(G)$ and the size of a maximum stable set $\alpha(G)$. The lower bound on $\chi(G)$ (due to A. Hoffman [67]), and the upper bound on $\chi(G)$ (due to H. Wilf [132]) are probably the earliest applications of spectral bounds to a graph optimization problem. On the other hand, theoretically most important is the Lovász's bound $\vartheta(G)$ on the maximum stable set size.

We recall that the *chromatic number* $\chi(G)$ is the minimum number of colors needed to color the vertices of G so that no two adjacent vertices have the same color. A set S is called *stable* or *independent* if no two vertices of S are adjacent. The maximum size of a stable set is denoted by $\alpha(G)$. We let $\overline{G}$ denote the complement of G. The coloring and the stable set problems make sense only for simple graphs. Therefore all graphs in this section are assumed to be simple.

4.1. Chromatic number. The first eigenvalue bounds on the chromatic number are due to H. Wilf [132] and A. Hoffman [67] who formulated an upper and a lower bound, respectively.

THEOREM 4.1. *([132,67]) Let $A = A(G)$ be the adjacency matrix of a simple graph G. Then*

(48)
$$1 + \frac{\lambda_{\max}(A)}{|\lambda_{\min}(A)|} \le \chi(G) \le 1 + \lambda_{\max}(A) .$$

Proof. (See [133].) Let G_{crit} be a maximal color-critical subgraph of G (i.e. a subgraph of G with the same chromatic number and such that deleting any edge decreases the chromatic number). Let d denote the minimum degree of G_{crit}. We have

$$\chi(G) - 1 \le d \le \lambda_{\max}(A(G_{crit})) \le \lambda_{\max}(A(G)) .$$

A short proof of the lower bound, based on the interlacing property of eigenvalues, can be found in [55] or [86]. □

Let us remark that the upper bound can be realized by an efficient algorithm, i.e., one can color any graph G by $1 + \lambda_{\max}(A(G))$ colors in polynomial time. (This question was raised in [8].) The algorithm follows from an easy lemma (see, e.g., [34]).

LEMMA 4.2. *The average degree $\overline{d}$ of G is less or equal to $\lambda_{\max}(A(G))$.* □

Let v_1 be a vertex of G whose degree is at most the average degree $\overline{d}$. Clearly, such a vertex can easily be found. By the interlacing property of eigenvalues, $\lambda_{\max}(A(G - v_1)) \le \lambda_{\max}(A(G))$. Assume that $G - v_1$ is already colored with at most $k = \lfloor 1 + \lambda_{\max}(A(G)) \rfloor$ colors. Since the degree of v_1 is at most $k - 1$, the coloring can be extended to G. This gives a polynomial time algorithm.

The following two theorems are due to Cvetković [32] and Hoffman [67].

THEOREM 4.3. *([32,34]) Let G be a simple graph of order n. Then*

$$\chi(G) \ge \frac{n}{n - \lambda_{\max}(A(G))} .$$

THEOREM 4.4. *([67]) Let G be a simple graph of order n distinct from the complete graph K_n, and let $\overline{G}$ denote the complement of G. Then*

$$\chi(\overline{G}) \ge \frac{n + \lambda_{n-1}(A(G)) - \lambda_n(A(G))}{1 + \lambda_{n-1}(A(G))} .$$

Eigenvectors can also be used to obtain a coloring of a graph. In fact, bipartite graphs are fully characterized by the following theorem.

THEOREM 4.5. *([34], Theorems 3.4 and 3.11) A graph G is bipartite if and only if*

$$\lambda_{\min}(A(G)) = -\lambda_{\max}(A(G)) .$$

The bipartition of a bipartite graph G is given by the sign-pattern of an eigenvector corresponding to $\lambda_{\min}(A(G))$. The idea has been extended by Apswall and Gilbert [8] to obtain a heuristic graph coloring algorithm using the following result as the basis of the heuristic. Let $x_1, \ldots, x_r$ be a collection of vectors from $\boldsymbol{R}^n$. The vectors $x_\ell = (x_{1\ell}, \ldots, x_{n\ell})$, $\ell = 1, \ldots, r$, determine a partition (i.e. coloring) of $\{1, \ldots, n\}$ as follows. Let i and j belong to the same partition class if and only if either $x_{i\ell} \geq 0$ and $x_{j\ell} \geq 0$ for $\ell = 1, \ldots, r$, or $x_{i\ell} < 0$ and $x_{j\ell} < 0$ for $\ell = 1, \ldots, r$. In other words, the partition classes are given by the sign-patterns of the collection (where the zero entries are considered as positive).

THEOREM 4.6. ([8]) Let $x_1, \ldots, x_n$ be pairwise orthogonal eigenvectors of $\lambda_1(A(G)), \ldots, \lambda_n(A(G))$, respectively. Then, for every pair i and j of adjacent vertices, there exits an eigenvector x_ℓ for which $x_{i\ell}$ and $x_{j\ell}$ have distinct signs.

In some cases, the minimum coloring can be obtained via sign-pattern of a collection of vectors. An explicit class of graphs is given in Theorem 4.7. A graph $G = (V, E)$ is called block regular k-partite, if (i) G is a k-partite graph with partition $V = V_1 \cup \ldots \cup V_k$, and (ii) for every $i, j, i \neq j$, the number of edges from a vertex $u \in V_i$ to V_j depends only on i and j.

THEOREM 4.7. ([8]) Let G be a block regular k-partite graph. Then there exists a set of at most $k - 1$ eigenvectors whose sign-pattern induces a proper k-coloring of G. □

For ideas of the similar flavor we also refer to [115,116].

4.2. Lower bounds on stable sets. H. S. Wilf [134] derived a spectral bound on the size of the maximum stable set using an earlier result of Motzkin and Straus [101].

THEOREM 4.8. ([134]) Let G be a simple graph. Then

$$\alpha(G) \geq \frac{s^2}{s^2 - \lambda_{\max}(A(\overline{G}))}$$

where s is the sum of entries of the normalized eigenvector corresponding to $\lambda_{\max}(A(\overline{G}))$.

Proof. Let $G = (V, E)$ be a graph with vertex set $V = \{1, \ldots, n\}$. We need a result of Motzkin and Straus [101], who proved that $\alpha(G)$ can be expressed as an optimum of a quadratic program as follows:

$$\max\left\{\sum_{ij \notin E} x_i x_j \mid \sum_{i=1}^n x_i = 1, x \geq 0\right\} = 1 - \frac{1}{\alpha(G)}.$$

Let $u = (u_i) \geq 0$ be the normalized eigenvector corresponding to the eigenvalue $\lambda_{\max}(A(\overline{G}))$, and let $s := \sum_{i=1}^n u_i$. Then $y = u_i/s$ satisfies $y \geq 0$ and $\sum_{i=1}^n y_i = 1$, and hence

$$1 - \frac{1}{\alpha(G)} \geq \sum_{ij \notin E} y_i y_j = y^T A(\overline{G}) y = \lambda_{\max}(A(\overline{G})) \|y\|^2 = \frac{1}{s^2} \lambda_{\max}(A(\overline{G})).$$

COROLLARY 4.9. *([134])* *For a simple graph G we have*

$$\alpha(G) \geq \frac{n}{n - \lambda_{\max}(A(\overline{G}))}.$$

Proof. It follows from the fact that $s^2 \leq n$.

Corollary 4.9 generalizes the bound of Cvetković given in Theorem 4.3, since

$$\chi(G) \geq \alpha(\overline{G}) \geq \frac{n}{n - \lambda_{\max}(A(G))}.$$

In fact, H. S. Wilf derived a hierarchy of spectral bounds. One of them is given in the next theorem.

THEOREM 4.10. *([134])* *Let G be a d-regular graph on n vertices. Then*

$$\alpha(G) \geq \frac{n}{d + 1 + (\lambda_{\min}(A(G)) + 1) \max(M_+^2, M_-^2)/n}$$

where $M_+ = \min_{u_i > 0} \frac{1}{u_i}$, $M_- = \min_{u_i < 0} \frac{1}{|u_i|}$, and u is the normalized eigenvector of the second largest eigenvalue of $A(G)$.

4.3. Upper bounds on stable sets. L. Lovász introduced an eigenvalue bound $\vartheta(G)$ on $\alpha(G)$ in the connection with his solution of the problem of Shannon capacity of the 5-cycle, see [85]. Given a graph $G = (V, E)$, the number $\vartheta(G)$ is defined by

$$(49) \qquad \vartheta(G) := \min_{A \in \mathcal{A}} \lambda_{\max}(A)$$

where $\mathcal{A}$ is the class of real $n \times n$ matrices $A = (a_{ij})$, where $n = |V|$, satisfying

$$(50) \qquad a_{ij} = \begin{cases} 1 & \text{for } ij \notin E, \text{ or } i = j, \\ \text{arbitrary} & \text{otherwise.} \end{cases}$$

THEOREM 4.11. *([85])* *For every graph G, we have $\alpha(G) \leq \vartheta(G)$.*

Proof. Let $k = \alpha(G)$ and let S be a stable set of size $k = \alpha(G)$. Assume that $A \in \mathcal{A}$ is a matrix for which $\lambda_{\max}(A) = \vartheta(G)$. (It is easy to see that the minimum in (49) is attained.) Define a vector $y = (y_i)$, $\|y\| = 1$, by

$$(51) \qquad y_i = \begin{cases} k^{-1/2} & \text{for } i \in S \\ 0 & \text{otherwise.} \end{cases}$$

Then

$$\alpha(G) = k = y^T A y \leq \max_{\|x\|=1} x^T A x = \lambda_{\max}(A) = \vartheta(G).$$

Important is the dual characterization of $\vartheta(G)$.

THEOREM 4.12. *([85]) Let $\mathcal{B}$ denote the class of positive semidefinite symmetric matrices $B = (b_{ij})$ satisfying $\sum_{i=1}^{n} b_{ii} = 1$ and $b_{ij} = 0$ for i and j adjacent. Then*

$$\vartheta(G) = \max_{B \in \mathcal{B}} \sum_{i,j} B_{i,j}.$$

□

Lovász gives also several other equivalent definitions of $\vartheta(G)$.

THEOREM 4.13. *([85]) We have*

$$\vartheta(G) = \max_{B \in \mathcal{B}} \left(1 - \frac{\lambda_{\max}(B)}{\lambda_{\min}(B)} \right)$$

where $B \in \mathcal{B}$ if and only if $b_{ij} = 0$ for $ij \in E$ and for $i = j$, and b_{ij} is arbitrary otherwise.

□

Lovász also proves that $\chi(G) \geq \vartheta(\overline{G})$, and hence Theorem 4.13 yields the Hoffman's lower bound on $\chi(G)$ presented in Theorem 4.1.

The number $\vartheta(G)$ is probably the best known efficiently computable estimate of $\alpha(G)$. Moreover, $\alpha(G) = \vartheta(G)$ for perfect graphs. However, there is a big gap between the performance of $\alpha(G)$ and $\vartheta(G)$ on random graphs. It is well known that $\alpha(G) \approx 2 \ln n$. On the other hand, Juhász [73] determined $\vartheta(G)$ for random graphs.

THEOREM 4.14. *([73]) Let G be a random graph with edge probability $p = 1/2$. Then, with probability $1 - o(1)$ for $n \to \infty$,*

$$\frac{1}{2}\sqrt{n} + O(n^{1/3} \log n) \leq \vartheta(G) \leq 2\sqrt{n} + O(n^{1/3} \log n).$$

□

4.4. k-colorable subgraphs. Narasimhan and Manber [102] recently extended the eigenvalue bound $\vartheta(G)$ to obtain a bound on the maximum size of a k-colorable subgraph.

Given a graph $G = (V, E)$, let $\alpha_k(G)$ denote the maximum size of a k-colorable subgraph of G, i.e.

$$\alpha_k(G) = \max\{|S_1 \cup \ldots \cup S_k| \; : \; S_i \text{ is stable}, i = 1, \ldots, k\}.$$

Let $\mathcal{A}$ be the class of matrices defined by (50). Let

$$\vartheta_k(G) := \min \left\{ \sum_{i=1}^{k} \lambda_{n-i+1}(A) \mid A \in \mathcal{A} \right\}$$

i.e., $\vartheta_k(G)$ is the sum of the k largest eigenvalues of $A \in \mathcal{A}$, where A is chosen so that the sum in minimized. The bound $\vartheta_k(G)$ improves the obvious bound $\alpha_k(G) \leq k\vartheta(G)$.

THEOREM 4.15. *([102]) We have $\alpha_k(G) \leq \vartheta_k(G)$ for every graph G.*

Proof. The proof is quite analogous to that of Theorem 4.11. Let $S_1 \cup \ldots \cup S_k$ be the maximum k-colorable subgraph with S_t stable sets. For every $t = 1, \ldots, k$, let $y^{(t)}$

be the vector defined by (51) for S_t. We may assume that S_t are pairwise disjoint. Then the vectors $y^{(t)}$ are pairwise orthogonal. The result follows by using the Fan's Theorem [44] (cf. (8)). □

COROLLARY 4.16. *If $\vartheta_k(G) < |V(G)|$ then $\chi(G) > k$.*

5. Routing Problems. In this section we will mention a number of results relating eigenvalues of graphs (in particular $\lambda_2(G)$) to metric (distance) parameters and walks in graphs. A great variety of such results have been discovered recently. For example, eigenvalues are related to the diameter, mean distance, forwarding indices, routings, several properties of random walks on graphs, etc. We do not intend to be complete, so some parts will be covered with not too many details.

5.1. Diameter and the mean distance. There is a great interest in obtaining good upper bounds on the diameter of graphs. The diameter can be efficiently computed but in practice graphs with several hundreds of thousands of vertices are dealt with and eigenvalue bounds, which can sometimes be estimated "by construction" (analytically), come into play.

There is a lower bound

$$(52) \qquad \mathrm{diam}(G) \geq \frac{4}{n\lambda_2(G)}$$

which was obtained by Brendan McKay (private communication, cf. [97] for a proof). More important is that there are upper bounds on the diameter of a graph G in terms of $\lambda_2(G)$ [97]:

$$(53) \qquad \mathrm{diam}(G) \leq 2\left\lceil \sqrt{\frac{\lambda_n(G)}{\lambda_2(G)}} \sqrt{\frac{\alpha^2-1}{4\alpha}} + 1 \right\rceil \left\lceil \log_\alpha \frac{n}{2} \right\rceil$$

where α is any real number which is > 1. For any particular choice of n, λ_n, and λ_2 one can find the value of α which imposes the lowest upper bound on the diameter of the graph. See [97] for details. A good general choice is $\alpha = 7$.

In [97] another upper bound on the diameter of a graph is obtained

$$(54) \qquad \mathrm{diam}(G) \leq 2\left\lceil \frac{\Delta + \lambda_2(G)}{4\lambda_2(G)} \ln(n-1) \right\rceil.$$

This improves a bound obtained previously by Alon and Milman [6], and usually also supersedes the bound obtained by F. Chung [28]. Another very nice improvement was obtained by Chung, Faber, and Manteuffel [29]:

$$(55) \qquad \mathrm{diam}(G) \leq \left\lceil \frac{\cosh^{-1}(n-1)}{\cosh^{-1}((\lambda_n + \lambda_2)/(\lambda_n - \lambda_2))} \right\rceil + 1.$$

where $\lambda_2 = \lambda_2(G)$ and $\lambda_n = \lambda_n(G)$.

A relation between the diameter and the spectral properties is also observed in [39]. Another diameter bound was obtained by Nilli [103]. If G contains two edges whose endvertices are at distance at least κ in G then

$$(56) \qquad \lambda_2(G) \leq \Delta - \sqrt{\Delta - 1} + \frac{4\sqrt{\Delta - 1} - 2}{\kappa}$$

where Δ is the maximal vertex degree in the graph G. This bound holds also for weighted graphs, which is not the case with other bounds mentioned above. On the other hand, (56) makes sense only for the so-called *Ramanujan graphs* [89,90] where $\lambda_2(G) \geq \Delta - \sqrt{\Delta - 1}$.

In [97], some bounds on the mean distance $\bar{\rho}(G)$ are derived. Recall that the mean distance is equal to the average of all distances between distinct vertices of the graph. A lower bound is

$$(n - 1)\bar{\rho}(G) \geq \frac{2}{\lambda_2(G)} + \frac{n - 2}{2}$$

and an upper bound, similar to (54), is

$$\bar{\rho}(G) \leq \frac{n}{n - 1} \left\lceil \frac{\Delta + \lambda_2(G)}{4\lambda_2(G)} \ln(n - 1) \right\rceil.$$

There is also an upper bound on $\bar{\rho}(G)$ related to the inequality (53). Cf. [97].

There is an interesting formula for the mean distance of a tree (due to B. D. McKay). See [92] or [97] for a proof.

THEOREM 5.1. *Let T be a tree of order n and $\lambda_2, \lambda_3, \ldots, \lambda_n$ the non-zero Laplacian eigenvalues of T. Then the mean distance $\bar{\rho}(T)$ is equal to:*

$$(n - 1)\bar{\rho}(T) = 2 \sum_{i=2}^{n} \frac{1}{\lambda_i}.$$

Some related results can also be found in [93].

5.2. Routings. Eigenvalues of graphs can also be used in relation to several problems which involve a system of shortest paths between vertices of a graph. Roughly speaking, one is interested in a set of shortest paths between all pairs of distinct vertices, $\mathcal{P} = \{P_{uv} \mid u, v \in V(G), u \neq v, P_{uv} \text{ is a geodesic path}\}$. Such a system of paths is called a *routing* on G. The goal is to find a routing which minimizes the maximal number of occurrences of any edge in the routing. There is also a vertex version of the problem. The obtained minimum is called the *forwarding index* (respectively, the *vertex forwarding index*) of the graph. There are some simple inequalities that relate the forwarding parameters with expanding properties of graphs, and these relations give rise to eigenvalue applications in these problems. We refer to [125] for details.

5.3. Random walks. It has been known for a long time that the properties of a random walk on a graph G and the spectral properties of its transition matrix $P(G)$ are closely related (see, e.g., [82]). Let us recall that the entries p_{uv} of the transition matrix are defined by $p_{uv} = a_{uv}/\deg(u)$, where a_{uv} is the element of the (weighted) adjacency matrix of G. We will only refer to some recent papers in this area. There is a close relationship between the random walk properties and combinatorial optimization, for example in the area of randomized algorithms. See, e.g., [87].

The *cover time* of a graph G is the maximum over all vertices $v \in V(G)$ of the expected time required in a random walk starting at v to visit all vertices of G. Results related to the cover time of graphs appear in [3,20,38,39,80,109,119]. A closely related notion is the *hitting time* of the random walk [2,19,107,109].

The *mixing rate* of a random walk is a measure which measures how fast we approach the stationary distribution by starting the random walk at an arbitrary vertex. Related works are [1,87,124]. A random walk is *rapidly mixing* if it approaches stationary distribution "very fast" (in the logarithmic number of steps). For a random walk on a graph G this notion is closely related to $\lambda_2(G)$. Jerrum and Sinclair [70] obtained a fully polynomial approximation scheme for the permanent evaluation (and thus for counting the number of perfect matchings in graphs) by using the rapid mixing property.

Some other related results appeared in [4,39,107,108,128,130].

6. Embedding Problems. The eigenvalue approach is also useful in the study of embedding problems, where it is used to guarantee existence or nonexistence of certain graph embeddings. We include this topic in our survey, because determining the minimum dimension of an embedding has a flavour of an optimization problem. Moreover, the proofs rely on a different property of eigenvalues – on the Sylvester's law of inertia of symmetric quadratic forms.

Let G be a graph. The distance $\text{dist}_G(u,v)$ of two vertices u and v is the length of a shortest path between u and v. The *distance* matrix $D = D(G)$ is the matrix of pairwise distances between the vertices of G.

A mapping $f : V(G) \to V(H)$ is called an *isometric embedding* of a graph G into a graph H if $\text{dist}_H(f(u), f(v)) = \text{dist}_G(u,v)$ for every pair $u, v \in V(G)$. A recent survey of results on isometric graph embedding problems can be found in [59]. We recall only two results, namely isometric embedding of graphs in cubes and squashed cubes, where eigenvalue technique was applied.

The *squashed cube* Q_r^* is the graph with vertex set $\{0, 1, *\}^r$, and the edge set formed by the pairs $(u_1, \ldots, u_r)$ and $(v_1, \ldots, v_r)$ such that there is exactly one coordinate i for which $\{u_i, v_i\} = \{0, 1\}$. P. Winkler [135] proved that every connected graph G can be isometrically embedded in a squashed cube Q_r^* of dimension $r \leq |V(G)| - 1$. A lower bound on the minimum r was known earlier, and the result is reported to Witsenhausen (see [60]).

THEOREM 6.1. *Assume that a graph G can be isometrically embedded in the squashed cube Q_r^*. Then*

$$r \geq \max(n^+, n^-)$$

where n^+ and n^- denote the number of positive and negative eigenvalues of the distance matrix $D(G)$ of G.

Proof. Let d_{ij} denote the distance $\text{dist}_G(i, j)$. We will consider the quadratic form

(57) $$\frac{1}{2} x^T D(G) x = \sum_{i<j} d_{ij} x_i x_j$$

in variables $x = (x_1, \ldots, x_n)^T$, $n = |V(G)|$.

Let $f : V(G) \to \{0, 1, *\}^r$ be an isometric embedding of G in the squashed cube Q_r^*. For every $k = 1, \ldots, r$, let

$$X_k = \{i \mid f(i)_k = 0\} \quad \text{and} \quad Y_k = \{i \mid f(i)_k = 1\} \, .$$

The sets X_k, Y_k, $k = 1, \ldots, r$, are used to express the quadratic form (57) as

(58)
$$\sum_{i<j} d_{ij} x_i x_j = \sum_{k=1}^{r} \left(\sum_{i \in X_k} x_i \right) \left(\sum_{j \in Y_k} x_j \right) .$$

Using the identity $ab = \frac{1}{4}((a + b)^2 - (a - b)^2)$, this form can be written as a sum and difference of squares

(59)
$$\sum_{i<j} d_{ij} x_i x_j = \frac{1}{4} \sum_{k=1}^{r} \left\{ \left(\sum_{i \in X_k} x_i + \sum_{j \in Y_k} x_j \right)^2 - \left(\sum_{i \in X_k} x_i - \sum_{j \in Y_k} x_j \right)^2 \right\} .$$

By the Sylvester's law of inertia, the number of positive and negative squares must be at least n^+ and n^-, respectively. $\square$

The *hypercube* Q_r is the graph with vertex set $\{0, 1\}^r$, and two vertices $(u_1, \ldots, u_r)$ and $(v_1, \ldots, v_r)$ form an edge if $\sum_{i=1}^{r} |u_i - v_i| = 1$. The graphs G isometrically embeddable into a hypercube have been first characterized by Djoković [40]. From our point of view, the following characterization is interesting.

THEOREM 6.2. *([120]) A graph G is isometrically embeddable in a hypercube Q_r if and only if G is bipartite and the distance matrix $D(G)$ has exactly one positive eigenvalue.*

The dimension of the minimum embedding was determined by Graham and Pollack [60] as the number of negative eigenvalues of $D(G)$.

THEOREM 6.3. *([60]) Let a graph G be isometrically embeddable a hypercube Q_r. Then the minimum r (the dimension of the embedding) is equal to the number of negative eigenvalues of the distance matrix $D(G)$.* $\square$

A. Appendix: Computational aspects. In this Appendix we briefly discuss several ingredients of the computation and complexity of the problems arising by evaluating the eigenvalue bounds.

Complexity. The most commonly used model of complexity in the combinatorial optimization is the model based on the time complexity. From this point of view, problems are classified as 'easy' or 'difficult' depending on whether they are polynomial time solvable or NP-hard. A problem is said to be *polynomial time solvable*, if it admits a solution by an algorithm whose running time is bounded by a polynomial in the size of the input, measured by the length of the binary encoding. The opposite pole form the NP-hard problems, which are at least as difficult as the most difficult problems in the class NP (the class of nondeterministically polynomial problems). The formal definition of the complexity classes can be found in the book [53]. It is widely believed that the NP-complete problems cannot be solved in polynomial time, and the P=NP problem is definitely the most important open problem of the theoretical computer science.

There are only a few combinatorial optimization problems (like the graph isomorphism) whose complexity status is open. All the other most important problems are known to be either polynomial time solvable or NP-hard.

The combinatorial optimization problems surveyed in this article, with the exception of the edge- and vertex-connectivity studied in Section 2.2, belong to the NP-hard problems. Hence, the existence of good estimates is very important. Eigenvalue bounds provide in many cases quite good approximations. Moreover, as we show below, these estimates are efficiently computable.

Eigenvalue computation. Both in the direct formulas and the approximation by a continuous problem, we need to determine the eigenvalues numerically in order to get concrete bounds. The eigenvalues of a symmetric matrix are real but, in general, irrational numbers. Hence, given an integral (or rational) symmetric matrix M, we cannot compute its spectrum in polynomial time. However, it is known as a folklore result that the eigenvalues can be computed in polynomial time with an arbitrary prescribed precision. Unfortunately, we cannot refer to any detailed analysis of this question. Some partial discussions of this topic can be found in [8] and [36].

The polynomial time computability of eigenvalues has only theoretical importance. For applications, it is significant that the eigenvalues can be computed fast by a variety of numerical methods, see e.g., [58]. There also exist several software packages [31,123] for the eigenvalue computations. For example, the routine DNLASO from [123] was used by Poljak and Rendl [113] in the practical computations related to the max-cut problem.

Convexity. Convexity is a crucial property, which enables efficient solution of the relaxed continuous optimization problems.

Given a symmetric matrix M of size $n \times n$, an integer k, $1 \leq k \leq n$, and nonnegative coefficients c_i, $i = k, k+1, \ldots, n$, satisfying $c_k \leq c_{k+1} \leq \cdots \leq c_n$, let us consider the function $f(u), u \in \mathbf{R}^n$, defined by

$$(60) \qquad f(u) = \sum_{i=k}^{n} c_i \, \lambda_i(M + \mathrm{diag}(u)).$$

The following result is obtained by an easy application of the Rayleigh's principle (4) and the Fan's theorem (8).

THEOREM A.1. *[30] The function $f(u)$ defined by (60) is convex.* $\qquad \square$

Hence the functions given by formulas (13), (14), (26), (35), and (36) are convex (resp. concave), and their minimum (resp. maximum) is efficiently computable, as mentioned in the next paragraph. It is also easy to see (by an application of the Rayleigh's principle) that the ϑ-function defined by (49) is obtained by the minimization of a convex function.

As a consequence of the convexity, dual characterizations are possible (cf. Theorems 2.11 and 4.12). A general duality theorem of that type is proved also in [105].

Global optimization. Another important fact is that the problem of the minimization of a convex function f over a convex region K in $\mathbf{R}^n$ is, under some technical assumptions on f and K, polynomial time solvable in the following sense. Given an $\varepsilon > 0$, the algorithm finds a rational number $\bar{z}$ and a rational vector $\bar{u} \in K$ such that $|\bar{z} - f(\bar{u})| \leq \varepsilon$, and

(61) $f(\bar{u}) \leq f(u) + \varepsilon$ for every u such that $\{u' \in \mathbf{R}^n \mid \|u - u'\| < \varepsilon\} \subset K$.

A detailed analysis of the ellipsoid algorithm for the minimization of f is given in the book by Grötschel, Lovász, and Schrijver [62].

The general theory can be applied to the convex function f given by (60) and the convex set $K = \{u \in \mathbf{R}^n \mid \sum_{i=1}^n u_i = 0\}$. In this concrete case, the minimum $\bar{u}$ can be approximated in a stronger sense, namely so that

$$f(\bar{u}) \leq f(u) + \varepsilon \quad \text{for all} \quad u \in K$$

holds (instead of (61)).

Again, the polynomial time computability is mainly of theoretical interest, because it indicates that the relaxed problems are easy. For the concrete computation, other algorithms than the ellipsoid method are more suitable. The problem of practical minimization of (60) is slightly complicated by the fact that f is not differentiable at all points. In particular, it is typically not differentiable at its minimum, because several eigenvalues are often identified as a result of the minimization.

Several methods of minimization of (60) were proposed in [30,105,122].

Branch and bound algorithms. The eigenvalue relaxations can also be used to compute exact optima in the combination with the branch and bound method. For that purpose, it is important to show that the relaxation is monotone with respect to the branching. Computational experiments for the max-cut problem were done by Poljak and Rendl [113]. In particular, the computing was speeded up by a suitable initiation of parameters in solving the subproblems. This was possible due to the combinatorial properties of the eigenvalue bound.

B. Appendix: Eigenvalues of random graphs. The usual model for random graphs is the following. For a positive integer n we consider (labelled) graphs on n vertices in which each edge appears with probability $p = p(n)$. The adjacency matrix of a random graph is a random symmetric 01-matrix with zeros on the main diagonal. For a (random) graph G with the adjacency matrix A let us denote by $\bar{d}(A)$ the density of edges of G, more precisely, $\bar{d}(A) = 2|E(G)|/n^2$. The following bound is easy to verify:

PROPOSITION B.1. *([71]) Let G be a graph of order n and A its adjacency matrix. Then*

$$n\bar{d}(A) \leq \lambda_{\max}(A) \leq n\sqrt{\bar{d}(A)} .$$

Proof. The first inequality follows by (1.5) by taking $x = \mathbf{1}$,

$$\lambda_{\max}(A) \geq \frac{(A\mathbf{1}, \mathbf{1})}{(\mathbf{1}, \mathbf{1})} = n\bar{d}(A) .$$

The other inequality follows from the fact that

$$\sum_{i=1}^{n} \lambda_i(A)^2 = tr\, A^2 = \sum_{v \in V(G)} deg(v) = n^2 \overline{d}(A)\ .$$

□

Note that for random graphs the expected value of $\overline{d}(A)$ is $\frac{n-1}{n}p$ which is approximately equal to p. The bounds of Proposition B.1 hold in general. They can be improved for random graphs.

THEOREM B.2. *([71]) For almost all graphs of order n (with fixed edge probability p) the maximal eigenvalue of the adjacency matrix A is equal to*

$$\lambda_{\max}(A) = np(1 + o(1))\ .$$

By the above result almost all graphs have the maximal eigenvalue close to np. This is to be expected. Much more surprising is the result for the second largest eigenvalue.

THEOREM B.3. *([71]) For almost all graphs G of order n with fixed edge probability p $(0 < p < 1)$ and an arbitrary $\varepsilon > 0$, the second largest adjacency matrix eigenvalue is*

$$\lambda_{n-1}(A) = O(n^{1/2+\varepsilon})\ .$$

These results were upgraded by Füredi and Komlós [52]:

THEOREM B.4. *([52]) Let $A = (a_{ij})$ be an $n \times n$ random symmetric matrix in which $a_{ii} = 0$ and a_{ij} $(i < j)$ are independent, identically distributed bounded random variables with distribution function H. Denote the moments of H by $\mu = \int x\,dH(x)$ and $\sigma^2 = \int (x - \mu)^2 dH(x)$. Then:*

(a) If $\mu > 0$ then $\lambda_{\max}(A) = \mu n + O(1)$ in measure, and $\max_{1 \le i < n} |\lambda_i(A)| = 2\sigma\sqrt{n} + O(n^{1/3} \log n)$ in probability.

(b) If $\mu = 0$ then $\max_{1 \le i \le n} |\lambda_i(A)| = 2\sigma\sqrt{n} + O(n^{1/3} \log n)$ in probability. □

The special case of Theorem B.4(a) where H is a discrete random variable with values 0 and 1, the latter with probability p, gives Theorem B.3. The additional step is a result of Wigner [131] which determines the overall eigenvalue distribution of random graphs.

THEOREM B.5. *([131]) Let $A = (a_{ij})$ be an $n \times n$ random symmetric matrix in which $a_{ii} = 0$ and a_{ij} $(i < j)$ are independent, identically distributed bounded random variables with distribution function H. Denote the moments of H by $\mu = \int x\,dH(x)$ and $\sigma^2 = \int (x - \mu)^2 dH(x)$, and let $F_n(x)$ be the cumulative distribution function of the eigenvalues of A. Then we have for an arbitrary x:*

$$\lim_{n \to \infty} F_n(x) = \int_{-\infty}^{x} f(x)\,dx \quad in\ probability$$

where

$$f(x) = \begin{cases} \frac{1}{2\pi\sigma^2 n}\sqrt{4\sigma^2 n - x^2}, & |x| < 2\sigma\sqrt{n} \\ 0, & otherwise \end{cases} .$$

□

It should be mentioned that Theorem C.4 does not follow from Theorem B.5. The latter result only implies that at most $o(n)$ eigenvalues are in absolute value larger than $2\sigma\sqrt{n}(1+o(1))$. The non-symmetric random matrices were considered by Juhász [72].

The Laplacian spectrum of random graphs was considered by Juvan and Mohar [79]. (A weaker result was obtained independently by Juhász [74].) The distribution of eigenvalues follows easily from the Wigner's result (a version where the diagonal entries need not to be identically 0; cf. [71]). The important is the estimation of $\lambda_2(G)$ for a random graph G.

THEOREM B.6. *([79]) For a fixed edge probability p ($0 < p < 1$) and any $\varepsilon > 0$ almost all graphs have their Laplace eigenvalues $\lambda_2(G)$ and $\lambda_{\max}(G)$ bounded by:*

$$pn - f_\varepsilon^+(n) < \lambda_2(G) < pn - f_\varepsilon^-(n)$$

and

$$pn + f_\varepsilon^+(n) > \lambda_{\max}(G) > pn + f_\varepsilon^-(n)$$

where

$$f_\varepsilon^+(n) = \sqrt{(2+\varepsilon)p(1-p)n\log n} \quad and \quad f_\varepsilon^-(n) = \sqrt{(2-\varepsilon)p(1-p)n\log n}.$$

□

There were also serious attacks on the eigenvalue problem for random regular graphs. For an integer $d \geq 2$ consider random d-regular graphs. McKay [91] determined the expected eigenvalue distribution of large d-regular graphs. Moreover, his result approximates eigenvalue distribution for arbitrary large regular graphs which do not have too many short cycles (which is the case with random regular graphs in any sensible model).

THEOREM B.7. *([91]) Let $d \geq 2$ be a fixed integer and let $G_1, G_2, G_3, \ldots$ be a sequence of simple d-regular graphs with increasing orders and such that for each $k \geq 3$ the number $c_k(G_n)$ of cycles of length k in G_n satisfies*

$$\lim_{n\to\infty} c_k(G_n)/|V(G_n)| = 0 .$$

Then the cumulative eigenvalue distribution functions $F(G_n, x)$ of graphs G_n converge to $F(x)$ for every x, where $F(x) = \int_{-\infty}^{x} f(t)dt$ and

$$f(t) = \begin{cases} \frac{d\sqrt{4(d-1)-t^2}}{2\pi(d^2-t^2)}, & |t| \leq 2\sqrt{d-1} \\ 0, & otherwise \end{cases} .$$

□

Godsil and Mohar [56] determined the expected eigenvalue distribution of random semiregular bipartite graphs (and some other families). Recall that a bipartite graph G is (d_1, d_2)-semiregular if the vertices in one bipartition class all have degree d_1, and the vertices in the other class have degree d_2.

THEOREM B.8. *([56]) Let $d_1, d_2 \geq 2$ be a integers, $p = \sqrt{(d_1 - 1)(d_2 - 1)}$, and let $G_1, G_2, G_3, \ldots$ be a sequence of simple (d_1, d_2)-semiregular bipartite graphs with increasing orders and such that for each $k \geq 3$ the number $c_k(G_n)$ of cycles of length k in G_n satisfies*

$$\lim_{n \to \infty} c_k(G_n)/|V(G_n)| = 0 .$$

Then the cumulative eigenvalue distribution functions $F(G_n, x)$ of graphs G_n converge to $F(x)$ for every x, where

$$F(x) = \int_{-\infty}^{x} f(t)dt + \frac{1}{2}\frac{|d_1 - d_2|}{d_1 + d_2}\delta(0)$$

where $\delta(0)$ is the discrete distribution with a unit point mass at the point 0, and

$$f(t) = \begin{cases} \dfrac{d_1 d_2 \sqrt{-(t^2 - d_1 d_2 + (p-1)^2)(t^2 - d_1 d_2 + (p+1)^2)}}{\pi(d_1 + d_2)(d_1 d_2 - t^2)|t|}, \\ \qquad \text{if } |\sqrt{d_1 - 1} - \sqrt{d_2 - 1}| \leq |t| \leq \sqrt{d_1 - 1} + \sqrt{d_2 - 1} \\ 0, \qquad \text{otherwise} . \end{cases}$$

□

The above result shows that all but $o(n)$ eigenvalues of a random d-regular graph are in absolute value smaller than $2\sqrt{d-1}$. However, due to the importance of the second eigenvalue of graphs, this is not a sufficient result for applications surveyed in this paper. Broder and Shamir [21] were able to show that random d-regular graphs (for d an even integer) have

$$\rho_2(G) = O(d^{3/4})$$

where

$$\rho_2(G) = \max\{|\lambda| \, ; \, \lambda \text{ an eigenvalue of } A(G), \lambda \neq d\} .$$

The above estimate holds for almost all d-regular graphs according to the following model. Let d be an even integer. For a random d-regular graph on n vertices choose $\frac{d}{2}$ random permutations $\sigma_1, \sigma_2, \ldots, \sigma_{d/2}$ of $V = \{1, 2, \ldots, n\}$, and define the d-regular graph corresponding to these permutations as the graph on the vertex set V with the vertex i $(1 \leq i \leq n)$ adjacent to vertices $\sigma_j(i), \sigma_j^{-1}(i), j = 1, 2, \ldots, \frac{d}{2}$. (Note that some of the obtained graphs contain loops or parallel edges. If we do not want loops we may consider random fixed point free permutations σ_j. It turns out, however, that a positive portion of these graphs are simple, so the properties holding for almost all graphs in this model also hold for almost all simple graphs among them.)

The $O(d^{3/4})$ bound for $\rho_2(G)$ was improved to $O(\sqrt{d})$ by Friedman (using the same model) [50]. A weaker version than Friedman's was obtained independently by Kahn and Szemerédi (cf. [51]). They proved the following results:

THEOREM B.9. *([51]) For a fixed even integer d, a random d-regular graph G of order n has*

$$\rho_2(G) = O(\sqrt{d})$$

with probability $1 - n^{-\Omega(\sqrt{d})}$ *as n tends to infinity.* □

It is not known if the methods used by Kahn and Szemerédi yield the expected bound $(2 + o(1))\sqrt{d-1}$ with probability $1 - o(1)$.

THEOREM B.10. *([50]) For a fixed even integer d, a random d-regular graph G of order n has the expectation*

$$E(\rho_2(G)^m) \le \left(2\sqrt{d-1} + 2\log d + O(1) + O\left(\frac{d^{3/2}\log\log n}{\log n}\right)\right)^m$$

for any integer $m \le 2\lfloor\log n\lfloor\sqrt{d-1}/2\rfloor/\log(d/2)\rfloor$ *(with an absolute constant in the O-notation), where all logarithms are base e.* □

The following result is even more useful:

THEOREM B.11. *([50]) For a fixed even integer d and an arbitrary* $\varepsilon > 0$, *a random d-regular graph G of order n has*

$$\rho_2(G) \le \left(2\sqrt{d-1} + 2\log d + O(1) + O\left(\frac{d^{3/2}\log\log n}{\log n}\right)\right)(1+\varepsilon)$$

with probability at least $1 - (1+\varepsilon)^2 n^{-2\lfloor\sqrt{d-1}/2\rfloor\log(1+\varepsilon)/\log(d/2)}$. □

On the other hand, Alon and Boppana (cf. [5]) proved that

$$\rho_2(G) \ge 2\sqrt{d-1} - O\left(\frac{\log d}{\log n}\right)$$

hold for any d-regular graph G, as far as $n \ge d^2$. Friedman's results are close to this bound but still far from the bound $\rho_2(G) \le 2\sqrt{d-1} + \varepsilon$ conjectured for random d-regular graphs by Alon [5].

REFERENCES

[1] D. Aldous, On the time taken by random walks on finite groups to visit every state, ZW 62 (1983) 361–374.
[2] D. Aldous, Hitting times for random walks on vertex-transitive graphs, Math. Proc. Camb. Phil. Soc. 106 (1989) 179–191.
[3] D. Aldous, Lower bounds for covering times for reversible Markov chains and random walks on graphs, J. Theoret. Probab. 2 (1989) 91–100.
[4] R. Aleliunas, R. M. Karp, R. J. Lipton, L. Lovász, and C. Rakoff, Random walks, universal traversal sequences, and the complexity of maze problems, in "20th FOCS," IEEE, 1979, 218–223.
[5] N. Alon, Eigenvalues and expanders, Combinatorica 6 (1986) 83–96.
[6] N. Alon, V. D. Milman, λ_1, isoperimetric inequalities for graphs and superconcentrators, J. Combin. Theory, Ser. B 38 (1985) 73–88.

[7] W. N. Anderson and T. D. Morley, Eigenvalues of a Laplacian of a graph, Lin. Multilin. Alg. 18 (1985) 141–145.

[8] B. Aspvall and J. R. Gilbert, Graph coloring using eigenvalue decomposition, SIAM J. Alg. Disc. Meth. 5 (1984) 526–538.

[9] F. Barahona and A. R. Mahjoub, On the cut polytope, Math. Prog. 36 (1986) 157–173.

[10] E. R. Barnes, An algorithm for partitioning the nodes of a graph, SIAM J. Alg. Disc. Meth. 3 (1982) 541–550.

[11] E. R. Barnes, Partitioning the nodes of a graph, in "Graph Theory and its Applications to Algorithms and Computer Science," Ed. Y. Alavi, Wiley, 1985, pp. 57–72.

[12] E. R. Barnes and A. J. Hoffman, Partitioning, spectra, and linear programming, in: "Progress in Combinatorial Optimization" (W. Pulleyblank, ed.), Academic Press, 1984, pp. 13–25.

[13] E. R. Barnes and A. J. Hoffman, On transportation problems with upper bounds on leading rectangles, SIAM J. Alg. Discr. Meth. 6 (1985) 487–496.

[14] C. Berge, Graphs, North-Holland, Amsterdam, 1985.

[15] F. Bien, Constructions of telephone networks by group representations, Notices Amer. Math. Soc. 36 (1989) 5–22.

[16] N. L. Biggs, Algebraic graph theory, Cambridge Univ. Press, Cambridge, 1974.

[17] M. Bolla, Relations between spectral and classification properties of multigraphs, DIMACS Technical Report 91-27, 1991.

[18] R. B. Boppana, Eigenvalues and graph bisection: An average case analysis, 28th Annual Symp. Found. Comp. Sci., IEEE, 1987, pp. 280–285.

[19] G. Brightwell, P. Winkler, Maximum hitting time for random walks on graphs, Random Str. Algor. 1 (1990).

[20] A. Z. Broder, A. R. Karlin, Bounds on the cover time, J. Theoret. Probab. 2 (1989) 101–120.

[21] A. Broder, E. Shamir, On the second eigenvalue of random regular graphs, 28th Annual Symp. Found. Comp. Sci., IEEEE, 1987, pp. 286–294.

[22] R. Brooks, Combinatorial problems in spectral geometry, in "Curvature and topology of Riemannian manifolds", Lecture Notes in Math. 1201, Springer, 1986, pp. 14–32.

[23] R. E. Burkard, Quadratic assignment problem, Europ. J. Oper. Res. 15 (1984) 283–289.

[24] P. Buser, On the bipartition of graphs, Discrete Appl. Math. 10 (1984) 105–109.

[25] J. Cheeger, A lower bound for the smallest eigenvalue of the Laplacian, in "Problems in analysis," (R.C. Gunnig, ed.), Princeton Univ. Press, 1970, pp. 195–199.

[26] P. Z. Chinn, J. Chvátalová, A. K. Dewdney, N. E. Gibbs, The bandwidth problem for graphs and matrices — a survey, J. Graph Theory 6 (1982) 223–254.

[27] F. R. K. Chung, Labelings of graphs, in "Selected Topics in Graph Theory 3," Academic Press, 1988, pp. 151–168.

[28] F. R. K. Chung, Diameter and eigenvalues, J. Amer. Math. Soc. 1 (1989) 187–196.

[29] F. R. K. Chung, V. Faber, and T. Manteuffel, An upper bound on the diameter of a graph from eigenvalues associated with its Laplacian, preprint, 1989.

[30] J. Cullum, W. E. Donath and P. Wolfe, The minimization of certain nondifferentiable sums of eigenvalues of symmetric matrices, Math. Prog. Study 3 (1976) 55–69.

[31] J. K. Cullum and R. A. Willoughby, Lanczos methods for large symmetric eigenvalue computations, Volume 1 and 2, Birkhäuser, Basel, 1985.

[32] D. M. Cvetković, Chromatic number and the spectrum of a graph, Publ. Inst. Math. (Beograd) 14 (1972) 25–38.

[33] D. M. Cvetković, M. Doob, I. Gutman, and A. Torgašev, Recent results in the theory of graph spectra, Ann. Discr. Math. 36, North-Holland, 1988.

[34] D. M. Cvetković, M. Doob and H. Sachs, Spectra of graphs, Academic Press, New York, 1979.

[35] C. Delorme and S. Poljak, Laplacian eigenvalues and the maximum cut problem, Technical Report 599, Université de Paris–Sud, Centre d'Orsay, 1990.

[36] C. Delorme and S. Poljak, Combinatorial properties and the complexity of a max–cut approximation, Technical Report 91687, Institut für Diskrete Mathematik, Universität Bonn, 1991.

[37] C. Delorme and S. Poljak, The performance of an eigenvalue bound on the max–cut problem in some classes of graphs, in "Colloque Marseille," 1990.

[38] C. Delorme, P. Solé, Diameter, covering radius and eigenvalues, Europ. J. Combin. 12 (1991) 95–108.

[39] P. Diaconis, D. Stroock, Geometric bounds for eigenvalues of Markov chains, preprint, 1989.

[40] D. Z. Djoković, Distance preserving subgraphs of hypercubes, J. Combin. Theory, Ser. B 14 (1973) 263–267.

[41] J. Dodziuk, Difference equations, isoperimetric inequality and transience of certain random walks, Trans. Amer. Math. Soc. 24 (1984) 787–794.

[42] W. E. Donath and A. J. Hoffman, Lower bounds for the partitioning of graphs, IBM J. Res. Develop. 17 (1973) 420–425.

[43] M. Dyer, A. Frieze, and R. Kannan, A random polynomial-time algorithm for approximating the volume of convex bodies J. Assoc. Comput. Mach. 38 (1991) 1-17.

[44] K. Fan, On a theorem of Weyl concerning eigenvalues of linear transformations. I, Proc. Nat. Acad. Sci. U.S.A. 35 (1949) 652–655.

[45] M. Fiedler, Algebraic connectivity of graphs, Czech. Math. J. 23 (98) (1973) 298–305.

[46] M. Fiedler, A property of eigenvectors of nonnegative symmetric matrices and its application to graph theory, Czech. Math. J. 25 (100) (1975) 619–633.

[47] M. Fiedler, Laplacian of graphs and algebraic connectivity, in "Combinatorics and graph theory", Banach Center Publ. 25, Warsaw, 1989, pp. 57–70.

[48] G. Finke, R. E. Burkard and F. Rendl, Quadratic assignment problem, Annals of Discrete Mathematics 31 (1987) 61-82.

[49] S. Friedland, Lower bounds for the first eigenvalue of certain M-matrices associated with graphs, preprint, 1991.

[50] J. Friedman, On the second eigenvalue and random walks in random d-regular graphs, Combinatorica 11 (1991) 331–362.

[51] J. Friedman, J. Kahn, and E. Szemerédi, On the second eigenvalue in random regular graphs, Proc. 21st Annual ACM Symp. Theory Comput., Seattle, 1989 (ACM, New York, 1989) pp. 587–598.

[52] Z. Füredi and J. Komlós, The eigenvalues of random symmetric matrices, Combinatorica 1 (1981) 233–241.

[53] M. R. Garey and D. S. Johnson, Computers and Intractability: A guide to the theory of NP–completeness, San Francisco, Freeman, 1979.

[54] M. R. Garey, D. S. Johnson, R. L. Stockmeyer, Some simplified NP-complete problems, Proc. 6th ACM Symposium on Theory of Computing, 1974, pp. 47–63.

[55] C. D. Godsil, Tools from linear algebra, Research Rep. CORR 89–35, University of Waterloo, 1989.

[56] C. D. Godsil, B. Mohar, Walk–generating functions and spectral measures of infinite graphs, Linear Algebra Appl. 107 (1988) 191–206.

[57] M. K. Goldberg, R. Gardner, On the minimal cut problem, in "Progress in Graph Theory", Eds. J. A. Bondy and U. S. R. Murty, Academic Press, Toronto, 1984, pp. 295–305.

[58] G. H. Golub and C. F. van Loan, Matrix Computations, Johns Hopkins Series in the Mathematical Sciences, Johns Hopkins Univ. Press, Second edition, 1989.

[59] R. L. Graham, Isometric embeddings of graphs, in: "Selected Topics in Graph Theory 3" (L. W. Beineke, R. J. Wilson, eds.), Academic Press, 1988.

[60] R. L. Graham and H. O. Pollack, On the addressing problem for loop switching, Bell Syst. Tech. J. 50 (1971) 2495–2519.

[61] M. Gromov, V. D. Milman, A topological application of the isoperimetric inequality, Amer. J. Math. 105 (1983) 843–854.

[62] M. Grötschel, L. Lovász, and A. Schrijver, Geometric algorithms and combinatorial optimization, Springer-Verlag, Berlin, 1988.

[63] S. W. Hadley, F. Rendl, and H. Wolkowicz, Symmetrization of nonsymmetric quadratic assignment problems and the Hoffman-Wielandt inequality, Linear Algebra Appl. 167 (1992) 53–64.

[64] S. W. Hadley, F. Rendl, and H. Wolkowicz, Bounds for the quadratic assignment problem using continuous optimization techniques, Proc. "Combinatorial Optimization", Waterloo, 1990, pp. 237–248.

[65] S. W. Hadley, F. Rendl, and H. Wolkowicz, A new lower bound via projection for the quadratic assignment problem, preprint 1991.

[66] F. O. Hadlock, Finding a maximum cut of a planar graph in polynomial time, SIAM J. Comput. 4 (1975) 221–225.

[67] A. J. Hoffman, On eigenvalues and colorings of graphs, in "Graph Theory and Its Applications" (B. Harris, ed.), Acad. Press, 1970, pp. 79–91.

[68] A. J. Hoffman and H. W. Wielandt, The variation of the spectrum of a normal matrix, Duke Math. J. 20 (1953) 37–39.

[69] R. A. Horn and C. R. Johnson, Matrix Analysis, Cambridge University Press, 1985.

[70] M. Jerrum, A. Sinclair, Approximating the permanent, SIAM J. Comput. 18 (1989) 1149–1178.

[71] F. Juhász, On the spectrum of a random graph, in: "Algebraic Methods in Graph Theory" (L. Lovász, V. T. Sós, eds.), Colloq. Math. Soc. J. Bolyai 25, North–Holland, Amsterdam, 1982, pp. 313–316.

[72] F. Juhász, On the asymptotic behaviour of the spectra of non-symmetric random (0,1) matrices, Discrete Math. 41 (1982) 161–165.

[73] F. Juhász, The asymptotic behaviour of Lovász' ϑ function for random graphs, Combinatorica 2 (1982) 153–155.

[74] F. Juhász, The asymptotic behaviour of Fiedler's algebraic connectivity for random graphs, Discrete Math. 96 (1991) 59–63.

[75] F. Juhász, On a method of cluster analysis, ZAMM 64 (1984) T335–T336.

[76] F. Juhász, On the theoretical backgrounds of cluster analysis based on the eigenvalue problem of the association matrix, Statistics 20 (1989) 573–581.

[77] F. Juhász and K. Mályusz, Problems of cluster analysis from the viewpoint of numerical analysis, in: "Numerical Methods", Colloq. Math. Soc. J. Bolyai 22, North-

Holland, Amsterdam, 1977, pp. 405–415.

[78] M. Juvan and B. Mohar, Optimal linear labelings and eigenvalues of graphs, to appear in Discr. Appl. Math.

[79] M. Juvan and B. Mohar, Laplace eigenvalues and bandwidth-type invariants of graphs, preprint, 1990.

[80] J. N. Kahn, N. Linial, N. Nisan, M. E. Saks, On the cover time of random walks on graphs, J. Theoret. Probab. 2 (1989) 121–128.

[81] R. M. Karp, Reducibility among combinatorial problems, in: "Complexity of Computer Computation" (R. E. Miller, J. W. Thather, eds), Plenum Press, New York, 1972, pp. 85–103.

[82] J. G. Kemeny, J. L. Snell, Finite Markov chains, Van Nostrand, 1960.

[83] P. Lancaster, Theory of matrices. Academic Press, 1969.

[84] T. Lengauer, Combinatorial Algorithms for Integrated Circuit Layout, J. Wiley, New York, 1990.

[85] L. Lovász, On the Shannon capacity of a graph, IEEE Trans. Inform. Theory IT–25 (1979) 1–7.

[86] L. Lovász, Combinatorial Problems and Exercises, North–Holland, Amsterdam, 1979.

[87] L. Lovász, M. Simonovits, The mixing rate of Markov chains, an isoperimetric inequality, and computing the volume, preprint, 1990.

[88] A. Lubotzky, Discrete groups, expanding graphs and invariant measures, manuscript, 1989.

[89] A. Lubotzky, R. Phillips and P. Sarnak, Ramanujan graphs, Combinatorica 8 (1988) 261–277.

[90] G. A. Margulis, Explicit group–theoretical constructions of combinatorial schemes and their application to the design of expanders and superconcentrators, Problemy Pered. Inform. 24 (1988) 51–60 (in Russian); Engl. transl. Problems Inform. Transm. 24 (1988) 39–46.

[91] B. D. McKay, The expected eigenvalue distribution of a large regular graph, Lin. Algebra Appl. 40 (1981) 203–216.

[92] R. Merris, An edge version of the Matrix-Tree Theorem and the Wiener index, Lin. Multilin. Alg. 25 (1989) 291–296.

[93] R. Merris, The distance spectrum of a tree, J. Graph Theory 14 (1990) 365–369.

[94] B. Mohar, Isoperimetric inequalities, growth, and the spectrum of graphs, Linear Algebra Appl. 103 (1988) 119–131.

[95] B. Mohar, Isoperimetric numbers of graphs, J. Combin. Theory, Ser. B 47 (1989) 274–291.

[96] B. Mohar, The Laplacian spectrum of graphs, in: "Graph Theory, Combinatorics, and Applications," (Y. Alavi et al., eds.), J. Wiley, New York, 1991, pp. 871–898.

[97] B. Mohar, Eigenvalues, diameter, and mean distance in graphs, Graphs Comb. 7 (1991) 53–64.

[98] B. Mohar, Some algebraic methods in graph theory and combinatorial optimization, Discrete Math., to appear.

[99] B. Mohar and S. Poljak, Eigenvalues and the max–cut problem, Czech. Math. J. 40 (115) (1990) 343–352.

[100] C. Moler and D. Morrison, Singular value analysis of cryptograms, Amer. Math. Monthly 90 (1983) 78–87.

[101] T. Motzkin and E. G. Straus, Maxima for graphs and new proof of a theorem of Turán, Canad. J. Math. 17 (1965) 533–540.

[102] G. Narasimhan and R. Manber, A generalization of Lovász Θ function, DIMACS Series in Discrete Math. and Comp. Sci. 1, 1990, pp. 19–27.

[103] A. Nilli, On the second eigenvalue of a graph, Discrete Math. 91 (1991) 207–210.

[104] G. I. Orlova and Y. G. Dorfman, Finding the maximal cut in a graph, Engrg. Cybernetics 10 (1972) 502–506.

[105] M. L. Overton, On minimizing the maximum eigenvalue of a symmetric matrix, SIAM J. Matrix Anal. 9 (1988) 256–268.

[106] M. L. Overton and R. S. Womersley, On the sum of the largest eigenvalues of a symmetric matrix, SIAM J. Matrix Anal. 13 (1992) 41–45.

[107] J. L. Palacios, Bounds on expected hitting times for a random walk on a connected graph, Linear Algebra Appl. 141 (1990) 241–252.

[108] J. L. Palacios, On a result of Aleliunas et al. concerning random walks on graphs, Prob. Eng. Info. Sci. 4 (1990) 489–492.

[109] J. L. Palacios, Expected hitting and cover times of random walks on some special graphs, preprint, 1991.

[110] C. H. Papadimitriou, The NP-completeness of the bandwidth minimization problem, Computing 16 (1976) 263–270.

[111] P. M. Pardalos and G. P. Rodgers, Computational aspects of a branch and bound algorithm for quadratic zero-one programming, Computing 40 (1990) 131–144.

[112] S. Poljak, Polyhedral and eigenvalue approximations of the max–cut problem, Technical Report 91691, Institut für Diskrete Mathematik, Universität Bonn, 1991. Submitted to Proc. Conf. 'Sets, Graphs and Numbers' (Budapest 1991).

[113] S. Poljak and F. Rendl, Computing the max–cut by eigenvalues, Report No. 91735-OR, Institut für Diskrete Mathematik, Universität Bonn, 1991.

[114] A. Pothen, H. D. Simon and K.-P. Liou, Partitioning Sparse Matrices with Eigenvectors of Graphs, SIAM J. Matrix Anal. Appl. 11 (1990) 430–452.

[115] D. L. Powers, Structure of a matrix according to its second eigenvalue, in: "Current Trends in Matrix Theory" (F. Uhlig and R. Grone, eds.), Elsevier, 1987, pp. 121–133.

[116] D. L. Powers, Graph partitioning by eigenvectors, Linear Algebra Appl. 101 (1988) 121–133.

[117] F. Rendl and H. Wolkowicz, Applications of parametric programming and eigenvalue maximization to the quadratic assignment problem, Math. Progr. 53 (1992) 63-78.

[118] F. Rendl and H. Wolkowicz, A projection technique for partitioning the nodes of a graph, Technical Report, University of Technology, Graz, 1990.

[119] R. Rubinfeld, The cover time of a regular expander is $O(n \log n)$, Inform. Proc. Lett. 35 (1990) 49–51.

[120] R. L. Roth and P. M. Winkler, Collapse of the metric hierarchy for bipartite graphs, Europ. J. Combin. 7 (1986) 371–375.

[121] T. L. Saaty, A scalling method for priorities in hierarchical structures, J. Math. Psych. 15 (1977) 234–281.

[122] H. Schramm and J. Zowe, A combination of the bundle approach and the trust region concept, in: "Advances in Mathematical Optimization" (J. Guddat et al., ed.), Akademie Verlag, Berlin, 1988, pp. 196–209.

[123] D. S. Scott, Block Lanczos software for symmetric eigenvalue problems, Technical Report ORNL/CSD-48, Oak Ridge National Laboratory, 1979.

[124] A. Sinclair, M. Jerrum, Approximate counting, uniform generation and rapidly mixing Markov chains, Inform. and Comput. 82 (1989) 93–133.

[125] P. Solé, Expanding and forwarding, submitted.

[126] A. Srivastav and P. Stangier, A provably good algorithm for the graph partitioning problem, Preprint of Institute of Discrete Math., Univ. Bonn, 1991.

[127] P. D. Straffin, Jr., Linear algebra in geography: Eigenvectors of networks, Math. Mag. 53 (1980) 269–276.

[128] V. S. Sunderam, P. Winkler, Fast information sharing in a distributed system, preprint, 1988.

[129] R. M. Tanner, Explicit concentrators from generalized n-gons, SIAM J. Alg. Discr. Meth. 5 (1984) 287–293.

[130] N. Th. Varopoulos, Isoperimetric inequalities and Markov chains, J. Funct. Anal. 63 (1985) 215–239.

[131] E. P. Wigner, Characteristic vectors of bordered matrices with infinite dimensions, Ann. Math. 62 (1955) 548–564.

[132] H. S. Wilf, The eigenvalues of a graph and its chromatic number, J. London Math. Soc. 42 (1967) 330–332.

[133] H. S. Wilf, Graphs and their spectra: Old and new results, Congr. Numer. 50 (1985) 37–42.

[134] H. S. Wilf, Spectral bounds for the clique and independence numbers of graphs, J. Combin. Theory, Ser. B 40 (1986) 113–117.

[135] P. M. Winkler, Proof of the squashed cube conjecture, Combinatorica 3 (1983) 135–139.

EUTACTIC STARS AND GRAPH SPECTRA

PETER ROWLINSON*

Abstract. The eutactic stars to be considered arise naturally from the spectral decomposition of an adjacency matrix of a graph. The role of these stars is discussed in relation to (i) local modifications of graphs, (ii) the construction of star partitions of a graph. Some exploratory results relate these star partitions to the structure of the underlying graph.

AMS(MOS) subject classification: 05C50

1. Introduction. A *eutactic star*, in a subspace $\mathcal{U}$ of an inner product space $\mathcal{V}$, is (to within a scalar multiple) the orthogonal projection onto $\mathcal{U}$ of an orthonormal set of vectors in $\mathcal{V}$ [S]. In what follows, $\mathcal{V} = \mathbb{R}^n$ and the inner product of two column vectors $\mathbf{u}, \mathbf{v}$ is just the scalar product $\mathbf{u}^T \mathbf{v}$. The stars to be considered are obtained by projecting the standard orthonormal basis $\{\mathbf{e}_1, \mathbf{e}_2, \ldots, \mathbf{e}_n\}$ of $\mathbb{R}^n$ onto the eigenspaces of a symmetric matrix A with real entries. Thus if A has spectral decomposition

$$(1.1) \qquad A = \mu_1 P_1 + \mu_2 P_2 + \cdots + \mu_m P_m$$

then we define $\mathcal{S}_i$ as the eutactic star consisting of the vectors $P_i \mathbf{e}_1, P_i \mathbf{e}_2, \ldots, P_i \mathbf{e}_n$. These n vectors, called the *arms* of $\mathcal{S}_i$, span the eigenspace $\mathcal{E}(\mu_i)$ corresponding to the eigenvalue μ_i. Fig. 1 depicts $\mathcal{S}_i$ in a situation where $n = 3$ and $\mathcal{E}(\mu_i)$ has dimension 2. We shall use implicitly the facts that $P_i^2 = P_i = P_i^T$ $(i = 1, \ldots, m)$; $P_i P_j = 0$ $(i \neq j)$; $\sum_{i=1}^m P_i = I$; and each P_i is a polynomial in A.

For the time being we shall take A to be the $(0,1)$ adjacency matrix of a simple graph G whose vertices are labelled $1, 2, \ldots, n$. The eigenvalues of A are independent of this vertex-ordering and so we refer to them as the eigenvalues of G, comprising the *spectrum* of G. If G is connected then A is irreducible and so the largest eigenvalue of G is simple [G,Ch. XIII]: in this case, the arms of the corresponding eutactic star all lie along a single line. In the special case that G is strongly regular, when $m = 3$, Bier [B] has used the remaining two stars to distinguish such graphs with identical parameters, essentially by considering the largest number of arms on one side of a hyperplane of an eigenspace. Here we explore the role of the stars $\mathcal{S}_1, \ldots, \mathcal{S}_m$ in the general context of arbitrary graphs, with particular reference to (i) the change in spectrum resulting from various local modifications of a graph, (ii) the construction of star partitions of the vertex-set of a graph. In §4 we present some exploratory results relating the structure of a graph to its star partitions. The results of §2 extend those of [PR1, Section 2], and the results of §§3,4 extend those of [CRS, Sections 2,3,4].

*Department of Mathematics, University of Stirling, Stirling FK9 4LA, Scotland

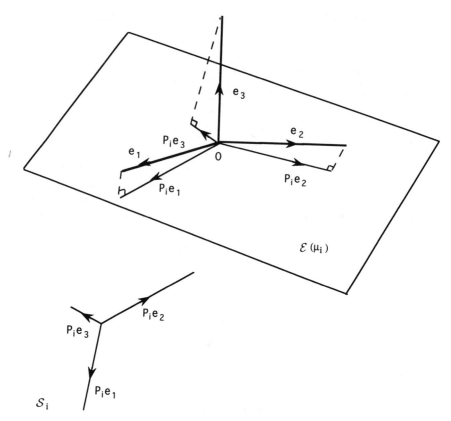

FIGURE 1

Before discussing local modifications we note the relation of the stars $S_1, \ldots, S_m$ to the graph angles introduced by Cvetković [C]. The length of the arm $P_i \mathbf{e}_j$ ($i = 1, \ldots, m; j = 1, \ldots, n$) is just α_{ij}, where $\cos^{-1}(\alpha_{ij})$ is the angle between $\mathbf{e}_j$ and $\mathcal{E}(\mu_i)$. The α_{ij} themselves are customarily called the *angles* of G. From the spectrum of G and the angles α_{iu} ($i = 1, \ldots, m$) one can determine the spectrum of $G - u$, the graph obtained from G by deleting vertex u and all edges incident with u [CD, Section 5]; see equation (2.3). The next section is devoted to analogous results for other local modifications of G.

Finally, we recall that, given the spectrum of G, knowledge of the angles at u is equivalent to knowledge of the number of $u - u$ walks in G of length k for each $k = 0, 1, \ldots, m - 1$. This is proved by equating $(u, u)-$ entries in the spectral decomposition of A^k. It follows that the spectrum and angles of G determine the number of 4-cycles and 5-cycles in G [CR], the spectrum alone being sufficient to determine only the number of 3-cycles in G.

2. Local modifications of graphs. With the notation of the previous sections, $\det(xI - A)$ is called the *characteristic polynomial* of G, denoted by $\varphi_G(x)$. In order to investigate $\varphi_{G-U}(x)$, where now U is an arbitrary subset of vertices of G, we use an argument of Jacobi and write $xI - A = \begin{pmatrix} P & Q \\ R & S \end{pmatrix}, (xI - A)^{-1} = \begin{pmatrix} P^* & Q^* \\ R^* & S^* \end{pmatrix}$, where the rows and columns of P and P^* are indexed by U. Then $\begin{pmatrix} I & O \\ R & S \end{pmatrix} = \begin{pmatrix} P^* & Q^* \\ O & I \end{pmatrix}\begin{pmatrix} P & Q \\ R & S \end{pmatrix}$, whence

$$(2.1) \qquad \det(S) = \det(P^*)\det(xI - A),$$

that is, $\varphi_{G-U}(x) = \det(P^*)\varphi_G(x)$, as noted by Godsil in [Go, Theorem 3.2]. Since $(xI - A)^{-1} = \sum_{i=1}^{m}(x - \mu_i)^{-1}P_i$, we have $\det(P^*) = \sum_{i=1}^{m}(x - \mu_i)^{-1}P_i^U$ where P_i^U denotes the submatrix of P_i whose rows and columns are indexed by U. Thus equation (2.1) yields

$$(2.2) \qquad \varphi_{G-U}(x) = \varphi_G(x)\det\left\{\sum_{i=1}^{m}(x - \mu_i)^{-1}P_i^U\right\}.$$

When $U = \{u\}$ we recover the relation

$$(2.3) \qquad \varphi_{G-u}(x) = \varphi_G(x)\sum_{i=1}^{m}\frac{\alpha_{iu}^2}{x - \mu_i} .$$

In general, $\varphi_{G-U}(x)$ is determined by $\varphi_G(x)$ and the m Gram matrices of the vectors $P_i e_u (u \in U)$ $(i = 1, \ldots, m)$. The foregoing results appear in [CRS,§4], while the case in which U consists of two non-adjacent vertices is treated explicitly in [PR1, §2.5]. The latter article treats also the addition of an edge and the addition of a bridging vertex (i.e. a new vertex adjacent to two existing non-adjacent vertices). The results are described briefly here as a preliminary to the discussion of some further modifications local to two prescribed vertices u and v. We describe a polynomial as *known* if it is determined by $\varphi_G(x)$ and the matrices $P_i^{\{u,v\}}$ $(i = 1, \ldots, m)$. Note that knowledge of $P_i^{\{u,v\}}$ is equivalent to knowledge of $|P_i e_u|, |P_i e_v|$ and $|P_i e_u + P_i e_v|$, essentially the dimensions of the (possibly degenerate) parallelogram determined by $P_i e_u, P_i e_v$. Given $\varphi_G(x)$, we may characterize known polynomials in terms of the structure of G rather than the structure of the stars $S_1, \ldots S_m$. For let $p_{uv}^{[i]}$ be the (u, v) - entry of P_i and let $a_{uv}^{(k)}$ be the number of $u - v$ walks of length k in G. Since $A^k = \sum_{i=1}^{m} \mu^k P_i$ we have

$$\begin{pmatrix} a_{uv}^{(0)} \\ a_{uv}^{(1)} \\ \vdots \\ a_{uv}^{(m-1)} \end{pmatrix} = \begin{pmatrix} 1 & 1 & 1 & \cdots & 1 \\ \mu_1 & \mu_2 & \mu_3 & \cdots & \mu_m \\ \vdots & & & & \\ \mu_1^{m-1} & \mu_2^{m-1} & \mu_3^{m-1} & \cdots & \mu_m^{m-1} \end{pmatrix} \begin{pmatrix} p_{uv}^{[1]} \\ p_{uv}^{[2]} \\ \vdots \\ p_{uv}^{[m]} \end{pmatrix}.$$

Thus knowledge of $\varphi_G(x)$ and $P_i^{\{u,v\}}$ $(i = 1, \ldots, m)$ is equivalent to knowledge of $\varphi_G(x)$ and $a_{uu}^{(k)}, a_{vv}^{(k)}, a_{uv}^{(k)}$ $(k = 0, 1, \ldots, m - 1)$.

In view of our initial observations, $\varphi_{G-u}(x)$ and $\varphi_{G-v}(x)$ are known; and $\varphi_{G-u-v}(x)$ is known, whether or not u and v are adjacent. Now suppose that u and v are non-adjacent and consider addition of the edge uv. We take $u = 1$, $v = 2$ without loss of generality, and we let

$$A = \begin{bmatrix} 0 & 0 & \mathbf{r}^T \\ 0 & 0 & \mathbf{s}^T \\ \mathbf{r} & \mathbf{s} & A' \end{bmatrix}.$$

Using columnwise linearity of determinants, we obtain [PR2, Lemma 1.1]:

$$(2.4) \qquad \varphi_{G+uv}(x) = \varphi_G(x) - \varphi_{G-u-v}(x) + \begin{vmatrix} 0 & -\mathbf{r}^T \\ -\mathbf{s} & A' \end{vmatrix} + \begin{vmatrix} 0 & -\mathbf{s}^T \\ -\mathbf{r} & A' \end{vmatrix}.$$

Equivalently,

$$(2.5) \qquad \varphi_{G+uv}(x) = \varphi_G(x) - \varphi_{G-u-v}(x) - 2\psi_{uv}(x),$$

where $\psi_{uv}(x)$ is the (u, v)-entry of $\mathrm{adj}(xI - A)$, i.e. $\psi_{uv}(x) = \varphi_G(x) \sum_{i=1}^{m} (x - \mu_i)^{-1} p_{uv}^{[i]}$. Thus $\psi_{uv}(x)$, and hence $\varphi_{G+uv}(x)$, is known. In fact, $\psi_{uv}(x) = \sqrt{\varphi_{G-u}(x)\varphi_{G-v}(x) - \varphi_G(x)\varphi_{G-u-v}(x)}$ [Go, Corollary 3.3], a result which follows from equation (2.1) by writing $\det(P^*)$ as $\varphi_G(x)^{-2}\{\psi_{uu}(x)\psi_{vv}(x) - \psi_{uv}(x)^2\}$ when $U = \{u, v\}$. Here there is no ambiguity in the sign of the square root because $x\psi_{uv}(x)/\varphi_G(x)$ is the (u, v)-entry of $(I - x^{-1}A)^{-1}$, i.e. the (u, v)-entry of the walk generating function $\sum_{i=0}^{\infty} x^{-i} A^i$.

Let G^* be the multigraph obtained from G by amalgamating the vertices u and v. In this case, columnwise linearity of determinants yields [PR2, Lemma 1.2]

$$(2.6) \qquad \varphi_{G^*}(x) = \varphi_{G-u}(x) + \varphi_{G-v}(x) - x\varphi_{G-u-v}(x) - 2\psi_{uv}(x)$$

and so $\varphi_{G^*}(x)$ is known also. Moreover, on eliminating $\psi_{uv}(x)$ from equations (2.5) and (2.6) we obtain:

$$(2.7) \quad \varphi_{G+uv}(x) = \varphi_G(x) + \varphi_{G^*}(x) + (x - 1)\varphi_{G-u-v}(x) - \varphi_{G-u}(x) - \varphi_{G-v}(x).$$

We turn next to the case in which u and v are adjacent vertices. With $u = 1, v = 2$ and $A = \begin{bmatrix} 0 & 1 & \mathbf{r}^T \\ 1 & 0 & \mathbf{s}^T \\ \mathbf{r} & \mathbf{s} & A' \end{bmatrix}$, equation (2.4) becomes:

$$(2.8) \qquad \varphi_G(x) = \varphi_{G-uv}(x) - \varphi_{G-u-v}(x) + \begin{vmatrix} 0 & -\mathbf{r}^T \\ -\mathbf{s} & A' \end{vmatrix} + \begin{vmatrix} 0 & -\mathbf{s}^T \\ -\mathbf{r} & A' \end{vmatrix}.$$

Now $\begin{vmatrix} 0 & -\mathbf{r}^T \\ -\mathbf{s} & A' \end{vmatrix} = \begin{vmatrix} -1 & -\mathbf{r}^T \\ -\mathbf{s} & A' \end{vmatrix} + \begin{vmatrix} 1 & -\mathbf{r}^T \\ \mathbf{0} & A' \end{vmatrix} = -\psi_{uv}(x) + \varphi_{G-u-v}(x)$ and so

$$(2.9) \qquad \varphi_{G-uv}(x) = \varphi_G(x) - \varphi_{G-u-v}(x) + 2\psi_{uv}(x).$$

In particular, $\varphi_{G-uv}(x)$ is known.

In the case of adjacent vertices u, v, equation (2.7) may be rewritten in the form of the Deletion–Contraction Algorithm [PR2, Theorem 1.3]:

$$(2.10) \quad \varphi_G(x) = \varphi_{G-uv}(x) + \varphi_{(G-uv)^*}(x) + (x-1)\varphi_{G-u-v}(x) - \varphi_{G-u}(x) - \varphi_{G-v}(x).$$

From this we deduce that $\varphi_{(G-uv)^*}(x)$ is known.

Next let G_{uv} be the graph obtained from G by subdividing the edge uv, and let w be the vertex of subdivision. On applying equation (2.10) to G_{uv} and the edge uw we find [PR2, Proposition 1.7]:

$$(2.11) \quad \varphi_{G_{uv}}(x) = \varphi_G(x) + (x-1)\varphi_{G-uv}(x) - \varphi_{G-u}(x) - \varphi_{G-v}(x) + \varphi_{G-u-v}(x)$$

and so $\varphi_{G_{uv}}(x)$ is known.

For completeness we note that we can now deal with the addition of a bridging vertex between non-adjacent vertices u and v. Let $G(u, v)$ denote the graph obtained in this way. On applying equation (2.11) to $G + uv$ we obtain

$$(2.12) \quad \varphi_{G(u,v)}(x) = \varphi_{G+uv}(x) + (x-1)\varphi_G(x) - \varphi_{G-u}(x) - \varphi_{G-v}(x) + \varphi_{G-u-v}(x)$$

and so $\varphi_{G(u,v)}(x)$ is known. In fact, Lowe and Soto [LS] had already noted that $\varphi_{G(u,v)}(x)$ is determined just by $\varphi_G(x)$ and $|P_i\mathbf{e}_u + P_i\mathbf{e}_v|$ $(i = 1, \ldots, m)$, and the following explicit expression was derived in [PR3, §2]:

$$(2.13) \qquad \varphi_{G(u,v)}(x) = x\varphi_G(x) - \varphi_G(x) \sum_{i=1}^{m} \frac{|P_i\mathbf{e}_u + P_i\mathbf{e}_v|^2}{x - \mu_i}.$$

This follows from equations (2.3), (2.5) and (2.12).

We summarize our results as follows, where (as above) an asterisk denotes amalgamation of non-adjacent vertices u and v.

THEOREM 2.1. *Let u, v be vertices of the graph G. If u and v are non-adjacent then the spectra of $G - u - v, G + uv, G^*, G(u, v)$ are determined by the spectrum of G and the lengths $|P_i \mathbf{e}_u|, |P_i \mathbf{e}_v|, |P_i \mathbf{e}_u + P_i \mathbf{e}_v|$ $(i = 1, \ldots, m)$. If u and v are adjacent then the same is true of $G - u - v, G - uv, (G - uv)^*, G_{uv}$.*

$\square$

This result completes the answer to questions raised in [PR1,§2.7]. Although these questions were posed in terms of eutactic stars, we may (in view of our earlier remarks) replace $|P_i \mathbf{e}_u|, |P_i \mathbf{e}_v|, |P_i \mathbf{e}_u + P_i \mathbf{e}_v|$ $(i = 1, \ldots, m)$ in Theorem 2.1 with $a_{uu}^{(k)}, a_{vv}^{(k)}, a_{uv}^{(k)}$ $(k = 0, \ldots, m-1)$ to obtain a result whose statement and proof are free of references to stars.

3. Star bases. We retain the notation of previous sections, and we write k_i for the multiplicity of μ_i as an eigenvalue of G $(i = 1, \ldots, m)$. If μ_i is not an eigenvalue of $G - U$ then $|U| \geq k_i$; more precisely, if $|U| \leq k_i$ then by consideration of $\mathcal{E}(\mu_i) \cap \langle \mathbf{e}_j : j \notin U \rangle$, the multiplicity of μ_i as an eigenvalue of $G - U$ is at least $k_i - |U|$. Simić asked whether $V(G)$, the vertex set of G, has a partition $X_1 \dot\cup \ldots \dot\cup X_m$ such that μ_i is not an eigenvalue of $G - X_i$ $(i = 1, \ldots, m)$. In this situation, $|X_i| = k_i$ $(i = 1, \ldots, m)$, while if $U \subset X_i$ then μ_i is an eigenvalue of $G - U$ with multiplicity $k_i - |U|$. For the time being we call such a partition a *polynomial partition*, in view of the relations $\varphi_{G - X_i}(\mu_i) \neq 0$ $(i = 1, \ldots, m)$.

We now turn to what will prove to be a related question. If μ_i is a simple eigenvalue then a basis for $\mathcal{E}(\mu_i)$ is unique to within a nonzero scalar multiple. If however $k_i > 1$ then the existence of infinitely many essentially different bases for $\mathcal{E}(\mu_i)$ makes it desirable to restrict the choice of bases in a natural way. Cvetković asked whether $V(G)$ always has a partition $X_1 \dot\cup \ldots \dot\cup X_m$ such that $\{P_i \mathbf{e}_j : j \in X_i\}$ is a basis for $\mathcal{E}(\mu_i)$ $(i = 1, \ldots, m)$. In this situation, stringing together such bases for the m eigenspaces we obtain a basis for $\mathbf{R}^n$. Such a basis is called a *star basis* for $\mathbf{R}^n$ corresponding to G, and the underlying partition of $V(G)$ is called a *star partition*.

In fact it is easy to see that star partitions always exist by considering a multiple Laplacian development of the determinant of an appropriate transition matrix. Let $\{\mathbf{x}_1, \ldots, \mathbf{x}_n\}$ be a basis of $\mathbf{R}^n$ obtained by stringing together arbitrary fixed bases of $\mathcal{E}(\mu_1), \ldots, \mathcal{E}(\mu_m)$; say $\mathcal{E}(\mu_i)$ has basis $\{\mathbf{x}_h : h \in R_i\}$ where $R_1 \dot\cup \ldots \dot\cup R_m$ is a fixed partition of $\{1, 2, \ldots, n\}$. Let $\mathbf{e}_j = \sum_{h=1}^{n} t_{hj} \mathbf{x}_h$, so that

$$(3.1) \qquad P_i \mathbf{e}_j = \sum_{h \in R_i} t_{hj} \mathbf{x}_h;$$

and let $T = (t_{hj})$. Let $C_1 \dot\cup \ldots \dot\cup C_m$ be any partition of $\{1, 2, \ldots, n\}$ such that $|C_i| = k_i$ $(i = 1, \ldots, m)$, and let M_i be the $k_i \times k_i$ matrix (t_{hj}) whose rows are indexed by R_i and whose columns are indexed by C_i. To within sign, $\prod_{i=1}^{m} \det(M_i)$ accounts for $k_1! \ldots k_m!$ of the summands in $\det(T)$, and it follows that $\det(T) =$

$$\sum \left\{ \pm \prod_{i=1}^{m} \det(M_i) \right\}$$ where the sum is taken over all $n!/k_1! \ldots k_m!$ partitions

$C_1 \dot\cup \ldots \dot\cup C_m$. Since T is invertible, some $\prod_{i=1}^{m} \det(M_i)$ is non-zero, say that determined by the partition $X_1 \dot\cup \ldots \dot\cup X_m$. This partition is a star partition because in view of equation (3.1), the invertibility of M_i guarantees that each $\mathbf{x}_h$ ($h \in R_i$) is a linear combination of the vectors $P_i \mathbf{e}_j$ ($j \in X_i$).

It turns out that the star partitions of $V(G)$ are precisely the polynomial partitions [CRS, Section 3]. To prove this we first establish another characterization of star partitions.

LEMMA 3.1. *Let* $X_1 \dot\cup \ldots \dot\cup X_m$ *be a partition of* $V(G)$ *and let* $\mathcal{V}_i = \langle \mathbf{e}_h : h \notin X_i \rangle$ *($i = 1, \ldots, m$). Then* $X_1 \dot\cup \ldots \dot\cup X_m$ *is a star partition if and only if* $\mathbb{R}^n = \mathcal{E}(\mu_i) \oplus \mathcal{V}_i$ *for each* $i \in \{1, \ldots, m\}$.

Proof. If $\mathbb{R}^n = \mathcal{E}(\mu_i) \oplus \mathcal{V}_i$ then $\mathbb{R}^n = \mathcal{E}(\mu_i)^\perp \oplus \mathcal{V}_i^\perp$ and so $\mathcal{E}(\mu_i) = P_i(\mathbb{R}^n) = P_i(\mathcal{V}_i^\perp) = \langle P_i \mathbf{e}_h : h \in X_i \rangle$. Hence $X_1 \dot\cup \ldots \dot\cup X_m$ is a star partition if $\mathbb{R}^n = \mathcal{E}(\mu_i) \oplus \mathcal{V}_i$ for each i. Conversely suppose that $X_1 \dot\cup \ldots \dot\cup X_m$ is a star partition, and let $\mathbf{x} \in \mathcal{E}(\mu_i) \cap \mathcal{V}_i$. Then $P_i \mathbf{x} = \mathbf{x}$ and so $(P_i \mathbf{x})^T \mathbf{e}_h = 0$ for all $h \in X_i$, i.e. $\mathbf{x}^T (P_i \mathbf{e}_h) = 0$ for all $h \in X_i$. Hence $\mathbf{x} \in \mathcal{E}(\mu_i)^\perp$ and so $\mathbf{x} = 0$. Thus $\mathcal{E}(\mu_i) \cap \mathcal{V}_i = 0$ and by a comparison of dimensions, $\mathbb{R}^n = \mathcal{E}(\mu_i) \oplus \mathcal{V}_i$.

$\square$

THEOREM 3.2. *The partition* $X_1 \dot\cup \ldots \dot\cup X_m$ *of* $V(G)$ *is a star partition if and only if it is a polynomial partition.*

Proof. Suppose that $X_1 \dot\cup \ldots \dot\cup X_m$ is a star partition. We have to show that μ_i is not an eigenvalue of $G - X_i$ ($i = 1, \ldots, m$). We take $i = 1$ without loss of generality and we let A' be the adjacency matrix of $G - X_1$. It suffices to show that if $A'\mathbf{x}' = \mu_1 \mathbf{x}'$ then $\mathbf{x}' = \mathbf{0}$. Now A has the form $\begin{bmatrix} * & * \\ * & A' \end{bmatrix}$ and so if $\mathbf{x} = \begin{pmatrix} \mathbf{0} \\ \mathbf{x}' \end{pmatrix}$ then $A\mathbf{x} - \mu_1 \mathbf{x} \in \mathcal{V}_1^\perp$. But $A\mathbf{x} - \mu_1 \mathbf{x}$ also lies in $\mathcal{E}(\mu_1)^\perp$, and so by Lemma 3.1, $\mathbf{x} \in \mathcal{E}(\mu_1)$. But $\mathbf{x}$ also lies in $\mathcal{V}_1$ and so $\mathbf{x} = \mathbf{0}, \mathbf{x}' = \mathbf{0}$. Hence μ_1 is not an eigenvalue of $G - X_1$.

For the converse, suppose by way of contradiction that the polynomial partition $X_1 \dot\cup \ldots \dot\cup X_m$ is not a star partition. Then by Lemma 3.1, $\mathcal{E}(\mu_i) \cap \mathcal{V}_i \neq 0$ for some i, say for $i = 1$. A non-zero vector in $\mathcal{E}(\mu_1) \cap \mathcal{V}_1$ has the form $\begin{pmatrix} \mathbf{0} \\ \mathbf{x}' \end{pmatrix}$ where $\mathbf{x}' \neq \mathbf{0}$. Then $\mathbf{x}'$ is an eigenvector of $G - X_1$ corresponding to μ_1, a contradiction.

$\square$

Theorem 3.2 may be used to verify that the following are star partitions. In examples (ii) and (iii), the vertices in X_i are labelled with the corresponding eigenvalue μ_i.

Examples. (i) For a complete bipartite graph $K_{r,s}$ with vertex-set V, a star partition has the form $X_1 = \{u\}, X_2 = \{v\}, X_3 = V - \{u, v\}$ where $u \sim v$, $\{\mu_1, \mu_2\} = \{-\sqrt{rs}, \sqrt{rs}\}$ and $\mu_3 = 0$.

(ii)　　　　　　　　　　　　　　　　　(iii)

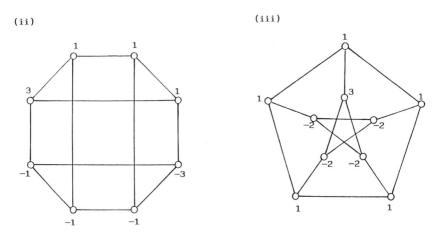

Yet a third characterization of star partitions, suggested by Simić, is established below.

PROPOSITION 3.3. *The partition* $X_1 \dot{\cup} \ldots \dot{\cup} X_m$ *of $V(G)$ is a star partition if and only if for each $i \in \{1, \ldots, m\}, \mathcal{E}(\mu_i)$ has a basis of eigenvectors $\{\mathbf{s}_h : h \in X_i\}$ such that* $\mathbf{s}_h^T \mathbf{e}_k = \delta_{hk}$ *whenever* $h, k \in X_i$.

Proof. If $X_1 \dot{\cup} \ldots \dot{\cup} X_m$ is a star partition, if $\mathcal{V}_i = \langle \mathbf{e}_h : h \notin X_i \rangle$ and if Q_i denotes the orthogonal projection $\mathbf{R}^n \to \mathcal{V}_i^\perp$ then by Lemma 3.1 we have $\langle \mathbf{e}_h : h \in X_i \rangle = \mathcal{V}_i^\perp = Q_i(\mathbf{R}^n) = Q_i(\mathcal{E}(\mu_i) \oplus V_i) = Q_i(\mathcal{E}(\mu_i))$. Hence for each $h \in X_i$ there exists a vector $\mathbf{s}_h \in \mathcal{E}(\mu_i)$ such that $Q_i(\mathbf{s}_h) = \mathbf{e}_h$.

Conversely if $\mathcal{E}(\mu_i)$ has a basis $\{\mathbf{s}_h : h \in X_i\}$ with the property specified then the vectors $P_i \mathbf{e}_j$ $(j \in X_i)$ are linearly independent and hence form a basis for $\mathcal{E}(\mu_i)$. For if $\sum_{j \in X_i} \alpha_j P_i \mathbf{e}_j = \mathbf{0}$ then $\sum_{j \in X_i} \alpha_j \mathbf{e}_j \in \ker(P_i) = \mathcal{E}(\mu_i)^\perp$; it follows that $\sum_{j \in X_i} \alpha_j \mathbf{s}_h^T \mathbf{e}_j = 0$ for each $h \in X_i$, hence that $\alpha_j = 0$ for each $j \in X_i$.

$\square$

Remark. In the situation of Proposition 3.3, the vertices of G may be ordered so that the matrix $(\mathbf{s}_1 | \mathbf{s}_2 | \ldots | \mathbf{s}_n)$ has the form $\begin{bmatrix} I & * & \ldots & * \\ * & I & \ldots & * \\ \multicolumn{4}{c}{\dotfill} \\ * & * & \ldots & I \end{bmatrix}$.

$\square$

Cvetković has proposed the use of star bases to define a total ordering of finite graphs with loops. Details are given in [CRS, Section 5] and we do not propose to repeat them here. We merely note the role of star bases in the construction, which is by induction on the number of vertices. In the nontrivial case that $m > 1$, the Gram matrix of the vectors $P_i \mathbf{e}_j$ ($j \in X_i$) is a linear combination of adjacency matrices of graphs with fewer vertices than G. The ordering of these graphs is used to order the m-tuples of Gram matrices and the corresponding star bases associated with G. The star bases which are minimal with respect to this ordering include a canonical basis of $\mathbf{R}^n$ corresponding to G, and the graphs with n vertices can now be ordered using a lexicographical ordering of the corresponding canonical bases.

4. Structural considerations. Throughout this section, $X_1 \dot{\cup} \ldots \dot{\cup} X_m$ denotes a star partition of an arbitrary simple graph G: the cell X_i corresponds to an eigenvalue μ_i of multiplicity k_i.

For $X \subseteq V(G)$, let $\overline{X}$ denote the complement of X in $V(G)$. The *coboundary* of X is the set of edges between X and $\overline{X}$, denoted by ∂X. We begin with an elementary observation which shows that $|\partial X_i| \geq |X_i|$ for each cell X_i of a star partition associated with a nontrivial connected graph.

PROPOSITION 4.1. *Let v be a non-isolated vertex of G. If $v \in X_i$ then v is adjacent to at least one vertex in $\overline{X}_i$.*

Proof. We have $\mu_i P_i \mathbf{e}_v = AP_i \mathbf{e}_v = P_i A \mathbf{e}_v = \sum_{u \sim v} P_i \mathbf{e}_u$. Thus if all vertices u adjacent to v lie in X_i then the vectors $P_i \mathbf{e}_j$ ($j \in X_i$) are linearly dependent, a contradiction.

$\square$

Remarks. (i) In the argument used in the above proof, we may replace A with any polynomial in A. For example if G is distance-regular then for each positive integer k not exceeding the diameter of G, there exists a polynomial f_k such that the (u, v)-entry of $f_k(A)$ is 0 or 1 according as $d(u, v) \neq k$, $d(u, v) = k$. Then $f_k(\mu_i)P_i \mathbf{e}_v$ is the sum of the vectors $P_i \mathbf{e}_u$ taken over all vertices u at distance k from v. The conclusion is that for each $v \in X_i, \overline{X}_i$ contains a vertex at distance k from v.
(ii) The conclusion of Proposition 4.1 remains valid if A is replaced by L, where L is the Lagrangian matrix $D - A$. Here $D = \text{diag}(d_1, \ldots, d_n)$, where d_i is the degree of vertex i.

$\square$

We write $\Delta(v) = \{u \in V(G) : u \sim v\}$, and for $v \in X_i$ we write $\Gamma(v) = \Delta(v) \cap X_i$, $\overline{\Gamma}(v) = \Delta(v) \cap \overline{X}_i$.

PROPOSITION 4.2. *Let h, k be distinct vertices in X_i. If $\overline{\Gamma}(h) = \overline{\Gamma}(k)$ then either*
(a) $h \sim k, \mu_i = -1$ and $\Delta(h) \cup \{h\} = \Delta(k) \cup \{k\}$, *or*
(b) $h \not\sim k, \mu_i = 0$ and $\Delta(h) = \Delta(k)$.

Proof. We have $\mu_i P_i \mathbf{e}_h = \sum_{j \in \Gamma(h)} P_i \mathbf{e}_j + \sum_{j \in \overline{\Gamma}(h)} P_i \mathbf{e}_j$ together with a similar expression for $\mu_i P_i \mathbf{e}_k$. It follows that if $\overline{\Gamma}(h) = \overline{\Gamma}(k)$ then $\mu_i P_i \mathbf{e}_h - \sum_{j \in \Gamma(h)} P_i \mathbf{e}_j =$

$\mu_i P_i \mathbf{e}_k - \sum_{j \in \Gamma(k)} P_i \mathbf{e}_j$. When $h \sim k$ this becomes $(\mu_i + 1)P_i\mathbf{e}_h - (\mu_i + 1)P_i\mathbf{e}_k - \sum_{j \in \Gamma(h)-\{k\}} P_i\mathbf{e}_j + \sum_{j \in \Gamma(k)-\{h\}} P_i\mathbf{e}_j = \mathbf{0}$. When $h \not\sim k$ we have instead

$$\mu_i P_i \mathbf{e}_h - \mu_i P_i \mathbf{e}_k - \sum_{j \in \Gamma(h)} P_i \mathbf{e}_j + \sum_{j \in \Gamma(k)} P_i \mathbf{e}_j = \mathbf{0}.$$

The respective conclusions (a), (b) now follow from linear independence of the vectors $P_i \mathbf{e}_j$ $(j \in X_i)$.

□

Remark. If we replace A by L, the conclusion of Proposition 4.2 becomes: either

(a) $h \sim k, \mu_i = d_h + 1 = d_k + 1$ and $\Delta(h) \cup \{h\} = \Delta(k) \cup \{k\}$; or

(b) $h \not\sim k, \mu_i = d_h = d_k$ and $\Delta(h) = \Delta(k)$.

□

COROLLARY 4.3. *If $\mu_i \notin \{-1,0\}$ then the sets $\overline{\Gamma}(h)$ $(h \in X_i)$ are distinct subsets of $\overline{X}_i$.*

□

Remark. For graphs of bounded degree, Corollary 4.3 can provide a crude bound on the multiplicity of an eigenvalue different from -1 or 0. For example if G is a cubic graph and $h \in X_i$ then (by Proposition 4.1 and Corollary 4.3) $\Gamma(h)$ is a subset of $\overline{X}_i$ of size 1, 2 or 3. Hence if we write r_i for $n - k_i$ then for $\mu_i \notin \{-1, 0\}$ we have

$$n - r_i \leq \binom{r_i}{1} + \binom{r_i}{2} + \binom{r_i}{3}, \text{ whence } n \leq \frac{1}{6} r_i(r_i^2 + 11).$$

□

In order to extend the foregoing observations we can consider the dimension of the subspace spanned by the vectors $\sum_{j \in \overline{\Gamma}(h)} P_i \mathbf{e}_j$ $(h \in X_i)$. The next result appears in [CRS, Section 3]. Here and subsequently we write G_i for the subgraph of G induced by X_i.

THEOREM 4.4. *If k_i' (possibly zero) is the multiplicity of μ_i as an eigenvalue of G_i then the number of vertices of $\overline{X}_i$ which are adjacent to some vertex of X_i is at least $k_i - k_i'$.*

Proof. We let $\overline{\Gamma}(X_i) = \bigcup\{\overline{\Gamma}(h) : h \in X_i\}$, and we write A_i for the adjacency matrix of G_i. As before we have $\mu_i P_i \mathbf{e}_h - \sum_{j \in \Gamma(h)} P_i \mathbf{e}_j = \sum_{j \in \overline{\Gamma}(h)} P_i \mathbf{e}_j$ for each $h \in X_i$. It follows that the linear transformation of $\mathcal{E}(\mu_i)$ defined by $P_i\mathbf{e}_h \longmapsto \sum_{j \in \overline{\Gamma}(h)} P_i \mathbf{e}_j$ has matrix $\mu_i I - A_i$ with respect to the basis $P_i\mathbf{e}_h (h \in X_i)$, hence has rank $k_i - k_i'$. Accordingly

$$k_i - k_i' \leq \dim\langle P_i\mathbf{e}_j : j \in \overline{\Gamma}(X_i)\rangle \leq |\overline{\Gamma}(X_i)|.$$

□

COROLLARY 4.5. *If μ_i is not an eigenvalue of G_i then there are at least k_i vertices of $\overline{X}_i$ which are adjacent to some vertex of X_i.*

$\square$

In view of Corollary 4.5, we should investigate the situation in which $k_i' \neq 0$.

PROPOSITION 4.6. *For $j \in X_i$ let $\mathbf{c}_j$ be the column of length $n - k_i$ consisting of the entries of $P_i \mathbf{e}_j$ indexed by $\overline{X}_i$. If the k_i columns $\mathbf{c}_j$ $(j \in X_i)$ are linearly dependent then μ_i is an eigenvalue of G_i.*

Proof. We take $i = 1$ and $X_1 = \{1, \ldots, k_1\}$ without loss of generality. Suppose that $\sum_{j \in X_1} \alpha_j \mathbf{c}_j = \mathbf{0}$ where not all of the α_j $(j \in X_1)$ are zero. Then $\sum_{j \in X_1} \alpha_j P_1 \mathbf{e}_j$ has the form $\begin{pmatrix} \mathbf{x} \\ \mathbf{0} \end{pmatrix}$, where $\mathbf{x} \neq 0$ because the vectors $P_1 \mathbf{e}_j$ $(j \in X_1)$ are linearly independent. Since $A \begin{pmatrix} \mathbf{x} \\ \mathbf{0} \end{pmatrix} = \mu_1 \begin{pmatrix} \mathbf{x} \\ \mathbf{0} \end{pmatrix}$, it follows that $\mathbf{x}$ is an eigenvector of A_1 corresponding to μ_1.

$\square$

As a partial converse of Proposition 4.6 we have the following result.

THEOREM 4.7. *If μ_i is an eigenvalue of G_i then either the k_i columns $\mathbf{c}_j$ $(j \in X_i)$ are linearly dependent or μ_i lies strictly between the smallest and largest eigenvalues of $G - X_i$.*

Proof. We take $i = 1$ and $X_1 = \{1, \ldots, k_1\}$ without loss of generality, and we write $A = \begin{bmatrix} A_1 & B_1^T \\ B_1 & C_1 \end{bmatrix}$ where A_1, C_1 are the adjacency matrices of $G_1, G - X_1$ respectively. Let $A_1 \mathbf{x} = \mu_1 \mathbf{x}$ where $\mathbf{x} \neq \mathbf{0}$. By Lemma 3.1, we have $\mathbf{R}^n = \mathcal{E}(\mu_1) \oplus \mathcal{V}_1$ and so there exists a unique vector $\begin{pmatrix} \mathbf{0} \\ \mathbf{y} \end{pmatrix} \in \mathcal{V}_1$ such that $\begin{pmatrix} \mathbf{x} \\ \mathbf{y} \end{pmatrix} \in \mathcal{E}(\mu_1)$. The equation $\begin{bmatrix} \mu_1 \mathbf{x} \\ \mu_1 \mathbf{y} \end{bmatrix} = \begin{bmatrix} A_1 & B_1^T \\ B_1 & C_1 \end{bmatrix} \begin{bmatrix} \mathbf{x} \\ \mathbf{y} \end{bmatrix}$ yields $B_1^T \mathbf{y} = \mathbf{0}$ and $\mu_1 \mathbf{y} = B_1 \mathbf{x} + C_1 \mathbf{y}$, whence $\mathbf{y}^T (\mu_1 I - C_1) \mathbf{y} = 0$.

If $\mathbf{y} = \mathbf{0}$ then $\begin{pmatrix} \mathbf{x} \\ \mathbf{0} \end{pmatrix} \in \langle P_1 \mathbf{e}_j : j \in X_1 \rangle$ and (since $\mathbf{x} \neq \mathbf{0}$) it follows that the columns $\mathbf{c}_j$ are linearly dependent. If $\mathbf{y} \neq \mathbf{0}$ then the eigenvalues of the symmetric matrix $\mu_1 I - C_1$ are neither all negative nor all positive, and so μ_1 lies in the spectral range of C_1. By Proposition 3.3, μ_1 is not an eigenvalue of $G - X_1$, and the result follows.

$\square$

We conclude with a reconstruction theorem from [CRS]. This result [CRS, Theorem 4.6] asserts that G_i is known if we know all other edges in G.

THEOREM 4.8. *G is reconstructible from the eigenvalue μ_i, the graph $G - X_i$ and the coboundary ∂X_i.*

Proof. We take $i = 1$ and $X_1 = \{1, \ldots, k_1\}$ without loss of generality. Let $\{\mathbf{s}_1, \ldots, \mathbf{s}_{k_1}\}$ be the basis of $\mathcal{E}(\mu_1)$ given by Proposition 3.3, so that the matrix with columns $s_1, \ldots s_{k_1}$ has the form $\begin{pmatrix} I \\ X \end{pmatrix}$. If $A = \begin{pmatrix} A_1 & B_1^T \\ B_1 & C_1 \end{pmatrix}$, where A_1 is the adjacency matrix of G_1, then $\begin{pmatrix} A_1 & B_1^T \\ B_1 & C_1 \end{pmatrix} \begin{pmatrix} I \\ X \end{pmatrix} = \begin{pmatrix} \mu_1 I \\ \mu_1 X \end{pmatrix}$. It follows that $A_1 + B_1^T X = \mu_1 I$ and $B_1 + C_1 X = \mu_1 X$. Now μ_1 is not an eigenvalue of $G - X_1$ and so $\mu_1 I - C_1$ is invertible. Hence $X = (\mu_1 I - C_1)^{-1} B_1$ and $A_1 = \mu_1 I - B_1^T (\mu_1 I - C_1)^{-1} B_1$. Accordingly, given μ_1, B_1, C_1 (equivalently $\mu_1, \partial X_1, G - X_1$) we can find A_1 (equivalently G_1).

$\square$

REFERENCES

[Go] C.D. GODSIL, *Walk-generating functions, Christoffel–Darboux identities and the adjacency matrix of a graph*, Combinatorics, Probability and Computing (to appear).

[G] F.R. GANTMACHER, *The Theory of Matrices*, Vol. II, Chelsea (New York), 1959.

[B] T. BIER, *A distribution invariant for association schemes and strongly regular graphs*, Lin. Alg. & Appl. 57 (1984), 105–113.

[CD] D. CVETKOVIĆ AND M. DOOB, *Developments in the theory of graph spectra*, Linear and Multilinear Algebra 18 (1985), 153–181.

[PR1] P. ROWLINSON, *Graph perturbations*, in: Surveys in Combinatorics 1991 (ed. A.D. Keedwell), Cambridge University Press (Cambridge), 1991.

[CRS] D. CVETKOVIĆ, P. ROWLINSON AND S.K. SIMIĆ, *A study of eigenspaces of graphs*, Lin. Alg. & Appl. (to appear).

[PR2] P. ROWLINSON, *A deletion-contraction algorithm for the characteristic polynomial of a multigraph*, Proc. Royal Soc. Edinburgh 105A (1987), 153–160.

[PR3] P. ROWLINSON, *Graph angles and isospectral molecules*, Univ. Beograd Publ. Elektrokehn. Fac. Sci. Ser. Mat. 2 (1991), 61–66.

[C] D. CVETKOVIĆ, *Some graph invariants based on the eigenvectors of the adjacency matrix*, Proc. 8th Yugoslav Seminar on Graph Theory 1987 (Inst. Math., Univ. Novi Sad, 1989), 31–42.

[S] J.J. SEIDEL, *Eutactic stars*, Colloq. Math. Soc. János Bolyai, 18 (1976), 983–999.

[CR] D. CVETKOVIĆ AND P. ROWLINSON, *Further properties of graph angles*, Scientia (Ser. A) 1 (1988), 41–51.

[LS] J.P. LOWE AND M.R. SOTO, *Isospectral graphs, symmetry and perturbation theory*, Match 20 (1986), 21–51.

SOME MATRIX PATTERNS ARISING IN QUEUING THEORY

CLARK JEFFRIES*

Abstract. A remarkably versatile dynamical system is given by

$$dp_i/dt = \sum_{\substack{j=1 \\ j \neq i}}^{n} a_{ij} p_j - \left(\sum_{\substack{j=1 \\ j \neq i}}^{n} a_{ji} \right) p_i$$

where $n \geq 2$, $\{p_1, p_2, \ldots, p_n\}$ are system variables, and $\{a_{ij}\}$ is a matrix of nonnegative real numbers. Diagonal entries in $\{a\}$ do not occur in the model in the sense that they are both added and subtracted in the given sum; for secretarial purposes we set $a_{ii} = 0$. The system variables themselves arise from probabilities, so we also assume throughout that at time $t = 0$, all p_i are nonnegative and the component sum $p_1 + p_2 + \cdots + p_n$ is equal to 1.

The goal of this paper is to describe the trajectories of the model in terms of a Lyapunov-like function. Doing so involves the use of theorems of Brualdi and Gersgorin, the graph-theoretic notion of balanced cycles, and the qualitative solvability of an equation involving a Hadamard product.

Key words. qualitative matrix theory, dynamical system, queuing theory, M-matrix

AMS(MOS) subject classifications. 47A20, 15A57, 93D99

1. Introduction. Suppose a system such as a queue has a finite number $n > 1$ of possible states and the probability of being in state i at time t is $p_i(t)$, $i = 1, 2, \ldots, n$. A linear model describing the rates of change of $p_i(t)$ is the dynamical system

$$(1) \qquad dp_i/dt = \sum_{\substack{j=1 \\ j \neq i}}^{n} a_{ij} p_j - \left(\sum_{\substack{j=1 \\ j \neq i}}^{n} a_{ji} \right) p_i$$

where a_{ij} is the average rate at which the system in state j becomes the system in state i due, say, to batch arrival or batch service in a queue. Without loss of generality, we set each $a_{ii} = 0$. Since

$$(2) \qquad \sum_{i=1}^{n} \dot{p}_i = 0$$

it follows that $p_1 + p_2 + \cdots + p_n$ is a constant (namely 1 for probabilities) for all time. In this paper the positive orthant (respectively nonnegative orthant) of n-dimensional real space refers to all n-tuples with each component positive (nonnegative).

---

*Department of Mathematical Sciences, Clemson University, Clemson, South Carolina 29634-1907.

We call $a_{ij}p_j - a_{ji}p_i$ the *net flow* from state j to state i. Clearly if a combination of probabilities p exists with every net flow zero, then p is a constant trajectory for (1).

An alternative way to write (1) is

(3)
$$dp_i/dt = \sum_{j=1}^{n} M_{ij}p_j$$

where $\{M_{ij}\}$ is the summed system matrix with $M_{ij} = a_{ij}$, $i \neq j$, and $M_{ii} = -(a_{1i} + a_{2i} + \cdots + a_{ni})$. Given any initial state $p(0)$ for (3), there is a unique, smooth subsequent trajectory [S]. The matrix with $-M_{ij}$ in row i, column j, is an M-matrix, a class of matrices which have been studied by several authors, notably [BNS chapter 2, section 4], [BP chapter 6], and [HJ2 chapter 2, section 5]. Many of the results in this paper are closely related to results in these references. However, the fact that the column sums of M are 0 does distinguish the topic from general M-matrix theory.

THEOREM 1. *No trajectory for (3) which starts in the positive orthant has* $p_j(T) = 0$ *in any component j at any finite future time T.*

Proof. Considering (2), the only way a trajectory could reach the boundary of the positive orthant in future time T is with each $p_j(T)$ finite. Suppose we have such a trajectory with, say, $p_1(t) > 0$ for $t_0 \leq t < T$ and $p_1(T) = 0$. Consider the function $y(t)$ defined for $t_0 \leq t < T$ by $y(t) = ln(p_1(t))$. It follows that $y(t)$ is continuous and differentiable for $t_0 \leq t < T$ with $dy/dt \geq M_{11}$. Thus y approaches $-\infty$ as $t \to T$ yet dy/dt is bounded below by M_{11}, a contradiction. Thus no such trajectory can exist. ▯

The fact that the sum of every column of M is 0 implies 0 is an eigenvalue of M. It is known that all other eigenvalues of such M have nonpositive real part.

THEOREM 2. [HJ2 119] *The real part of any eigenvalue of M is nonpositive.*

Proof. For each $j = 1, 2, \ldots, n$, we define a *deleted column sum* of M as

$$C'_j = \sum_{\substack{i=1 \\ i \neq j}}^{n} M_{ij}.$$

According to a theorem of Gersgorin [HJ1 346] all the eigenvalues of M are located in the union of the n disks in the complex plane given by

$$|z - M_{jj}| \leq C'_j.$$

Since $-M_{jj} = C'_j$, the points in each such disk lie in the left half of the complex plane except for exactly one point, the origin. The same pertains to the union of such disks. ▯

THEOREM 3. *Every trajectory for (1) which starts in the positive orthant and which is not constant must asymptotically approach a constant trajectory in the nonnegative orthant.*

Proof. A diverging trajectory (having unbounded distance from the origin as $t \to +\infty$) exists for a linear dynamical system like (3) iff some eigenvalue of M has positive real part or some eigenvalue of M with zero real part has multiplicity > 1 in the minimal polynomial of M [S 56]. Considering theorems 1 and 2 and the fact that $p_1 + p_2 + \cdots + p_n$ is a constant, M must have no such trajectories. Using again the above theorem of Gersgorin, M can have no eigenvalues which are purely imaginary, hence no cyclic trajectories. Thus all nonconstant trajectories of (3) asymptotically approach constant trajectories. Thus theorem 1 implies that any trajectory starting in the positive orthant must be or must asymptotically approach a constant trajectory in the nonnegative orthant. □

Recall that the digraph associated with the matrix M has n labeled vertices and a directed edge from vertex j to vertex i iff $M_{ij} \neq 0$. (This is the causal convention from mathematical modeling, not the graph-theoretic convention associating the same edge with M_{ji}). Note that a digraph can be defined for any M arising from a constant matrix or a matrix with time-dependent entries of constant sign.

A p-cycle is a list of p distinct vertices, say, vertices $1, 2, \ldots, p$, and p distinct edges corresponding to a nonzero product $M_{21}M_{32} \ldots M_{pp-1}M_{1p}$. For each p-cycle with $p \geq 2$ we may consider the region of the complex plane given by

$$\prod_{\text{vertex } i \text{ in } p\text{-cycle}} |z - M_{ii}| \leq \prod_{\text{vertex } i \text{ in } p\text{-cycle}} C_i'$$

According to a theorem of Brualdi [BR 92], if every vertex of the digraph of a matrix M is in at least one p-cycle, $p \geq 2$, then all the eigenvalues of M are located in the union of all such regions. (The reference actually uses deleted row sums, an equivalent version.)

Deleting all the entries in row i and column i from M results in an $(n-1) \times (n-1)$ matrix denoted $M_{/i,i}$. The principal minors of M are the n real numbers $\{\det(M_{/i,i})\}$.

Of special interest are those instances of (1) or (3) in which every trajectory in the positive orthant which starts with component sum 1 asymptotically approaches the same constant trajectory (also in the nonnegative orthant with component sum 1). Considering the above theorems, it would be sufficient to show that the dimension of the kernel of M is one.

THEOREM 4. *Suppose for some i that in the digraph of $M_{/i,i}$ every p-cycle, $p \geq 2$, includes at least one vertex j with $|(M_{/i,i})_{j,j}| > C_j'$. Then there is a unique constant trajectory c in the nonnegative orthant with component sum 1; all other trajectories with component sum initially 1 must asymptotically approach c.*

Proof. Consider the characteristic polynomial $\det(zI - M) = z^n + k_{n-1}z^{n-1} + \cdots + k_1 z + 0$ of M. If the coefficient k_1 of z in $\det(zI - M)$ is nonzero, then the multiplicity of 0 as an eigenvalue is only 1. All other eigenvalues have negative real part. Considering the role of eigenvalues in linear dynamical systems with constant coefficients [S 55-65], any trajectory of (3) with component sum initially 1 but otherwise starting in any orthant must enter the nonnegative orthant. If two constant trajectories were to exist in the nonnegative orthant with component sums 1, then there would exist a third outside the nonnegative orthant. Thus if k_1 is nonzero, there is a unique constant trajectory in the nonnegative orthant with component sum 1.

Now k_1 is itself ± 1 times the sum of n determinants, the principal minors $\{\det M_{/i,i}\}$. Furthermore, each matrix $M_{/i,i}$, like M itself, has nonnegative off-diagonal entries and negative diagonal entries; each diagonal entry in magnitude is greater than or equal to the associated deleted column sum. Therefore the theorem of Gersgorin used in the proof of theorem 2 implies every eigenvalue of every $M_{/i,i}$ has negative real part except possibly the origin of the complex plane itself. Thus all nonzero $\{\det M_{/i,i}\}$ are of the same sign, namely the sign of $(-1)^{n-1}$. Hence k_1 is nonzero or every $\det M_{/i,i} = 0$. In the latter case, select i so that in the digraph of $M_{/i,i}$ every p-cycle, $p \geq 2$, has at least one vertex j with $|(M_{/i,i})_{jj}| > C_j'$. But $z = 0$ cannot be a solution of the inequality in the Brualdi theorem. Thus for each cycle containing vertex j the region defined in the Brualdi theorem cannot contain the origin of the complex plane. The same may be said of the union of such regions. Thus 0 cannot be an eigenvalue of $M_{/i,i}$. Thus k_1 is nonzero. □

THEOREM 5. *Suppose the off-diagonal entries in a matrix B are nonnegative and the column sums of B are nonpositive. Suppose no row of B is all zero (so every diagonal entry in B is negative). Suppose D is a nonpositive diagonal matrix not all zero. Then the real part of every eigenvalue of $B + D$ is negative.*

Proof. If B is 2×2, the conclusion of the theorem is obvious.

Suppose B is $n \times n$, $n > 2$, and is representative of the smallest size matrix for which the above statement is false. Using the Gersgorin theorem, the real part of every eigenvalue of B and $B + D$ is nonpositive. It suffices to show that $\det(B + D) \neq 0$.

Suppose D has exactly one nonzero entry, the negative $1, 1$ entry $-d_{11}$. By the definition of determinant, $\det(B + D) = \det(B) - d_{11} * \det(B_{/1,1})$; these summands cannot have opposite signs. Since row 1 of B has at least one positive entry, some column sum in $B_{/1,1}$ is negative. Then $B_{/1,1}$ itself can be written as a sum $B + D$ of size $(n-1) \times (n-1)$. Either $\det(B + D) \neq 0$ or n is not the smallest value possible; either contradiction establishes this case of the theorem.

Now suppose D has several negative diagonal entries, $-d_{11}, -d_{22}, \ldots$. It follows that $\det(B + D)$ is a multinomial of the form

$$\det(B+D) = \det(B) - d_{11} * \det(B_{/1,1}) - d_{22} * \det(B_{/2,2}) \ldots + d_{11} * d_{22} * \det([B_{/1,1}]_{/2,2}) + \ldots$$

Gersgorin's theorem can be used to establish that all nonzero summands are of the same sign. Furthermore, if all summands are zero, n is again not the smallest value for which the theorem does not hold, again a contradiction. □

THEOREM 6. *Suppose no row of M is all zero. Then there is a unique constant trajectory c in the nonnegative orthant with component sum 1; all other trajectories with component sum 1 must asymptotically approach c.*

Proof. As in the proof of Theorem 4, it suffices to show that the coefficient k_1 in $\det(zI - M) = z^n + k_{n-1}z^{n-1} + \cdots + k_1 z + 0$ is nonzero. Likewise k_1 is nonzero or every $\det M_{/i,i} = 0$. Theorem 5 implies every $\det M_{/i,i} \neq 0$. □

In certain cases the conclusion of theorem 6 can be obtained by applying the theorem of Brualdi. Given any row i, select column j so that $a_{ij} > 0$. Every diagonal entry in $M_{/i,i} < 0$ and the jj entry in $M_{/i,i}$ is less than -1 times the sum of all other entries in column j. In the latter case, suppose we can select i so that in the digraph of $M_{/i,i}$ every p-cycle, $p \geq 2$, has at least one vertex j with $|(M_{/i,i})_{jj}| > C'_j$. Here $z = 0$ cannot be a solution of the inequality in the Brualdi theorem. Thus for each cycle containing vertex j the region defined in the theorem of Brualdi cannot contain the origin of the complex plane. The same may be said of the union of such regions. Thus 0 cannot be an eigenvalue of $M_{/i,i}$. Thus k_1 is nonzero.

2. Queuing theory. This section is a sketch of the very basic nature of queue models and problems. The simplest model in queuing theory is that of a single-server queue with finite capacity n, random arrivals, and negative exponential service [KM1, KM2]. The model is

$$
\begin{aligned}
dp_0/dt &= & -\lambda p_0 &+ \mu p_1 \\
dp_1/dt &= \lambda p_0 - (\lambda + \mu)p_1 &+ \mu p_2 \\
dp_2/dt &= \lambda p_1 - (\lambda + \mu)p_2 &+ \mu p_3 \\
&\cdots \\
dp_n/dt &= \lambda p_{n-1} - (\lambda + \mu)p_n
\end{aligned}
$$

Here $p_i(t)$ is the probability of i customers standing in the queue at time t, and λ and μ are positive constants, the service and arrival coefficients. Note that the index of p runs here from 0 to n, a convention which is irritating from the algebraic viewpoint but compelling from the modeling viewpoint. Note also that in this model customers arrive and are served singly, not in batches.

A more general model uses positive constant coefficients $\{\lambda_i\}$ and $\{\mu_i\}$:

(4)
$$
\begin{aligned}
dp_0/dt &= & -\lambda_0 p_0 &+ \mu_1 p_1 4 \\
dp_1/dt &= \lambda_0 p_0 - (\lambda_1 + \mu_1)p_1 &+ \mu_2 p_2 \\
dp_2/dt &= \lambda_1 p_1 - (\lambda_2 + \mu_2)p_2 &+ \mu_3 p_3 \\
&\cdots \\
dp_n/dt &= \lambda_{n-1}p_{n-1} - (\lambda_n + \mu_n)p_n
\end{aligned}
$$

Theorem 6 readily applies to (4). Among all trajectories for any such system with component sum 1, there is one trajectory in the positive orthant which is constant and which is asymptotically approached by all others. The utility of theorem

6 can be seen in the case $n = 5$ (six dimensional) with M represented by

$$\begin{bmatrix} -1 & 1 & 0 & 0 & 0 & 0 \\ 1 & -2 & 1 & 0 & 0 & 0 \\ 0 & 1 & -2 & 1 & 0 & 0 \\ 0 & 0 & 1 & -2 & 1 & 0 \\ 0 & 0 & 0 & 1 & -2 & 1 \\ 0 & 0 & 0 & 0 & 1 & -1 \end{bmatrix}$$

This M has eigenvalues $-3.7321\ldots, -3, -2, -1, -.26795\ldots$, and 0. Note that the 2-cycle involving the last two vertices would not fulfill the conditions of the theorem of Brualdi if $M_{/1,1}$, $M_{/2,2}$, or $M_{/3,3}$ were selected. Likewise the 2-cycle involving the first two vertices would not do so if $M_{/4,4}$, $M_{/5,5}$, or $M_{/6,6}$ were selected.

In the generalization of (4) with variable μ and λ, one can define different measures of queue performance and try to respecify μ as some sort of function of $p(t_0)$, $\lambda(t)$, and p which optimizes performance [C]. Typical performance measures are the expected length of the queue $L(t)$ given by

$$L(t) = \sum_{i=1}^{n} i p_i(t)$$

and the expected departure rate $D(t)$ given by

$$D(t) = \sum_{i=1}^{n} \mu p_i(t)$$

All of the above can also be generalized in the obvious way if queue length is allowed to change by more than one customer per time step (batch arrival or batch service with nonzero entries in a occurring other than on the superdiagonal, diagonal, and subdiagonal of M).

3. Balanced cycles. In the remainder of the paper, let us use the algebraic convention that indices run from 1 to n. Suppose nonnegative constants $\{\lambda_{ij}\}$ make up a symmetric $n \times n$ matrix and $\lambda_{ij} > 0$ iff $a_{ij} > 0$ (so $\lambda_{ii} = 0$ and $a_{ij} > 0$ implies $a_{ji} > 0$). We call $\{\lambda_{ij}\}$ *stability multipliers* for a model (1) provided any two nonzero, nondiagonal entries in a given column of the Hadamard product of λ and a are equal, that is, $a_{ij} > 0$ and $a_{kj} > 0$ imply $\lambda_{ij} a_{ij} = \lambda_{kj} a_{kj} = \kappa_j > 0$. We shall make use of the weighted sum

(5) $$\Lambda = \frac{1}{2} \sum_{i<j} \lambda_{ij} (a_{ji} p_i - a_{ij} p_j)^2$$

This Λ is a weighted sum of the squares of the distinct net flows $a_{ji} p_i - a_{ij} p_j$ and is a type of measure of performance for some queues. If the dimension of the kernel of M is one, then each level set of Λ (where $\Lambda =$ a positive constant) is geometrically a cylinder, the direct product of a line and an $(n-2)$-dimensional hyperellipsoid; null vectors of M lie on the line.

THEOREM 7. *Suppose in the digraph of the matrix a in (1) there are no p-cycles with $p > 2$ and that every edge between two distinct vertices is in a 2-cycle (the system in (4) is an example). Then there exist stability multipliers for a.*

Proof. Insofar as connectivity by 2-cycles is concerned, the digraph of a is a tree. Choose an arbitrary vertex as a root of the tree, say, vertex 1. If vertex i is connected to vertex 1, let λ_{i1} be defined so that $\lambda_{i1}a_{i1} = 1 = \kappa_1$. Since λ is symmetric, this also defines λ_{1i}. If vertices j and k are connected to such a vertex i, define λ_{ji} and λ_{ki} so that $\lambda_{ji}a_{ji} = \lambda_{ki}a_{ki} = \kappa_i$; this is consistent since neither j nor k can be 1. In fact this process propagates through the digraph consistently precisely because the digraph is a tree. $\square$

THEOREM 8. *Suppose in (1) that for each index i at least one entry in $\{a_{ji}\}$ is positive. Suppose also that a matrix of stability multipliers λ exists for a. Then along every nonconstant trajectory which starts in the positive orthant with component sum 1, Λ decreases as the trajectory asymptotically approaches a constant trajectory.*

Proof. The rate of change of Λ in (5) along a trajectory is

$$\frac{d\Lambda}{dt} = \sum_{i<j} -\lambda_{ij}a_{ji}(a_{ij}p_j - a_{ji}p_i)\dot{p}_i - \lambda_{ij}a_{ij}(-a_{ij}p_j + a_{ji}p_i)\dot{p}_j$$

Using the symmetry of λ and interchanging the roles of i and j in the second terms in the summands lead to

$$\frac{d\Lambda}{dt} = \sum_{i \neq j} -\lambda_{ji}a_{ji}(a_{ji}p_i - a_{ij}p_j)\dot{p}_i.$$

Setting $\kappa_i = \lambda_{ji}a_{ji}$ for every j with $\lambda_{ji}a_{ji} > 0$, we find

(6)
$$\frac{d\Lambda}{dt} = \sum_{i=1}^{n} -\kappa_i \dot{p}_i^2$$

Thus Λ decreases along all trajectories except the unique constant trajectory in the positive orthant. To put things geometrically, trajectories proceed in the plane $p_1 + p_2 + \cdots + p_n = 1$ to puncture level sets of Λ inwardly as they asymptotically approach the constant trajectory. $\square$

We proceed to find additional conditions which suffice to guarantee the existence of stability multipliers.

If $a_{ij} > 0$ implies $a_{ji} > 0$ in (1), then clearly to every p-cycle, $p \geq 3$, is associated a second cycle which tours its p vertices in the reverse order. If the products of the two corresponding sets of p entries in a are equal, we say the p-cycle is *balanced*.

THEOREM 9. *If every p-cycle, $p \geq 3$, in (1) is balanced, then stability multipliers exist for a.*

Proof. Let us try to construct stability multipliers just as in the proof of theorem 6. If vertex i is connected to vertex 1, let λ_{i1} be defined so that $\lambda_{i1}a_{i1} = 1 = \kappa_1$.

Since λ is symmetric, this also defines λ_{1i}. If vertex j is connected to vertex i, define λ_{ji} so that $\lambda_{ji}a_{ji} = \lambda_{1i}a_{1i} = \kappa_i$. Suppose vertex j is also connected to vertex 1. Then $\lambda_{j1}a_{j1} = 1 = \kappa_1$, so $\lambda_{ji}a_{ji} = \lambda_{1i}a_{1i} = \kappa_i$. In fact this process propagates consistently through the digraph precisely because the digraph has balanced cycles. □

4. Nonlinear systems. Suppose next that in a two-dimensional queue a_{12} and a_{21} are continuous and positive functions of p and t. Regardless of these system functions, any two trajectories p and q for the queue asymptotically approach each other. That is, the rate of change of $(p_1-q_1)^2+(p_2-q_2)^2$ is $-4(p_1-q_1)^2(a_{12}+a_{21}) = -4(p_2-q_2)^2(a_{12}+a_{21})$. Given in addition a lower bound b satisfying $0 < b < a_{12}+a_{21}$, an initial separation of two trajectories, and any positive separation ε, one could calculate an upper bound on the time elapsed before separation becomes less than ε.

This two-dimensional nonlinear observation does not extend even to linear higher-dimensional models. For example, if M is the constant matrix

$$\begin{bmatrix} -1 & 10 & 0 \\ 1 & -10.1 & .1 \\ 0 & .1 & -.1 \end{bmatrix}$$

and if initially $p = (.5, .4, .1)$ and $q = (.3, .3, .4)$, then trajectories $p(t)$ and $q(t)$ initially move apart, the rate of change of $(p_1 - q_1)^2 + (p_2 - q_2)^2 + (p_3 - q_3)^2$ being $+.10424\ldots$. Put another way, the matrix

$$\begin{bmatrix} -2 & 11 & 0 \\ 11 & -20.2 & .2 \\ 0 & .2 & -.2 \end{bmatrix}$$

has a positive eigenvalue $(3.1784\ldots)$ and so is not negative semi-definite; nor, incidentally, is any product of a diagonal matrix with positive entries and this matrix negative semi-definite. Of course, theorem 5 shows that both $p(t)$ and $q(t)$ do eventually approach a common constant trajectory (and hence each other). Since this system has no 3-cycles, theorem 9 guarantees the existence of a matrix of stability multipliers; an example of λ is

$$\begin{bmatrix} 0 & 1 & 0 \\ 1 & 0 & 100 \\ 0 & 100 & 0 \end{bmatrix}.$$

Here the total derivative of $\Lambda = \frac{1}{2}(p_1 - 10p_2)^2 + 50(.1p_2 - .1p_3)^2$ is $-(dp_1/dt)^2 -10(dp_2/dt)^2 -10(dp_3/dt)^2$ and all nonconstant trajectories approach asymptotically (but not monotonically) the unique constant trajectory $(\frac{5}{6}, \frac{1}{12}, \frac{1}{12})$.

The purpose of the remainder of this section is to derive conditions on nonconstant $a_{ij}(p_j)$ which guarantee the existence of Λ as in (5) which decreases along all nonconstant trajectories. We consider

$$(7) \qquad dp_i/dt = \sum_{\substack{j=1 \\ j \neq i}}^{n} a_{ij}(p_j)p_j - \left(\sum_{\substack{j=1 \\ j \neq i}}^{n} a_{ji}(p_i) \right) p_i.$$

To illustrate what can "go wrong" with (7), suppose $n = 3$, $a_{21} = (9(1 + p_1^2)/10)^{-10}$, $a_{12} = a_{23} = a_{32} = 1$, and $a_{13} = a_{31} = 0$. This choice of a_{21} decreases from about 3 to 1 to about .003 as p_1 increase from 0 to $\frac{1}{3}$ to 1. One might regard this strange queue as a model of "mob psychology" in that if the system is near (but not at) the steady state $(\frac{1}{3}, \frac{1}{3}, \frac{1}{3})$ and a few customers drop out of line (so p_1 increases), then the rate at which others arrive in line decreases; likewise, if near $(\frac{1}{3}, \frac{1}{3}, \frac{1}{3})$ a few additional customers arrive in line, then the rate at which others arrive in line increases. The queue has three constant trajectories. Each has $p_2 = p_3 = (1 - p_1)/2$. The p_1 values are $.994\ldots$, $\frac{1}{3}$, and $.215\ldots$. Linear approximation and eigenvalue analysis at the constant trajectories shows the first and third are both stable and the second is unstable. Hence multiple attractors are possible with nonlinear queues. Note again that da_{21}/dp_1 is negative. Some representative trajectories for this system are shown in Figure 1.

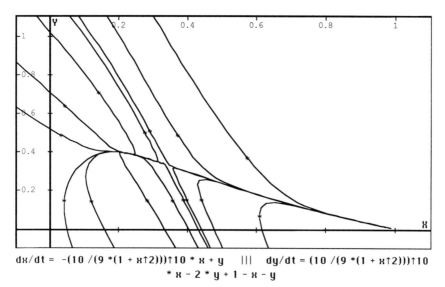

dɤ/dt = -(10 /(9 *(1 + ɤ↑2)))↑10 * ɤ + y ||| dy/dt = (10 /(9 *(1 + ɤ↑2)))↑10 * ɤ - 2 * y + 1 - ɤ - y

Figure 1. A nonlinear queue "mob psychology" model with (p_1, p_2) components of trajectories shown.

However, the stability of some nonlinear queues with da_{ji}/dp_i nonnegative can be clarified as follows.

THEOREM 10. *Suppose every p-cycle, $p \geq 3$, in (6) is balanced and all nondiagonal, nonzero entries in each column of a are proportional and bounded away from zero. Suppose also that each $da_{ij}/dp_j \geq 0$. Then stability multipliers exist for (7), and Λ in (5) still decreases along all nonconstant trajectories.*

Proof. We can construct stability multipliers just as in the proof of theorem

9. Proportionality of nonzero a_{ij} and a_{kj} and balanced loops together imply we can solve $\lambda_{ij} a_{ij} = \lambda_{kj} a_{kj}$ with constant, symmetric λ values. Likewise bounds on nondiagonal, nonzero entries imply each $\kappa_j(p_j) = \lambda_{ij} a_{ij}$ is positive and bounded away from zero. The rate of change of Λ in (6) can be shown to be

$$\frac{d\Lambda}{dt} = \sum_{i=1}^{n} - \left(\kappa_i + \frac{d\kappa_i}{dp_i} p_i \right) \dot{p}_i^2$$

Considering the signs of κ_i and $d\kappa_i/dp_i$, we see that Λ decreases along all nonconstant trajectories of (7). $\square$

Thus the algebraic existence of Λ still tells us much about the stability of (7) — the nonlinear generalization of (1).

REFERENCES

[BNS] A. BERMAN, M. NEUMAN, AND R. STERN, *Nonnegative Matrices in Dynamic Systems*, Wiley-Interscience, New York, 1989.

[BP] A. BERMAN AND R. PLEMMONS, *Nonnegative Matrices in the Mathematical Sciences*, Academic Press, New York, 1979.

[BR] R. BRUALDI AND H. RYSER, *Combinatorial Matrix Theory*, Cambridge University Press, 1991.

[C] J. CLEMENS, *A survey and extension of numerical techniques for modeling a computer network under nonstationary conditions*, M.S. thesis, Clemson University, 1991.

[HJ1] R. HORN AND C.R. JOHNSON, *Matrix Analysis*, Cambridge University Press, Cambridge, 1985.

[HJ2] R. HORN AND C.R. JOHNSON, *Topics in Matrix Analysis*, Cambridge University Press, Cambridge, 1991.

[KM1] S. KARLIN AND J. MCGREGOR, *The differential equations of birth and death processes and the Stieltjes moment problem*, Transactions of the American Mathematical Society, 85 (1957), pp. 489–546.

[KM2] S. KARLIN AND J. MCGREGOR, *The classification of birth and death processes*, Transactions of the American Mathematical Society, 86 (1957), pp. 366–400.

[S] D. SÁNCHEZ, *Ordinary Differential Equations and Stability Theory: An Introduction*, Freeman, San Francisco, 1968.

LAPLACIAN UNIMODULAR EQUIVALENCE OF GRAPHS

ROBERT GRONE*†, RUSSELL MERRIS**‡ AND WILLIAM WATKINS#

Abstract. Let G be a graph. Let $L(G)$ be the difference of the diagonal matrix of vertex degrees and the adjacency matrix. This note addresses the number of ones in the Smith Normal Form of the integer matrix $L(G)$.

AMS(MOS) subject classifications. 05C50, 15A36

Let $G = (V, E)$ be a graph with vertex set $V = V(G) = \{v_1, v_2, \ldots, v_n\}$ and edge set $E = E(G)$. Denote by $D(G)$ the diagonal matrix of vertex degrees and by $A(G)$ the $(0, 1)$-adjacency matrix. The *Laplacian matrix* is $L(G) = D(G) - A(G)$. Of course, G_1 and G_2 are isomorphic graphs if and only if there is a permutation matrix P such that $P^t L(G_1) P = L(G_2)$. One typically deduces that isomorphic graphs have similar Laplacian matrices, partially explaining interest in the Laplacian spectrum. (See, e.g., [3].)

Recall that an n-by-n integer matrix U is **unimodular** if $\det U = \pm 1$. So, the unimodular matrices are precisely those integer matrices with integer inverses. Two integer matrices A and B are said to be **congruent** if there is a unimodular matrix U such that $U^t A U = B$. Consequently, isomorphic graphs have congruent Laplacian matrices. Unlike its spectral counterpart, this observation has a partial converse: If G_1 is 3-connected, then G_1 is isomorphic to G_2 if and only if $L(G_1)$ and $L(G_2)$ are congruent [6]. The *computational* significance of this partial converse is problematic since there is no canonical form for congruence [2]. On the other hand, integer matrices cannot be congruent if they are not equivalent, and the question of unimodular equivalence is easily settled by the Smith Normal Form.

Recall that integer matrices A and B are **equivalent** if there exist unimodular matrices U_1 and U_2 such that $U_1 A U_2 = B$. Denote by $d_k(G)$ the k-th *determinantal divisor* of $L(G)$, i.e., the greatest common divisor (GCD) of all the k-by-k determinantal minors of $L(G)$. (It follows, e.g. from the Matrix-Tree Theorem, that $d_{n-1}(G) = t(G)$, the number of *spanning trees* in G; and $d_n(G) = 0$, because $L(G)$ is singular.) Of course, $d_k(G) | d_{k+1}(G)$, $1 \leq k < n$. The **invariant factors** of G are defined by $s_{k+1}(G) = d_{k+1}(G)/d_k(G)$, $0 \leq k < n$, where $d_0(G) = 1$. The Smith Normal Form of $L(G)$ (*Smith Normal Form of G*) is

---

*Department of Mathematical Sciences, San Diego State University, San Diego, CA 92182.

†The work of this author was supported by the National Science Foundation under grant DMS 9007048.

**Department of Mathematics and Computer Science, California State University, Hayward, CA 94542.

‡The work of this author was supported by a California State University RSCA award and by the National Security Agency under NSA/MSP grant MDA 90-H-4024.

#Department of Mathematics, California State University, Northridge, CA 91330.

$$S(G) = \mathrm{diag}(s_1(G), s_2(G), \ldots, s_n(G)).$$

(Previous work on $S(G)$ was reported in [4]. An application of $S(G)$ to graph "bicycles" appears in [1].)

If G_1 and G_2 are isomorphic, then $L(G_1)$ and $L(G_2)$ are equivalent, i.e., $S(G_1) = S(G_2)$. If this observation had a partial converse (e.g., for 3-connected graphs), it would have great computational significance because $S(G)$ can be obtained from $L(G)$ by a sequence of elementary row and column operations.

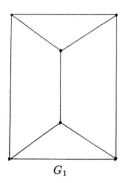 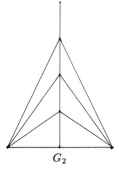

G_1 G_2

Figure 1. Two graphs with the same SNF.

Example 1. The graphs in Figure 1 share the Smith Normal Form, $\mathrm{diag}(1,1,1,5,15,0)$. Note that G_1 is 3-connected.

In spite of this discouraging example, the Smith Normal Form of G yields several bona fide graph theoretic invariants. In this note we are interested in one of them, namely the multiplicity of 1 in $S(G)$. We denote this multiplicity by $b(G)$; thus $b(G)$ is the largest value of k such that (either $d_k(G)$ or) $s_k(G) = 1$.

Suppose G is a connected graph with at least two vertices. Because G must have an edge, $b(G) \geq 1$. It is easy to see that $d_2(K_n) = n$, so equality holds for $G = K_n$. At the other extreme, $b(G) \leq n - 1$, with equality if and only if $t(G) = 1$, i.e., if and only if G is a tree.

THEOREM 1. *Let $S = \{v_1, v_2, \ldots, v_k\} \subseteq V(G)$. Suppose the induced subgraph $G[S]$ is a forest each of whose w components is a path on at least two vertices. Then $d_{k-w}(G) = 1$, i.e., $b(G) \geq k - w$.*

Proof. We may assume the numbering is chosen so that the vertices of S comprise the first k vertices of G, and that the numbering of the vertices of S respects its components, i.e., we may assume the leading k-by-k principal submatrix of $L(G)$ is a direct sum of w tridiagonal matrices of the form

$$A = \begin{pmatrix} a_1 & -1 & 0 & 0 & \cdots & 0 & 0 \\ -1 & a_2 & -1 & 0 & \cdots & 0 & 0 \\ 0 & -1 & a_3 & -1 & \cdots & 0 & 0 \\ & & \cdots & & & & \\ 0 & 0 & 0 & \cdots & & -1 & a_r \end{pmatrix}$$

where a_i is the degree in G of the i-th vertex in the component of $G[S]$ to which A corresponds. The $(r-1)$-square submatrix of A obtained by striking out its first row and last column is upper triangular with -1's down the main diagonal. Hence, its determinant is ± 1. Evidently, then, the leading k-by-k principal submatrix of $L(G)$ contains a $(k-w)$-square submatrix of determinant ± 1. □

COROLLARY 1. *Let G be a connected graph of diameter $\delta(G)$. Then $b(G) \geq \delta(G)$.*

Proof. Let $u, w \in V(G)$ be a distance $\delta = \delta(G)$ apart. Let $u = v_1, v_2, \ldots, v_{\delta+1} = w$ be a shortest path from u to w. If $\{v_i, v_j\} \in E$ for $i + 1 < j$, there would be a shorter path from u to w in G. Let $S = \{v_1, \ldots, v_{\delta+1}\}$. Then $G[S]$ is the path $P_{\delta+1}$. The result follows from Theorem 1. □

It has been shown that two connected graphs with the same blocks (maximal 2-connected subgraphs) have the same Smith Normal Form [4,5]. Therefore, Corollary 1 can be improved to the extent that $b(G)$ is an upper bound for the sum of the diameters of the blocks of G. Similarly, a certain "twist operation" (described in [6]) can be used to improve the result even further.

COROLLARY 2. *Let G be a connected graph on $n \geq 2$ vertices. Then $b(G) \geq 1$ with equality if and only if $G = K_n$.*

Proof. We have already seen that $b(K_n) = 1$. If $G \neq K_n$, then $\delta(G) \geq 2$. The result follows from Corollary 1. □

THEOREM 2. *Let $H = G - e$ be the edge subgraph of G obtained by deleting the (single) edge e. For $1 \leq k < n$,*

$$d_k(G) | d_{k+1}(H) \text{ and } d_k(H) | d_{k+1}(G).$$

Proof. Suppose edge $e = \{v_i, v_j\}$ is deleted from G to produce H. Let A be the matrix obtained from $L(G)$ by adding rows (and columns) $1, 2, \ldots, j-1, j+1, \ldots, n$ to row (and column) j (respectively). Note: (1) A is the matrix obtained from $L(G)$ by replacing each entry in its j-th row and column by 0; (2) there is a unimodular matrix U such that $A = U^t L(G) U$; (3) the matrix $B = U^t L(H) U$ is obtained from $L(H)$ by zeroing out its j-th row and column. Because equivalent matrices have the same sequence of determinantal divisors, $d_k(G) = d_k(A)$, the GCD of all the k-by-k determinantal minors of A, and $d_k(H) = d_k(B)$, $1 \leq k \leq n$. So, it suffices to prove that $d_k(A)|d_{k+1}(B)$ and $d_k(B)|d_{k+1}(A)$. Let $S,T \subseteq V(G) = V(H)$ be $(k+1)$-element sets of vertices. Denote by $A[S:T]$ and $B[S:T]$ the submatrices of A and B, respectively, lying in rows corresponding to S and columns corresponding to T. If $v_j \in S \cup T$, then $\det A[S:T] = \det B[S:T] = 0$. If $v_i \notin S \cup T$, then $A[S:T] = B[S:T]$. In either of these cases, $d_k(A)|\det B[S:T]$ and $d_k(B)|\det A[S:T]$. The only remaining possibility is that $v_i \in S \cup T$ but $v_j \notin S \cup T$. In this case, $A[S:T] = B[S:T] + F$, where F is the $(k+1)$-by-$(k+1)$ matrix whose only nonzero entry is a 1 in the row corresponding to vertex $v_i \in S$ and the column corresponding to vertex $v_i \in T$. Thus,

$$(1) \qquad \det A[S:T] = \det B[S:T] + \det B[S - v_i : T - v_i].$$

Since $d_k(B)$ divides both terms on the right-hand side of (1), $d_k(B)|\det A[S:T]$. But, $B[S - v_i : T - v_i] = A[S - v_i : T - v_i]$. It follows that $d_k(A)$ divides two of the terms in (1) so it must divide $\det B[S:T]$ as well. We have shown that $d_k(A)$ (respectively $d_k(B)$) divides every $(k+1)$-by-$(k+1)$ determinantal minor of B (respectively, A). Thus, it divides their GCD. □

Question 1. Let $H = G - e$. Does $s_k(G)|s_{k+1}(H)$, and/or does $s_k(H)|s_{k+1}(G)$?

COROLLARY 3. *Let $H = G - e$. Then $|b(G) - b(H)| \leq 1$.*

Proof. By definition, $1 = d_{b(G)}(G)$. By Theorem 2, $d_{b(G)-1}(H) = 1$, so $b(H) \geq b(G) - 1$. Similarly, $b(G) \geq b(H) - 1$. □

We now explore the relationship between $b(G)$ and the number of edges in G. For $n > k \geq 1$, let $F(n,k)$ denote the family of connected graphs G on n vertices satisfying $b(G) = k$. Then, for example, $F(n,1) = \{K_n\}$ and $F(n,n-1)$ is the set of trees on n vertices. Let $m(n,k)$ and $M(n,k)$ denote, respectively, the minimum and maximum of the number of edges in G, as G ranges over $F(n,k)$. Then, $m(n,k) \leq M(n,k) \leq C(n,2)$, *binomial coefficient n-choose-2*, with equality throughout for $k = 1$. On the other hand, $M(n,k) \geq m(n,k) \geq n-1$, with equality throughout for $k = n - 1$.

COROLLARY 4. *For $n > k > 1$, $m(n,k) = 2(n-1) - k$.*

Proof. Let $G \in F(n,k)$ be a graph with $m(n,k)$ edges. Since G is connected, it contains a spanning tree T. We may obtain T from G by successively removing the $m(n,k) - (n-1)$ edges of G not belonging to T. By Corollary 3, $b(T) - b(G) \leq m(n,k) - (n-1)$. Since $b(T) = n - 1$ and $b(G) = k$, we conclude that $m(n,k) \geq 2(n-1) - k$. It remains to produce a graph $G_{n,k} \in F(n,k)$ having $2(n-1) - k$ edges. For this, we need some additional notation.

If $G = (V(G), E(G))$ and $H = (V(H), E(H))$ are graphs on disjoint sets of vertices, let $G + H$ be their *union*, i.e., $V(G+H) = V(G) \cup V(H)$, and $E(G+H) =$

$E(G) \cup E(H)$. Suppose G has n_1 vertices and H has n_2 vertices. A *coalescence* of G and H is any of the $n_1 n_2$ graphs on $n_1 + n_2 - 1$ vertices that can be obtained from $G + H$ by identifying (i.e., coalescing into a single vertex) a vertex of G and a vertex of H. We denote any of these coalescences by $G * H$ (with the understanding that "$*$" is not a well defined binary operation).

For $n > k > 1$, let $G_{n,k} = K_{2,n-k} * P_{k-1}$, where $K_{2,n-k}$ is the complete bipartite graph and P_{k-1} is the path on $k - 1$ vertices. Then $G_{n,k}$ has n vertices and $m = 2(n - k) + (k - 2) = 2(n - 1) - k$ edges. By a direct computation (or by [5, Thm 1] or [4, Prop 1]), one obtains $b(G_{n,k}) = k$. $\square$

There is no 2-connected graph G with $b(G) = 3$ having $n = 6$ vertices and $m = m(6,3) = 7$ edges.

Question 2. What is the minimum number of edges in G, as G ranges over the 2-connected graphs in $F(n, k), k \geq 3$?

COROLLARY 5. *For $n - 1 > k \geq 1$, $M(n, k) = C(n, 2) - k + 1$.*

Proof. Let $G \in F(n, k)$ be a graph with $M(n, k)$ edges. We may obtain G by removing the $C(n, 2) - M(n, k)$ edges of K_n not belonging to G. By Corollary 3, $b(G) - b(K_n) \leq C(n, 2) - M(n, k)$. Since $b(G) = k$ and $b(K_n) = 1$, we see $M(n, k) \leq C(n, 2) - k + 1$. As in the proof of Corollary 4, it remains to describe a graph $G^{n,k} \in F(n, k)$ with $C(n, 2) - k + 1$ edges. For $k = 1$, there is only one possibility, namely, $G^{n,1} = K_n$. For $n - 1 > k > 1$, let $G^{n,k} = (P_k + K^c_{n-k})^c$, the complement of the union of P_k with the complement of K_{n-k}. Note that $G^{n,k}$ has $m = C(n, 2) - (k - 1)$ edges.

It remains to show that $b(G^{n,k}) = k$. A graph with $m = C(n, 2) - k + 1$ edges cannot have *more* than k ones in its Smith Normal Form because if it did, the addition of more edges would eventually lead to K_n. By Corollary 3, the addition of a single edge can change the number of ones in the Smith Normal Form by at most 1. So, $b(G^{n,k}) - b(K_n) \leq k - 1$ and, since $b(K_n) = 1$, we have $b(G^{n,k}) \leq k$.

Next we show that $b(G^{n,k}) \geq k$ by exhibiting a k-by-k submatrix of $L(G^{n,k})$ whose determinant is ± 1. We may suppose the vertices of $G^{n,k}$ have been numbered so that the k vertices associated with P_k come first. Thus, the leading $(k + 2)$-by-$(k + 2)$ principal submatrix of $L(G^{n,k})$ is

$$A = \begin{pmatrix} n-2 & 0 & -1 & -1 & \cdots & -1 & -1 & -1 & -1 \\ 0 & n-3 & 0 & -1 & \cdots & -1 & -1 & -1 & -1 \\ -1 & 0 & n-3 & 0 & \cdots & -1 & -1 & -1 & -1 \\ & & & & \cdots & & & & \\ -1 & -1 & -1 & -1 & \cdots & n-3 & 0 & -1 & -1 \\ -1 & -1 & -1 & -1 & \cdots & 0 & n-2 & -1 & -1 \\ -1 & -1 & -1 & -1 & \cdots & -1 & -1 & n-1 & -1 \\ -1 & -1 & -1 & -1 & \cdots & -1 & -1 & -1 & n-1 \end{pmatrix}.$$

Let B be the k-by-k submatrix of $L(G^{n,k})$ obtained by striking out rows k and $k+1$, and columns 1 and $k+2$ from A. We show that $\det B = \pm 1$ by a sequence of elementary row and column operations. Subtract row 1 (of B) from row k. If $k = 2$, the resulting matrix is

$$\begin{pmatrix} 0 & -1 \\ -1 & 0 \end{pmatrix},$$

and the proof is complete. Otherwise, the only nonzero entry in the k-th row of the resulting matrix is -1 in the $(k,1)$ position. We may use it to annihilate the other nonzero entries in column 1. In the resulting matrix, subtract row 2 from row 1. The only nonzero entry left in row 1 is -1 in column 2. Use it to annihilate any other nonzero entries in column 2. Then subtract row 3 from row 2. The only nonzero entry in row 2 is -1 in column 3. Use it to zero out any other nonzero entries in column 3. Continuing in this way, we eventually reduce B to

$$C = \begin{pmatrix} 0 & -1 & 0 & 0 & \cdots & 0 & 0 \\ 0 & 0 & -1 & 0 & \cdots & 0 & 0 \\ 0 & 0 & 0 & -1 & \cdots & 0 & 0 \\ 0 & 0 & \cdots & & & & \\ 0 & 0 & 0 & 0 & \cdots & 0 & -1 \\ -1 & 0 & 0 & 0 & \cdots & 0 & 0 \end{pmatrix}.$$

Since $\det B = \det C = \pm 1$, the proof is complete. □

REFERENCES

[1] K.A. BERMAN, *Bicycles and spanning trees*, SIAM J. Algebraic & Discrete Meth. 7 (1986), pp. 1–12.
[2] S. FRIEDLAND, *Quadratic forms and the graph isomorphism problem*, Linear Algebra Appl. 150 (1991), pp. 423–442.
[3] R. GRONE, R. MERRIS, AND V.S. SUNDER, *The Laplacian spectrum of a graph*, SIAM J. Matrix Anal. Appl. 11 (1990), pp. 218–238.
[4] R. MERRIS, *Unimodular equivalence of graphs*, Linear Algebra Appl., to appear.
[5] W. WATKINS, *The Laplacian matrix of a graph: unimodular congruence*, Linear & Multilinear Algebra, 28 (1990), pp. 35–43.
[6] W. WATKINS, *Unimodular congruence of the Laplacian matrix of a graph*, manuscript.

RANK INCREMENTATION VIA DIAGONAL PERTURBATIONS*

WAYNE BARRETT**, CHARLES R. JOHNSON†,
RAPHAEL LOEWY‡ AND TAMIR SHALOM#

Introduction. Let $M_n(F)$ denote the set of n-by-n matrices over the field F. We consider the following question: Among matrices $A \in M_n(F)$ with rank $A = k < n$, how many diagonal entries of A must be changed, at worst, in order to guarantee that the rank of A is increased. Our initial motivation arose from an error pointed out in [BOvdD], but we also view this problem as intrinsically important. The simplest example that shows that one entry does not suffice is the familiar Jordan block $\begin{bmatrix} 0 & 1 \\ 0 & 0 \end{bmatrix}$. Let

$$J_r = \begin{bmatrix} 0 & 1 & & & 0 \\ & \ddots & \ddots & & \\ & & \ddots & \ddots & \\ 0 & & & \ddots & 1 \\ & & & & 0 \end{bmatrix}$$

be the basic nilpotent r-by-r Jordan block. If $A = J_n$, rank $A = n - 1$, and all n diagonal entries of A must be changed in order to increase the rank of A to n. More generally, assume that A is a direct sum of such blocks. If rank $A = k$, A is the sum of $n - k$ blocks

$$(1) \qquad A = J_{r_1} \oplus J_{r_2} \oplus \cdots \oplus J_{r_{n-k}}$$

in which we take $r_1 \leq r_2 \leq \cdots \leq r_{n-k}$, and $r_1 + r_2 + \cdots + r_{n-k} = n$. To increase the rank of A it is necessary and sufficient to change all the diagonal entries in one block to nonzero entries. Thus, r_1 changes will do. Since

$$(n - k)r_1 \leq r_1 + \cdots + r_{n-k} = n$$

we have $r_1 \leq \frac{n}{n-k}$ and so $r_1 \leq \lfloor \frac{n}{n-k} \rfloor$, in which $\lfloor x \rfloor$, the "floor" function, denotes the greatest integer less than or equal to x. By taking $r_1 = r_2 = \cdots = r_{n-k-1} = \lfloor \frac{n}{n-k} \rfloor$ in equation (1) we see that we cannot hope, in general, to increase the rank of A by fewer than $\lfloor \frac{n}{n-k} \rfloor$ changes. We are therefore led to the statement of our first principal result.

---

*This manuscript was prepared while the first three authors were visitors at the Institute for Mathematics and its Applications, Minneapolis, Minnesota. The research of C.R. Johnson and R. Loewy was supported by grant No. 90-00471 from the United States–Israel Binational Science Foundation (BSF), Jerusalem, Israel.
**Dept. of Mathematics, Brigham Young University, Provo, Utah 84602.
†Dept. of Mathematics, The College of William and Mary, Williamsburg, Virginia 23185.
‡Dept. of Mathematics, Technion-Israel Institute of Technology, Haifa 32000, ISRAEL.
#Tecnomatix Technologies, Delta House, Herzliya 46733, ISRAEL.

THEOREM 1. *Let F be any field and let $A \in M_n(F)$ with rank $A = k < n$. Then it suffices to change at most $\lfloor \frac{n}{n-k} \rfloor$ diagonal entries of A in order to increase the rank of A.*

We note that $\lfloor \frac{n}{n-k} \rfloor = \lceil \frac{k+1}{n-k} \rceil = 1 + \lfloor \frac{2k}{n} \rfloor + \lfloor \frac{3k}{2n} \rfloor + \lfloor \frac{4k}{3n} \rfloor + \cdots$, each of which suggests an alternate understanding of the quantity calculated in theorem 1; here $\lceil \; \rceil$ denotes the "roof" function.

Although the argument given above shows that theorem 1 is sharp, it is unsatisfactory in one respect. Suppose, for example, that $A = B \oplus 0_{n-k}$, in which B is an invertible matrix in $M_k(F)$. Then rank $A = k$ and changing any diagonal entry in A outside B will increase the rank of A. In this case, the prediction of $\lfloor \frac{n}{n-k} \rfloor$ is overly pessimistic. We can make a slight adjustment to theorem 1 that accommodates such examples.

DEFINITION. For $A \in M_n(F)$, the *principal rank* of A is

$$p = p(A) \equiv \max_{\alpha \subseteq N} \{|\alpha| : A[\alpha] \text{ is invertible }\}$$

Here $N = \{1, 2, \ldots, n\}$, $|\alpha|$ denotes the cardinality of α, and $A[\alpha]$ denotes the principal submatrix of A whose rows and columns lie in α.

Of course, $p(A) \leq k(A) \equiv \text{rank}(A)$, and, by appeal to the presentation of the characteristic polynomial in terms of principal minor sums [HJ], the number of nonzero eigenvalues of A is $\leq p(A)$.

If we now consider $A = I_p \oplus J_{r_1} \oplus \cdots \oplus J_{r_{n-k}}$, so that $r_1 + r_2 + \cdots + r_{n-k} = n - p$, then we realize that we might have to change $\lfloor \frac{n-p}{n-k} \rfloor$ entries of A to increase its rank.

THEOREM 2. *Let $A \in M_n(F)$ with rank $A = k < n$ and principal rank $A = p$. Then it suffices to change at most $\lfloor \frac{n-p}{n-k} \rfloor$ diagonal entries of A in order to increase the rank of A.*

In the case $A = B \oplus 0_{n-k}$ cited above, $p = k$ and theorem 2 now indicates that it suffices to change one diagonal entry.

Note that all aspects of our problem are unchanged by permutation similarity of A, which we are free to use. Arbitrary similarities are not available, however.

We first note that theorem 2 is a consequence of theorem 1.

Proof of Theorem 2. If $p = 0$, there is nothing to prove.

Suppose $p > 0$ and let $\alpha \subseteq N$, $|\alpha| = p$ with $A[\alpha]$ invertible. Then by a permutation similarity, we may assume without loss of generality, that

$$A = \begin{bmatrix} A[\alpha] & B \\ C & D \end{bmatrix}.$$

We use the following elementary fact about the Schur complement of $A[\alpha]$ in A:

$$\text{rank } A = \text{rank } A[\alpha] + \text{rank}(D - C(A[\alpha])^{-1}B).$$

So $\text{rank}(D - C(A[\alpha])^{-1}B) = k - p < n - p$. Now, changing diagonal entries in D corresponds exactly to changing diagonal entries in $D - C(A[\alpha])^{-1}B$. Since $D - C(A[\alpha])^{-1}B \in M_{n-p}(F)$, by Theorem 1 we can increase its rank, and hence increase the rank of A, by changing at most $\lfloor \frac{n-p}{n-p-(k-p)} \rfloor = \lfloor \frac{n-p}{n-k} \rfloor$ diagonal entries. $\square$

COROLLARY. *Let $A \in M_n(F)$, and suppose that the algebraic multiplicity and the geometric multiplicity of zero as an eigenvalue of A are both $n - k > 0$. Then it suffices to change 1 diagonal entry of A in order to increase its rank.*

Proof. In this event $p = k$, as the number of nonzero eigenvalues is k. Thus, $\lfloor \frac{n-p}{n-k} \rfloor = 1$. $\square$

The corollary includes such special cases as diagonalizable, normal and Hermitian matrices. Each requires only one change of a diagonal entry.

We now proceed with the proof of theorem 1. Because of technical difficulties when F is a small finite field, we first prove the following weaker version.

THEOREM 3. *Let F be a field with $|F| \geq 2n$ and let $A \in M_n(F)$ with rank $A = k < n$. Then it suffices to change at most $\lfloor \frac{n}{n-k} \rfloor$ diagonal entries of A in order to increase the rank of A.*

The proof of theorem 3 relies upon a sequence of observations.

PROPOSITION 1. *Let $A \in M_n(F)$ with rank $A = k < n$. If there exist $\alpha, \beta \subseteq N$ with $\det A[\alpha|\beta] \neq 0$, $|\alpha| = |\beta| = k$ and $\alpha \cup \beta \neq N$, then the rank of A can be increased by changing one diagonal entry of A.*

Proof. Pick $i \notin \alpha \cup \beta$, let E_i be the matrix in $M_n(F)$ with 1 in the (i, i) position and 0 elsewhere, and let $A' = A + E_i$. Then $\det A'[\alpha \cup \{i\}|\beta \cup \{i\}]$
$= \pm \det A[\alpha|\beta] + \det A[\alpha \cup \{i\}|\beta \cup \{i\}]$. But $\det A[\alpha \cup \{i\}|\beta \cup \{i\}] = 0$ since rank $A = k$, so $\det A'[\alpha \cup \{i\}|\beta \cup \{i\}] \neq 0$ and the rank of A has been increased by changing one diagonal entry. $\square$

PROPOSITION 2. *Given a field F with $|F| > 2$, let $A \in M_n(F)$ with rank $A = k < n$, and let $A[\alpha|\beta]$, $|\alpha| = |\beta| = k$, be invertible. Assume $\alpha \cup \beta = N$. If there exist $i \in \alpha - \beta$ and $j \in \beta - \alpha$ such that $\det A[\alpha - \{i\}|\beta - \{j\}] \neq 0$, then the rank of A can be increased by changing at most two diagonal entries of A.*

Proof. Let $A'' = A + x(E_i + E_j)$. Assuming that the indices of A occur in the order $\alpha - \beta$, $\alpha \cap \beta$, $\beta - \alpha$, we have

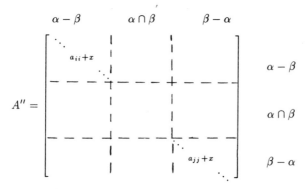

Then $\det A''[\alpha \cup \{j\}|\beta \cup \{i\}]$ is a polynomial of degree at most 2 in x. Since the coefficient of x^2 is $\pm \det A[\alpha - \{i\}|\beta - \{j\}]$, it has degree exactly 2 and hence has at most 2 roots in F. Since $|F| > 2$, we may choose $x \in F$ so that $\det A''[\alpha \cup \{j\}|\beta \cup \{i\}] \neq 0$. Hence rank $A'' \geq k+1$ and the rank of A can be increased by changing at most 2 diagonal entries. $\square$

LEMMA. *Let* $B \in M_m(F)$ *be invertible, and let* $\sigma, \tau \subseteq \{1, \ldots, m\}$. *Then*

$$\text{nullity } B[\sigma|\tau] = \text{nullity } B^{-1}[\tau^c|\sigma^c].$$

Proof. Various statements equivalent to this fact have arisen in a variety of places, e.g. [BGK], Corollary 3 in [FM], [G], Lemma 4 in [JL]. For completeness we give a simple proof.

Without loss of generality, via permutation equivalence, we may assume $B[\sigma|\tau] = B_{12}$ and $B^{-1}[\tau^c|\sigma^c] = A_{12}$ with

$$B = \begin{bmatrix} B_{11} & B_{12} \\ B_{21} & B_{22} \end{bmatrix} \text{ and } A = B^{-1} = \begin{bmatrix} A_{11} & A_{12} \\ A_{21} & A_{22} \end{bmatrix}.$$

If $x \in$ null space (B_{12}), then

$$\begin{bmatrix} 0 \\ x \end{bmatrix} = AB \begin{bmatrix} 0 \\ x \end{bmatrix} = A \begin{bmatrix} 0 \\ y \end{bmatrix} = \begin{bmatrix} A_{12}y \\ * \end{bmatrix}$$

exhibits an isomorphism between the null spaces of B_{12} and A_{12}. $\square$

Proof of Theorem 3.

Consider first the case $\frac{n}{n-k} < 2$, or equivalently, $2k < n$. Then $|\alpha \cup \beta| \leq |\alpha| + |\beta| = 2k$ so that $\alpha \cup \beta \neq N$. By proposition 1, it suffices to change $1 = \lfloor \frac{n}{n-k} \rfloor$ entry in order to increase rank A.

So from now on assume $\frac{n}{n-k} \geq 2$, or $2k \geq n$. In particular, $n \geq 2$ so $|F| \geq 4$. Also, assume that for any invertible submatrix $A[\alpha|\beta]$ with $|\alpha| = |\beta| = k$ that $\alpha \cup \beta = N$. Otherwise we are done by proposition 1. Since rank $A = k$, there is at least one invertible submatrix $A[\alpha|\beta]$ with $|\alpha| = |\beta| = k$. Let

(2)
$$B = A[\alpha|\beta],$$
$$C = A[\alpha \cap \beta|\alpha \cap \beta].$$

We call B the rank determining block of A. If the condition in Proposition 2 is met, we can increase the rank of A by changing at most two diagonal entries, and since $\frac{n}{n-k} \geq 2$ we are done.

We therefore assume that $\det A[\alpha - \{i\}|\beta - \{j\}] = 0$ for all $i \in \alpha - \beta$ and all $j \in \beta - \alpha$; equivalently, the cofactors B_{ij} in B are zero for $i \in \alpha - \beta$, $j \in \beta - \alpha$. Therefore,

$$(3) \qquad B^{-1}[\beta - \alpha | \alpha - \beta] = 0.$$

Now $|\alpha \cap \beta| = |\alpha| + |\beta| - |\alpha \cup \beta| = 2k - n$, so $|\alpha - \beta| = |\alpha| - |\alpha \cap \beta| = k - (2k - n) = n - k$, and $|\beta - \alpha| = n - k$ also. Since $B^{-1} \in M_k(F)$ is invertible, by the Frobenius-König Theorem (see [F] or p. 543 of the survey article [MP]), $|\alpha - \beta| + |\beta - \alpha| \leq k$, so we have $2n \leq 3k$. Since $k \geq \frac{2}{3}n$, $n - k \leq \frac{1}{3}n$, and now $\lfloor \frac{n}{n-k} \rfloor \geq 3$. The matrix $C \in M_{2k-n}(F)$ defined by (2) is a submatrix of B and by the lemma and equation (3)

$$\text{nullity } C = \text{nullity } B^{-1}[\beta - \alpha, \alpha - \beta] = n - k.$$

Therefore,

$$\text{rank } C = 2k - n - (n - k) = 3k - 2n$$

We now proceed by induction on n. Since $2n \leq 3k \leq 3(n-1)$, we have $n \geq 3$, so we know that the theorem holds for $n = 1, 2$. Let $n \geq 3$ be a fixed integer and assume the theorem is true for all positive integers $m < n$. Since $0 < 2k - n < n$, we can increase the rank of C by changing at most s diagonal entries of C where

$$s \leq \left\lfloor \frac{2k - n}{2k - n - (3k - 2n)} \right\rfloor = \left\lfloor \frac{2k - n}{n - k} \right\rfloor.$$

Let $i_1, \ldots, i_s \in \alpha \cap \beta$ be the indices corresponding to the entries which are changed, and let

$$\tilde{A}(x_1, \ldots, x_s) = A + x_1 E_{i_1} + x_2 E_{i_2} + \cdots + x_s E_{i_s}$$
$$\tilde{B}(x_1, \ldots, x_s) = \tilde{A}[\alpha | \beta]$$
$$\tilde{C}(x_1, \ldots, x_s) = \tilde{A}[\alpha \cap \beta | \alpha \cap \beta].$$

Then the fact that we can increase the rank of C by changing s diagonal entries is equivalent to the statement that there are nonzero $f_1, \ldots, f_s \in F$ such that $\text{rank } \tilde{C}(f_1, \ldots, f_s) > 3k - 2n$. However, although $\text{rank } B = \text{rank } \tilde{B}(0, \ldots, 0) = k$, it may now be the case that $\text{rank } \tilde{B}(f_1, \ldots, f_s) < k$. We overcome this obstruction by using the fact that $|F|$ is sufficiently large.

Let $\tilde{C}'(x_1, \ldots, x_s)$ be a submatrix of $\tilde{C}(x_1, \ldots, x_s)$ of order $3k - 2n + 1$ for which $\tilde{C}'(f_1, \ldots, f_s)$ is invertible. Since each x_j occurs at most once in $\tilde{C}'(x_1, \ldots, x_s)$ and $\det \tilde{C}'(f_1, \ldots, f_s) \neq 0$, then $\det \tilde{C}'(x_1, \ldots, x_s)$ is a nonzero polynomial in the integral domain $F[x_1, \ldots, x_s]$ with degree at most s.

Since $\det \tilde{B}(0, \ldots, 0) \neq 0$, a similar argument shows that $\det \tilde{B}(x_1, \ldots, x_s)$ is a nonzero polynomial in $F[x_1, \ldots, x_s]$ with degree at most s. Since $s \leq \left\lfloor \frac{2(n-1)-n}{n-(n-1)} \right\rfloor =$

$n-2$, $q(x_1,\ldots,x_s) = \det \widetilde{C}'(x_1,\ldots,x_s) \cdot \det \widetilde{B}(x_1,\ldots,x_s)$ is a nonzero polynomial of degree at most $2n-4$. But $|F| > 2n-4$, so we can choose $g_1,\ldots,g_s \in F$ such that $q(g_1,\ldots,g_s) \neq 0$. Then $\widetilde{C}'(g_1,\ldots g_s)$ and $\widetilde{B}(g_1,\ldots,g_s)$ are invertible so that

$$\text{rank}\,\widetilde{C}(g_1,\ldots,g_s) > 3k - 2n$$
$$\text{rank}\,\widetilde{B}(g_1,\ldots,g_s) = k.$$

For the remainder of the proof we abbreviate $\widetilde{A}(g_1,\ldots,g_s)$, $\widetilde{B}(g_1,\ldots,g_s)$, and $\widetilde{C}(g_1,\ldots,g_s)$ by $\widetilde{A}$, $\widetilde{B}$ and $\widetilde{C}$. Now $\widetilde{C} \in M_{2k-n}(F)$ so nullity $\widetilde{C} < n - k$. Applying Lemma 1 again, along with the invertibility of $\widetilde{B} = \widetilde{A}[\alpha|\beta]$,

$$\text{nullity}\,\widetilde{B}^{-1}[\beta - \alpha, \alpha - \beta] = \text{nullity}\,\widetilde{C} < n - k.$$

But $|\alpha - \beta| = n - k$, wo $\widetilde{B}^{-1}[\beta - \alpha, \alpha - \beta]$ is not the zero matrix. Thus, for some $i \in \alpha - \beta$ and some $j \in \beta - \alpha$, $\det \widetilde{B}[\alpha - \{i\}|\beta - \{j\}] \neq 0$ and since $\widetilde{B}$ is a submatrix of $\widetilde{A}$, $\det \widetilde{A}[\alpha - \{i\}|\beta - \{j\}] \neq 0$. By Proposition 2, the rank of $\widetilde{A}$ (which is at least k) can be increased by changing 2 diagonal entries. Therefore the rank of A has been increased by changing at most $s + 2 \leq \lfloor \frac{2k-n}{n-k} \rfloor + 2 = \lfloor \frac{2k-n}{n-k} + 2 \rfloor = \lfloor \frac{n}{n-k} \rfloor$ of its diagonal entries, which completes the proof of Theorem 3. $\square$

We give an example to show that the rank determining block $B = A[\alpha|\beta]$ may become singular as we increase the rank of $C = A[\alpha \cap \beta|\alpha \cap \beta]$, as mentioned in the proof. Suppose $F = GF(2)$, the field of two elements, that $n = 6$, $k = 4$ and that

$$A = \begin{bmatrix} 0 & 0 & 1 & 0 & 1 & 0 \\ 0 & 0 & 0 & 1 & 0 & 1 \\ 0 & 0 & 0 & 0 & 1 & 0 \\ 0 & 0 & 0 & 0 & 0 & 1 \\ 0 & 0 & 0 & 0 & 0 & 0 \\ 0 & 0 & 0 & 0 & 0 & 0 \end{bmatrix}$$

Then $B = A[\{1,2,3,4\}, \{3,4,5,6\}]$, $C = A[\{3,4\}, \{3,4\}]$ and rank $C = 3k - 2n = 0$. However, if any $1(= \lfloor \frac{2k-n}{n-k} \rfloor)$ diagonal entry of C is changed B becomes singular and rank A drops to 3. It is easy to construct similar examples for other values of n. This shows that no simple repair of the proof of Theorem 3 can yield a proof of Theorem 1. However, there is a construction along with Theorem 3 that gives the result.

Proof of Theorem 1. The only case to consider is $2 \leq |F| < 2n$. In this case, embed F in a field K with $|K| \geq 2n$. Then, by Theorem 3, there exist distinct indices $i_1,\ldots,i_s$, $s \leq \lfloor \frac{n}{n-k} \rfloor$, and nonzero elements $u_1,\ldots,u_s \in K$ such that $A + u_1 E_{i_1}, + u_2 E_{i_2} + \cdots + u_s E_{i_s}$ has rank greater than k. Let $\widetilde{A}(x_1,\ldots,x_s) = A + x_1 E_{i_1} + \cdots + x_s E_{i_s}$, $\widetilde{B}(x_1,\ldots,x_s)$ be a submatrix of $\widetilde{A}(x_1,\ldots,x_s)$ of order $k + 1$ such that $\det \widetilde{B}(u_1,\ldots,u_s) \neq 0$ and let

$$q(x_1,\ldots,x_s) = \det \widetilde{B}(x_1,\ldots,x_s).$$

Without loss of generality, assume that $x_1,\ldots,x_s$ all occur in $\widetilde{B}$.

Now by the definition of q, $p_1(x_1) = q(u_1 + x_1, u_2, \ldots, u_s)$ is either linear in x_1 or a nonzero element of K since $p_1(0) \neq 0$. Therefore p_1 has at most one zero in K. Choose x_1 such that $f_1 = u_1 + x_1 \in F$ and $p_1(x_1) = q(f_1, u_2, \ldots, u_s) \neq 0$. By considering $p_2(x_2) = q(f_1, u_2 + x_2, u_3, \ldots, u_s)$, we similarly obtain $f_2 \in F$ such that $q(f_1, f_2, u_3, \ldots u_s) \neq 0$. Continuing the process yields $q(f_1, f_2, \ldots, f_s) \neq 0$ for $f_1, f_2, \ldots, f_s \in F$. In other words $\widetilde{A}(f_1, \ldots, f_s)$ has a submatrix $\widetilde{B}(f_1, \ldots, f_s)$ of rank $k+1$, and we have increased the rank of A by replacing at most $\lfloor \frac{n}{n-k} \rfloor$ diagonal entries of A by elements of F. This completes the proof. $\square$

We end by stating the following refinement of theorem 2. It may be proven via the same type of argument just used in the proof of theorem 1.

THEOREM 4. *Let F be any field, let g be any nonzero element in F, and let $A \in M_n(F)$ with rank $A = k$ and principal rank $A = p$. Then it suffices to add g to at most $\lfloor \frac{n-p}{n-k} \rfloor$ diagonal entries of A in order to increase the rank of A.*

REFERENCES

[BGK]   H. BART, I. GOHBERG AND M.A. KAASHOEK, *The coupling method for solving integral equations*, in: Topics in Operator Theory, Systems and Networks, The Rehovot Workshop (Eds: H. Bart, I. Gohberg), Operator Theory: Adv. Appl. OT 12, Birkhäuser Verlag, Basel (1984), pp. 39–73.

[BOvdD] W.W. BARRETT, D.D. OLESKY AND P. VAN DEN DRIESSCHE, *A Note on Eigenvalues of Fixed Rank Perturbations of Diagonal Matrices*, Linear and Multilinear Algebra 30 (1991), pp. 13–16.

[F]   F.G. FROBENIUS, *Über zerlegbare Determinanten*, in: F.G. Frobenius, Gesammelte Abhandlungen, Band III, Springer-Verlag Berlin-Heidelberg (1968), pp. 701–702.

[FM]   M. FIEDLER, T.L. MARKHAM, *Completing a Matrix when Certain Entries of its Inverse are Specified*, Linear Algebra and its Applications, 74 (1986), pp. 225–237.

[G]   W.H. GUSTAFSON, *A Note on Matrix Inversion*, Linear Algebra and its Applications 57 (1984), pp. 71–73.

[HJ]   R.A. HORN AND C.R. JOHNSON, *Matrix Analysis*, Cambridge University Press, New York, NY, 1985.

[JL]   C.R. JOHNSON AND M. LUNDQUIST, *Operator Matrices with Chordal Inverse Patterns*, IMA Preprint Series # 885, October 1991.

[MP]   L. MIRSKY AND H. PERFECT, *Systems of Representatives*, Journal of Mathematical Analysis and Applications 15 (1966), pp. 520–568.

EIGENVALUES OF ALMOST SKEW SYMMETRIC MATRICES AND TOURNAMENT MATRICES

SHMUEL FRIEDLAND*

Abstract. A real square matrix C is called almost skew symmetric if $C = S + A$ where S is a rank one real symmetric matrix and A is a real skew symmetric matrix. We shall show that the real eigenvalues of almost skew symmetric matrices satisfy remarkable inequalities. We shall apply these and other inequalities to estimate the spectral radii of the tournament matrices.

1. Introduction. Let C be a real $n \times n$ matrix. Set

$$(1.1) \qquad C = S + A, \; S = \frac{C + C^T}{2}, \; A = \frac{C - C^T}{2}.$$

Thus, S and A are the symmetric and the skew (anti) symmetric components of C. We call C almost skew symmetric if S is a rank one matrix. In the terminology of [M-P] $C - \frac{I}{2}$ is a pseudo-tournament matrix. In §2 we discuss remarkable properties of the spectrum of almost skew symmetric matrices. Under certain conditions we estimate the largest real eigenvalue of an almost skew symmetric C from below and above when the nonzero eigenvalue of S is positive. Our estimate are quadratic in the entries of A. We also give conditions on A which ensure the existence of real eigenvalues for C in the case where the dimension of C is even. Our results are almost unrelated to the results discussed in [M-P]. A tournament matrix is a square $(0,1)$ matrix T such that $T + T^T = J - I$ where J is the all ones matrix. In §3 we note that a tournament matrix is an almost skew symmetric matrix minus $\frac{I}{2}$. We then apply our results to get some estimates on the spectral radius of tournament matrices. Our results can be viewed as a natural extension of the basic inequality due to Kirkland [Kir]. In §4 we bring a generalization of Ostrowski's inequality [Ost] for the spectral radius of nonnegative matrices which we apply to generalized tournament matrices. (A nonnegative matrix T is called a generalized tournament matrix if $T + T^T = J - I$.) In the last section we discuss the extremal tournament matrices of an even order n which maximize the spectral radius of the tournament matrices and the corresponding values of their spectral radii as a function of n. We were not able to prove or disprove the Brualdi - Li conjecture [B-L] on these extremal matrices. We did show that the Brualdi-Li matrix maximizes a function which is closely related to the spectral radius of a tournament matrix.

2. Perturbation of a rank one symmetric matrix by a skew symmetric matrix. As usual, denote by $M_n(\mathbf{F})$, $S_n(\mathbf{F})$, $A_n(\mathbf{F})$ be the algebra of $n \times n$ matrices, the space of $n \times n$ symmetric matrices and the space of $n \times n$ skew symmetric matrices with the entries in a field $\mathbf{F}$. Let $C \in M_n(\mathbf{R})$ and assume the decomposition (1.1). Let $\mathrm{spec}(C) = \{\lambda_1(C), ..., \lambda_n(C)\}$ be the spectrum of C. Then $\mathrm{spec}(S), \sqrt{-1}\,\mathrm{spec}(A) \subset \mathbf{R}$. Arrange the eigenvalues of S in the decreasing order:

$$\lambda_1(S) \geq \lambda_2(S) \geq ... \geq \lambda_n(S).$$

*University of Illinois at Chicago and Institute for Mathematics and its Applications.

As $x^T C x = x^T S x$, $x \in \mathbf{R}^n$ it follows that

$$\text{spec}(C) \cap \mathbf{R} \subset [\lambda_n(S), \lambda_1(S)].$$

Thus, the real eigenvalues of C can be viewed as a perturbation of the eigenvalues of S. More precisely, consider the coefficients of the characteristic polynomial of C as polynomials in the entries of S and A. As

$$\det(\lambda I - C) = \det(\lambda I - (S + A)) = \det(\lambda I - A^T) = \det(\lambda I - (S - A))$$

we deduce that the coefficients of the characteristic polynomial of C involve only even degree monomials in the entries of A. That is, the characteristic polynomial of C is a quadratic perturbation of the characteristic polynomial of S by the entries of A. The following theorem exhibits more precisely what happens to the spectrum of a real symmetric matrix S under the perturbation by a real skew symmetric matrix A:

THEOREM 2.1. *Let*

(2.2) $$C(z) = S + zA, \ S \in S_n(\mathbf{R}), \ A \in A_n(\mathbf{R}), \ z \in \mathbf{C}.$$

Then for a small enough disk $|z| < r$ the matrix $C(z)$ has n analytic eigenvalues $\lambda_1(z), ..., \lambda_n(z)$ with corresponding n independent analytic eigenvectors. Let $\lambda \in \text{spec}(S)$ be an eigenvalue of multiplicity k. Assume that the eigenvalues $\lambda_i(z), ..., \lambda_{i+k-1}(z)$ correspond to λ, i.e. $\lambda_j(0) = \lambda$, $j = i, ..., i + k - 1$. Suppose furthermore that k is odd. Then at least one of the eigenvalues $\lambda_j(z)$, $i \leq j \leq i + k - 1$ is real analytic. Furthermore, any real analytic eigenvalue $\lambda_l(z)$ is a function of z^2.

Proof. Let $z = \sqrt{-1}w$. Then $C(\sqrt{-1}w) = S + wA_1$, $A_1 = \sqrt{-1}A$. Thus, A_1 is a hermitian matrix. It then follows that $C(\sqrt{-1}w)$ is hermitian for real w. Rellich's theorem, e.g. [Kat], yields that for a small enough disk $|w| < r$ the matrix $C(\sqrt{-1}w)$ has n real analytic eigenvalues $\lambda_1(w), ..., \lambda_n(w)$ and corresponding n independent analytic eigenvectors. Replace $w = -\sqrt{-1}z$ to deduce the corresponding result for $C(z)$.

Assume that $\lambda_j(z)$ is real analytic. As $\lambda_j(\sqrt{-1}w)$ is real analytic we deduce that the power series of $\lambda_j(z)$ contain only even powers of z. Suppose finally that $\lambda \in \text{spec}(A)$ is of an odd multiplicity k. Assume that z is real. Then the eigenvalues of $C(z)$ are either real or come as conjugate complex pairs. As k is odd one of the eigenvalues $\lambda_i(z), ..., \lambda_{i+k-1}(z)$ is real analytic. ☐

In the space $\mathbf{R}^n$ we consider the standard inner product $(x, y) = x^T y$ and the l_2 norm $\|x\| = (x, x)^{\frac{1}{2}}$. The following lemma is well known, e.g. [G-V, §9.1.1, p'483] but we bring its proof for completeness:

LEMMA 2.3. *Let $A \in A_n(\mathbf{R})$. Assume that $e \in \mathbf{R}^n$, $\|e\| = 1$. Let $U = \text{span}\{e, Ae, ..., A^{k-1}e\} \subset \mathbf{R}^n$ be a cyclic subspace of A, that is $e, Ae, ..., A^{k-1}e$ are linearly independent and $A^k e \in U$. Assume that $k > 1$. Let $e^1 = e, ..., e^k$*

be the orthonormal vectors obtained by the Gram-Schmidt process from the vectors $e, Ae, ..., A^{k-1}e$:

(2.4) $\quad e^1 = e, \ e^2 = \dfrac{Ae}{\|Ae\|}, \ e^j = \dfrac{Ae^{j-1} - ((e^{j-2})^T Ae^{j-1})e^{j-2}}{\|Ae^{j-1} - ((e^{j-2})^T Ae^{j-1})e^{j-2}\|}, \ j = 3, ..., k.$

Then in the basis $e^1, ..., e^k$ the matrix A is represented by a skew symmetric $k \times k$ tridiagonal matrix $\hat{A} = (\hat{a}_{ij})_1^k$ whose lower nontrivial diagonal is given by:

(2.5)
$$\hat{a}_{21} = \alpha_1 = \|Ae\|, \ a_{(j+1)j} = \alpha_j = \|Ae^j + \alpha_{j-1}e^{j-1}\| = (e^{j+1})^T Ae^j, j = 2, ..., k-1.$$

Proof. As $x^T Ax = 0$ we deduce that Ae is orthogonal to e. Since $k > 1$ $Ae \neq 0$ hence $e_2 = \frac{Ae}{\|Ae\|}$. Assume by the induction hypothesis that (2.4) holds for $j = 1, ..., m-1$. Let e^m be given by (2.4). As $(e^{m-1})^T Ae^{m-1} = (e^{m-1})^T e^{m-2} = 0$ it follows that $(e^m)^T e^{m-1} = 0$. The formula for e^m yields that $(e^m)^T e^{m-2} = 0$. Consider $(e^m)^T e^j$ for $j < m - 2$. By the induction hypothesis $(e^{m-2})^T e^j = 0$. As $Ae^j \in \text{span}\{e^1, ..., e^{j+1}\}$ we deduce that $(e^{m-1})^T Ae^j = 0$. Hence, $(e^m)^T e^j = 0$. This shows that $e^1, ..., e^k$ form an orthonormal basis of U which is constructed by the Gramm-Schmidt process. The formula (2.4) yields that A is represented by a tridiagonal matrix $\hat{A}$ in the orthonormal basis $e^1, ..., e^k$. As A is skew symmetric so is $\hat{A}$. The equalities (2.5) follow straight forward from (2.4). $\square$

Let $\bar{\mathbf{R}} = \mathbf{R} \cup \{-\infty\}$. We adopt the natural convention $-\infty < x, \forall x \in \mathbf{R}$. Let $x, y \in \bar{\mathbf{R}}$. We then let $x \prec y$ ($y \succ x$) if either $x = y = -\infty$ or $y \in \mathbf{R}$ and $x < y$.

THEOREM 2.6. Let $S \in S_n(\mathbf{R})$, $A \in A_n(\mathbf{R})$. Assume furthermore that S is a rank one matrix with positive eigenvalue $\delta > 0$ and the corresponding eigenvector $Se = \delta e$, $\|e\| = 1$. If $Ae = 0$ then the $\text{spec}(S + A)$ is obtained from $\text{spec}(A)$ by removing one zero eigenvalue and replacing it by δ. Assume next that $Ae \neq 0$. Let $e^1 = e, ..., e^k$ be the orthonormal basis of the cyclic subspace of $U = \text{span}\{e, Ae, ..., A^{k-1}e\} \subset \mathbf{R}^n$ as in Lemma 2.3. Then in the basis $e^1, ..., e^k$ S is represented by the $k \times k$ diagonal matrix $D = (d_{ij})_1^k = \text{diag}\{\delta, 0, ..., 0\}$ and A by the $k \times k$ tridiagonal skew symmetric matrix $\hat{A} = (\hat{a}_{ij})_1^k$ given by (2.5). Furthermore, any real eigenvalue of $D + \hat{A}$ is nonzero. Moreover, all nonzero real eigenvalues of $S + A$ are the eigenvalues of $D + \hat{A}$. Set:

(2.7) $\qquad q_l(\lambda) = \det(\lambda I - (d_{ij} + \hat{a}_{ij})_1^l), \ l = 1, ..., k.$

Denote by ρ_l the maximal real root of $q_l(\lambda)$ for $l = 1, ..., k$. If l is even and $q_l(\lambda)$ does not have a real root we then let $\rho_l = -\infty$. Then

(2.8) $\qquad \rho_i > \rho_j, \ j = i + 1, ..., k; \ i = 1, 3, \ \rho_j \succ \rho_2, \ j = 3, ..., k.$

Suppose that $q_{2i-1}(\lambda)$, $i = 1, ..., \lfloor \frac{k+1}{2} \rfloor$ have exactly one real simple root ρ_{2i-1}, $i = 1, ..., \lfloor \frac{k+1}{2} \rfloor$. Then

(2.9)
$$\rho_{2i-1} > \rho_{2i+1}, \ i = 1, ..., \lfloor \frac{k-1}{2} \rfloor,$$
$$\rho_{2i} \prec \rho_{2i+2}, \ i = 1, ..., \lfloor \frac{k-2}{2} \rfloor,$$
$$\rho_{2i-1} > \rho_{2j}, \ i = 1, ..., \lfloor \frac{k+1}{2} \rfloor, \ j = 1, ..., \lfloor \frac{k}{2} \rfloor.$$

Proof. Let $U^\perp \subset \mathbf{R}^n$ be the orthogonal complement of U. As A is skew symmetric we deduce that $AU^\perp \subset U^\perp$. Clearly, $S \operatorname{span}(e)^\perp = 0$. Hence $SU \subset U$, $SU^\perp = 0$ and

$$\operatorname{spec}(S + A) = \operatorname{spec}(S + A|_U) \cup \operatorname{spec}(A|_{U^\perp}).$$

Assume first that $Ae = 0$. Then $U = \operatorname{span}(e)$ and the above equality yield that the $\operatorname{spec}(S + A)$ is obtained from the $\operatorname{spec}(A)$ by replacing a zero eigenvalue of A by δ.

We now assume that $Ae \neq 0$, that is, $k = \dim(U) > 1$. Suppose that $(S+A)x = 0$, $0 \neq x \in U$. As $0 = x^T(S + A)x = x^T Sx$ we deduce that $Sx = 0$. Thus $Ax = 0$. As U is a cyclic subspace of A we deduce that $U = \operatorname{span}(x)$ contrary to our assumption. Thus, the matrix $D + \hat{A}$ does not have a zero eigenvalue. We now show the inequalities (2.8).

Let

$$(2.10) \qquad p_l(\lambda) = \det(\lambda I - (d_{ij} + \hat{a}_{ij})_{k-l+1}^k), \ l = 1, ..., k.$$

Set $p_0(\lambda) = q_0(\lambda) = 1$. The following recursive formulas are established in a straightforward manner:

$$(2.11) \qquad \begin{aligned} p_k(\lambda) &= q_k(\lambda) = q_i(\lambda)p_{k-i}(\lambda) + q_{i-1}(\lambda)\alpha_i^2 p_{k-i-1}, \ i = 1, ..., k - 1, \\ p_i(\lambda) &= \lambda p_{i-1}(\lambda) + \alpha_{k-i+1}^2 p_{i-2}, \ i = 2, ..., k - 1. \end{aligned}$$

Note that $p_i(\lambda) > 0$, $\lambda > 0$, $i = 1, ..., k - 1$. The formulas (2.11) yield

$$q_k(\lambda) = (\lambda - \delta)p_{k-1}(\lambda) + \alpha_1^2 p_{k-2}(\lambda).$$

Hence, $q_k(\lambda) \geq \alpha_1^2 p_{k-2}(\lambda) > 0$ for $\lambda \geq \delta (> 0)$. Therefore, $\rho_k < \rho_1 = \delta$. By considering the matrix $(d_{ij} + \hat{a}_{ij})_1^l$ we deduce that $\rho_l < \rho_1$, $l = 2, ..., k$. Consider ρ_2. If $\rho_2 = -\infty$ we trivially have that $\rho_j \succ \rho_2$, $j = 3, ..., k$. Assume now that $\rho_2 \in \mathbf{R}$. Our arguments yield that $0 < \rho_2 < \delta$. The first set of the equalities of (2.11) imply:

$$q_k(\lambda) = q_2(\lambda)p_{k-2}(\lambda) + (\lambda - \delta)\alpha_2^2 p_{k-3}(\lambda).$$

Thus $q_k(\rho_2) = (\delta - \rho_2)\alpha_2^2 p_{k-3}(\rho_2) < 0$. Hence, q_k has a real root and $\rho_k > \rho_2$. The arguments above show that each q_l, $l = 3, ..., k$ has a real root and $\rho_l > \rho_2$, $l = 3, ..., k$. Assume finally that all q_{2i-1} have one real simple root. We prove the inequalities (2.9) by the induction on k. Suppose (2.9) holds for $k = l$. Let $k = l + 1$. The first set of equalities of (2.11) yield that

$$(2.12) \qquad q_{l+1} = \lambda q_l + \alpha_l^2 q_{l-1}.$$

Suppose first that l is even. Then the induction hypothesis yield that $\rho_l < \rho_{l-1}$. It then follows that $q_{l+1}(\lambda) > 0$ for $\lambda \geq \rho_{l-1}$. Hence, $\rho_{l+1} < \rho_{l-1}$. If $\rho_l = -\infty$ we obviously have the inequality $\rho_l < \rho_{l+1}$. If $0 < \rho_l$ then (2.12) yields $q_{l+1}(\rho_l) = \alpha_l^2 q_{l-1}(\rho_l) < 0$. The last inequality follows from the inequality $\rho_l < \rho_{l-1}$ and the assumption that ρ_{l-1} is the only real root of q_{l-1} which is also simple. Therefore $\rho_{l+1} > \rho_l$. Similar arguments apply in the case l is odd. $\square$

The two roots of:

$$(2.13) \qquad q_2(\lambda) = \lambda^2 - \delta\lambda + \alpha_1^2$$

are given by

$$(2.14) \qquad \tau_1 = \frac{\delta - \sqrt{\delta^2 - 4\alpha_1^2}}{2}, \ \tau_2 = \frac{\delta + \sqrt{\delta^2 - 4\alpha_1^2}}{2}.$$

Thus, if $\delta^2 \geq 4\alpha_1^2$ then $\rho_2 = \tau_2 \geq \frac{\delta}{2}$. Then Theorem 2.6 yields:

$$(2.15) \qquad \rho_i > \rho_2 = \frac{\delta + \sqrt{\delta^2 - 4\alpha_1^2}}{2}, \ i = 3, ...k, \ \delta^2 \geq 4\alpha_1^2.$$

In particular, the matrix $S + A$ has a positive real eigenvalue which is not less than ρ_2. This last statement can be deduced from the recent inequality of [Kir] for hypertournament matrices. We adopt Kirkland's proof to state a more precise result about the spectrum of almost skew symmetric matrices in the case $\delta^2 \geq 4\alpha_1^2$.

THEOREM 2.16. *Let $S \in S_n(\mathbf{R})$, $A \in A_n(\mathbf{R})$. Assume furthermore that S is a rank one matrix with positive eigenvalue $\delta > 0$ and the corresponding eigenvector $Se = \delta e$, $\|e\| = 1$. Set $\alpha_1 = \|Ae\|$. Assume that $\delta^2 \geq 4\alpha_1^2$. Then $S + A$ has one real eigenvalue λ satisfying the inequality:*

$$(2.17) \qquad \lambda \geq \frac{\delta + \sqrt{\delta^2 - 4\alpha_1^2}}{2}.$$

Furthermore, all other $n - 1$ eigenvalues of $A + S$ satisfy the inequality:

$$(2.18) \qquad \Re(\lambda) \leq \frac{\delta - \sqrt{\delta^2 - 4\alpha_1^2}}{2}.$$

Proof. Using a continuity argument it is enough to consider the case $\delta^2 > 4\alpha_1^2$. Let

$$(2.19) \qquad \begin{aligned} &H = S - \frac{\delta}{2}I, \ M_t = S + A - tI, \\ &P_t = M_t H + H M_t^T, \ \frac{\delta - \sqrt{\delta^2 - 4\alpha_1^2}}{2} < t < \frac{\delta + \sqrt{\delta^2 - 4\alpha_1^2}}{2}. \end{aligned}$$

Then $P_t = N_t \oplus \frac{t\delta}{2} I_{n-2}$. A straightforward calculation shows that the 2×2 matrix N_t is positive definite for the values of t prescribed in (2.19). Thus, P_t is positive definite. The result of Ostrowski and Schneider [O-S] claims that M_t and H have the same inertia. That is, M_t has exactly one eigenvalue with positive real part and all other $n-1$ eigenvalues have negative real part. As M_t is a real matrix we deduce that the unique eigenvalue with a positive real part must be real. To deduce (2.17) let t tend to $\rho_2 = \tau_2$ from below and to deduce (2.18) let t tend to τ_1 from above. $\square$

THEOREM 2.20. *For $\delta > 0$, $\alpha_1^2, \alpha_2^2 \geq 0$ Let $\rho_3(\alpha_1^2, \alpha_2^2)$ be the largest real root of*

$$(2.21) \qquad q_3(\lambda) = \lambda^3 - \delta\lambda^2 + \lambda(\alpha_1^2 + \alpha_2^2) - \delta\alpha_2^2.$$

For a fixed value $\alpha_1^2 > 0$, $4\alpha_1^2 \leq \delta^2$, ρ_3 is a strictly increasing function of α_2^2 such that

$$(2.22) \qquad \rho(\alpha_1^2, 0) = \rho_2(\alpha_1^2) = \frac{\delta + \sqrt{\delta^2 - 4\alpha_1^2}}{2}, \ \rho_2(\alpha_2^2, \infty) = \delta, \ \delta^2 \geq 4\alpha_1^2 > 0.$$

For a fixed value of $\alpha_2 > 0$ ρ_3 is a strictly decreasing function of α_1 on the interval $[0, \frac{\delta}{2}]$ such that $\rho_3(0, \alpha_2^2) = \delta$. For a fixed value $\alpha_2^2 \geq \frac{\delta^2}{3}$ ρ_3 is a strictly decreasing function of α_1^2 such that:

$$(2.23) \qquad \rho_3(0, \alpha_2^2) = \delta, \ \rho_3(\infty, \alpha_2^2) = 0,$$

Assume finally that $\alpha_1^2, \alpha_2^2 > 0$ and $4\alpha_1^2 \leq \delta^2$. Then

$$(2.24) \qquad 0 < \rho_3 - \rho_2 < \frac{\alpha_2^2(\delta - \rho_2)}{(2\rho_2 - \delta)\rho_2 + \alpha_2^2}.$$

Let $S \in S_n(\mathbf{R})$, $A \in A_n(\mathbf{R})$. Assume furthermore that S is a rank one matrix with positive eigenvalue $\delta > 0$ and the corresponding eigenvector $Se = \delta e$, $\|e\| = 1$. Assume that $A^2 e \neq 0$. Set

$$(2.25) \qquad \alpha_1^2 = \|Ae\|^2, \ \alpha_2^2 = \|\frac{A^2 e}{\alpha_1} + Ae\|^2 = \frac{\|A^2 e\|^2}{\|Ae\|^2} - \|Ae\|^2.$$

Then $S + A$ has exactly one positive simple eigenvalue λ satisfying

$$(2.26) \qquad \rho_2(\alpha_1^2) < \lambda \leq \rho_3(\alpha_1^2, \alpha_2^2).$$

Proof. Using the definitions (2.13) and (2.2) we deduce in a straightforward manner the equality

$$(2.27) \qquad q_3(\lambda) = \lambda q_2(\lambda) + \alpha_2^2(\lambda - \delta),$$

which is a special case of (2.12). Then ρ_3 is the maximal real solution of

$$(2.28) \qquad \frac{\lambda q_2(\lambda)}{\delta - \lambda} = \alpha_2^2.$$

Note that $q_2(\lambda)$ is an increasing function of λ for $\lambda \geq \frac{\delta}{2}$. Also, for $\delta^2 \geq 4\alpha_1^2 > 0, \alpha_2^2 > 0$ the argument in the proof of Theorem 2.6 yields that

$$(2.29) \qquad \delta > \rho_3(\alpha_1^2, \alpha_2^2) > \rho_2(\alpha_1^2) \geq \frac{\delta}{2}, \ 0 < 4\alpha_1^2 \leq \delta^2, \ 0 < \alpha_2^2.$$

It then follows that for a fixed α_1, $\delta^2 \geq 4\alpha_1^2 > 0$ ρ_3 is an increasing function in $\alpha_2^2 > 0$. Thus, when $\alpha_2^2 = 0$ we obtain the first equality in (2.23). As $\rho_3 < \delta$ we deduce that if $\alpha_2^2 \to \infty$ then $\rho_3 \to \delta$. This proves the second equality of (2.23).

Fix the value of $\alpha_2 > 0$. Then for $\alpha_1 \in (0, \frac{\delta}{2})$ we know that $\rho_3 > \rho_2 > \frac{\delta}{2}$. Note that the equalities (2.27) and (2.13) directly yield that for a fixed $\lambda > 0$, q_3 is an increasing function of α_1^2. Fix the values of α_1^2, α_2^2. Then q_3 is a strictly increasing function of λ for $\lambda > \frac{\delta}{2}$. These arguments imply straightforward the $\rho_3(\alpha_1, \alpha_2)$ is a strictly decreasing function of $\alpha_1 \in (0, \frac{\delta}{2})$. Assume that $\alpha_2^2 \geq \frac{\delta^2}{3}$. A straightforward calculation shows that $q_3' \geq 0$ for all λ. The previous arguments yield that ρ_3 is a strictly decreasing function of $\alpha_1 > 0$. The equalities (2.23) are straightforward.

We now prove (2.24). Assume that $0 < \alpha_1 \leq \frac{\delta}{2}$, $0 < \alpha_2$. Then the inequality $\rho_2 < \rho_3$ is a consequence of (2.8). Clearly

$$q_2(\rho_3) = q_2'(\rho_2)(\rho_3 - \rho_2) + (\rho_3 - \rho_2)^2 > q_2'(\rho_2)(\rho_3 - \rho_2).$$

Since ρ_3 satisfies the equation (2.28) from the above inequality and the fact $q_2'(\rho_2) \geq 0$ we deduce:

$$\alpha_2^2 > \frac{\rho_3 q_2'(\rho_2)(\rho_3 - \rho_2)}{\delta - \rho_3} \geq \frac{\rho_2 q_2'(\rho_2)(\rho_3 - \rho_2)}{\delta - \rho_3}.$$

The inequality (2.24) is equivalent to the observation that α_2^2 is greater than the last term in the above inequality. □

THEOREM 2.30. Let $S \in S_n(\mathbf{R})$, $A \in A_n(\mathbf{R})$. Assume furthermore that S is a rank one matrix with positive eigenvalue $\delta > 0$ and the corresponding eigenvector $Se = \delta e$, $\|e\| = 1$. Assume that that $Ae \neq 0$. Let

$$e^1 = e, ..., e^k, D, \hat{A}, q_1, ..., q_k, \rho_1, ..., \rho_k$$

be defined as in Theorem 2.6. Assume that $\alpha_1 = \|Ae\| \leq \frac{\delta}{2}$. Then all $\rho_1, ..., \rho_k$ are positive numbers which satisfy the strict inequalities:

(2.31)
$$\rho_{2i-1} > \rho_{2i+1}, \ i = 1, ..., \left\lfloor \frac{k-1}{2} \right\rfloor,$$

$$\rho_{2i} < \rho_{2i+2}, \ i = 1, ..., \left\lfloor \frac{k-2}{2} \right\rfloor,$$

$$\rho_{2i-1} > \rho_{2j}, \ i = 1, ..., \left\lfloor \frac{k+1}{2} \right\rfloor, \ j = 1, ..., \left\lfloor \frac{k}{2} \right\rfloor.$$

Proof. We prove (2.31) by the induction on k. For $k = 1, 2, 3$ we showed the validity of (2.31). Assume that (2.31) holds for $k = l \geq 3$. Let $k = l + 1$. Suppose first that l is even. The induction hypothesis yields that $\rho_l < \rho_{l-1}$. Use (2.12) to deduce that $q_{l+1}(\lambda) > 0$ for $\lambda \geq \rho_{l-1}$. Hence $\rho_{l+1} < \rho_{l-1}$. Combine Theorem 2.16, (2.12) and the induction hypothesis $\rho_{l-1} > \rho_l > \rho_2$ to deduce that $q_{l+1}(\rho_l) = \alpha_l^2 q_{l-1}(\rho_l) < 0$. Therefore $\rho_{l+1} > \rho_l$. Similar arguments apply in the case l is odd. □

Note that under the assumptions of Theorem 2.30 ρ_k is the maximal real eigenvalue of the matrix $S + A$.

3. Tournament matrices. A *tournament matrix* is a square $(0,1)$ matrix T such that $T + T^T = J - I$ where J is the all ones matrix. These matrices have many interesting properties, see [Moo] and the references therein. The spectral properties of these matrices were studied by quite a few numbers of authors, see for example [M-P] and [Kir] and the references therein. We first note that any tournament matrix is an almost skew symmetric matrix minus $\frac{I}{2}$

$$(3.1) \qquad T = C - \frac{I}{2}, \; C = \frac{J}{2} + A, \; A = \frac{T - T^T}{2}, \; J = uu^T, \; u = (1, ..., 1)^T.$$

To estimate the spectral radius $\rho(T)$ of T it is enough to estimate the maximal real eigenvalue of an almost skew symmetric matrix C. In the notation of Section 2

$$(3.2) \qquad \delta = \frac{n}{2}, \; e = \frac{u}{\sqrt{n}} = \frac{(1, ..., 1)^T}{\sqrt{n}}.$$

Set

$$(3.3) \qquad b = (b_1, ..., b_n)^T = Tu - T^T u \in \mathbf{Z}^n.$$

That is b_i is the number of wins minus the number of losses of the player number i. We call b the balance vector. Use (3.1), (2.4) and (2.25) to deduce:

$$(3.4) \qquad \alpha_1^2 = \frac{b^T b}{4n}, \; \alpha_2^2 = -\frac{b^T A^2 b}{b^T b} - \frac{b^T b}{4n}.$$

We assume that $\alpha_1 = \alpha_2 = 0$ if $b = 0$. Theorems 2.6 and 2.30 yield:

THEOREM 3.5. *Let T be an $n \times n$ tournament matrix. Assume $(3.1) - (3.5)$. Let $\rho_3(\alpha_1^2, \alpha_2^2)$ be the biggest real root of (2.21). Then*

$$(3.6) \qquad \rho(T) \le \rho_3(\alpha_1^2, \alpha_2^2) - \frac{1}{2}.$$

Suppose furthermore that $4b^T b \le n^3$. Let $\rho_2(\alpha_1^2)$ be the maximal root of (2.13). Then

$$(3.7) \qquad \rho(T) \ge \rho_2(\alpha_1^2) - \frac{1}{2}$$

and equality holds iff either $b = 0$ or $b \ne 0, \alpha_2 = 0$.

The inequality (3.7) is due to Kirkland [Kir]. He also gives the condition for equality stated in a different form.

The difficulty in using the inequality (3.6) is due to the fact that ρ_3 is a solution to the cubic equation. We now bring an upper bound on the spectral tournament matrices using Ostrowski's theorem. Let $B = (b_{ij})_1^n$ be a nonnegative matrix. Set

$$(3.8)$$
$$r_i(B) = \sum_{j=1}^n b_{ij}, c_i(B) = r_i(B^T), \; \hat{r}_i(B) = r_i(B) - b_{ii}, \; \hat{c}_i(B) = \hat{r}_i(B^T) \; i = 1, ..., n.$$

Ostrowski's theorem [Ost], see also [Bru] for generalization of Ostrowski's theorem, claims that $\mathrm{spec}(B)$ - the spectrum of B is contained in the following complex disks:

$$|z - b_{ii}| \leq \hat{r}_i(B)^t \hat{c}_i(B)^{1-t}, \; i = 1, ..., n$$

for any $0 \leq t \leq 1$. In particular:

(3.9) $\qquad \rho(B) \leq \mu_t(B), \; \mu_t(B) = \max_{1 \leq i \leq n} b_{ii} + \hat{r}_i(B)^t \hat{c}_i(B)^{1-t}, \; 0 \leq t \leq 1.$

COROLLARY 3.10. *Let T be an $n \times n$ tournament matrix. Then*

(3.11) $$\rho(T) \leq \mu_{\frac{1}{2}}(T).$$

Assume that n is even. Let ρ_n^* be the maximal spectral radius of all $n \times n$ tournament matrices. By abusing the notation we let $\rho^* = \rho_n^*$. A conjecture due to Brualdi-Li [B-L] gives a "natural" candidade for the tournament matrix T^*, $\rho(T^*) = \rho^*$.

(3.12)
$$\begin{aligned} & Q = (Q_{ij})_1^2, \; Q_{11} = Q_{22} = V, \; Q_{12} = I + V^T, \; Q_{21} = V^T, \\ & V = (v_{ij})_1^{\frac{n}{2}}, \; v_{ij} = 1 \; for \; 1 \leq i < j \leq \frac{n}{2}, \; v_{ij} = 0 \; for \; 1 \leq j \leq i \leq \frac{n}{2}. \end{aligned}$$

In particular, Q is almost regular. Recall that a tournament matrix T is called almost regular if $|r_i(T) - \frac{n-1}{2}| = \frac{1}{2}, \; i = 1, ..., n$. A weaker version of Brualdi-Li conjecture is that any extremal matrix T^* is almost regular. We now study these conjectures.

THEOREM 3.13. *Let T be a tournament matrix of an even order n. Then*

(3.14) $$\rho(T) \leq \sqrt{\frac{n}{2}(\frac{n}{2} - 1)}.$$

Assume furthermore that T satisfies the condition

(3.15) $$|r_i(T) - \frac{n-1}{2}| > \frac{1}{2}, \; i = 1, ..., n.$$

Then

(3.16) $$\rho(T) \leq \sqrt{(\frac{n}{2} + 1)(\frac{n}{2} - 2)}.$$

In particular, any tournament matrix which satisfies (3.15) satisfies the inequality

(3.17) $$\rho(T) < \frac{n-1}{2} - \frac{1}{n + \sqrt{n^2 - 4}}.$$

Proof. Note that for any tournament matrix of an even order n we have the inequality $r_i(T)c_i(T) \leq \frac{n}{2}(\frac{n}{2} - 1)$. Hence, (3.11) yields (3.14). The assumption (3.15) implies that $r_i(T)c_i(T) \leq (\frac{n}{2} + 1)(\frac{n}{2} - 2), i = 1, ..., n$. Thus, (3.11) implies (3.16). A straightforward calculation shows that the right-hand side of (3.16) is strictly less than the right-hand side of (3.17). □

The following lemma is due to Kirkland:

LEMMA 3.18. *Let $T = (T_{ij})_1^{2m}$, $T_{ij} \in M_m(\mathbf{R})$, $i, j = 1, 2$ be an almost regular tournament matrix of an even order. Assume furthermore that the row sum of each of the first m rows of T is m and $T_{11} = T_{22} = V$ where V is given by (3.12). Then T is equal to the Brualdi-Li matrix Q given by (3.12).*

Proof. As T is an almost regular tournament matrix with $r_i(T) = m$, $i = 1, ..., m$, we deduce that $c_{m+i}(T) = 2m - 1 - r_{m+i}(T) = m$, $i = 1, ..., m$. Clearly, $r_i(V) = c_{m-i+1}(V) = m - i$, $i = 1, ..., m$. The equalities $r_i(T) = c_{m+i}(T) = m$, $i = 1, ..., m$, yield

(3.19) $r_i(T_{12}) = m - r_i(V) = i$, $c_i(T_{12}) = m - c_i(V) = m - i + 1$, $i = 1, ..., m$.

We claim that the above equations have a unique $0-1$ matrix solution $T_{12} = I + V^T$. For $m = 1, 2$ the claim is trivial. Assume that the claim holds for $m \leq k - 1$. Let $m = k$. As T_{12} is a $0 - 1$ matrix it follows that the first column and the last row of T_{12} consists of all ones. The equality $r_1(T_{12}) = 1$ yields that the only nonzero entry on the first row of T_{12} is the entry $(1, 1)$. As $c_k(T_{12}) = 1$ we deduce that the only nonzero entry on the last column of T_{12} is the entry (k, k). Delete the first and the last rows and columns of T_{12} and use the induction hypothesis to show that the equalities (3.19) have a unique $0 - 1$ matrix solution $T_{12} = I + V^T$. As T is a tournament we deduce that $T_{21} = V^T$. □

THEOREM 3.20. *Let T be an almost regular tournament matrix of an even order n. Assume (3.1) − (3.4). Then $\alpha_1 = \frac{1}{2}$. In particular, Kirkland's inequality holds [Kir]:*

(3.21) $$\rho(T) \geq \rho_2(\frac{1}{4}) - \frac{1}{2} = \frac{n-1}{2} - \frac{1}{n + \sqrt{n^2 - 4}}.$$

Furthermore

(3.22) $$\alpha_2^2 \leq \frac{n^2 - 4}{12}.$$

Equality holds iff T is permutationally similar to the Brualdi-Li matrix given by (3.12). In particular, any almost regular tournament matrix T of an even order satisfies the inequality

(3.23) $$\rho(T) \leq \rho_3(\frac{1}{4}, \frac{n^2 - 4}{12}) - \frac{1}{2}$$

which implies

(3.24) $$\rho(T) < \frac{n-1}{2} - \frac{3}{8n}.$$

Proof. Let $n = 2m$. Then the balance vector b has m entries equal to 1 and m entries equal to -1. Use the first part of (3.4) to deduce the equality $\alpha_1^2 = \frac{1}{4}$. According to Theorem 2.20 we have that $\rho(\frac{1}{4}, \alpha_2^2)$ is an increasing function function of α_2^2. (Here $\delta = \frac{n}{2}$.) Use (3.4) to deduce that the maximal value of α_2^2 is achieved for the almost regular tournament matrices for which $\|Ab\|$ is maximal. Since $Jb = 0$ we get:

$$
\begin{aligned}
\text{(3.25)} \quad \|Ab\|^2 &= \|(A + \frac{J}{2})b\|^2 = \|(T + \frac{I}{2})b\|^2 = b^T T^T T b + \frac{b^T b}{4} + \frac{b^T(T + T^T)b}{2} = \\
& b^T T^T T b + \frac{n}{4} + \frac{b^T(J - I)b}{2} = b^T T^T T b - \frac{n}{4}.
\end{aligned}
$$

Thus, it is enough to consider almost regular tournament matrices which maximize $\|Tb\|^2$. Rename the indices $1, ..., 2m$ so that the first m coordinates of b are equal to 1 and the last m coordinates of b are equal to -1. Let $T = (t_{ij})_1^n = (T_{ij})_1^2$, $T_{ij} \in M_m(\mathbf{R})$. Assume that s_i, t_i are the $i - th$ row sum of T_{11}, T_{22} respectively, for $i = 1, ...m$. It then follows that

$$
b^T T^T T b = \sum_1^m (2s_i - m)^2 + (2t_i - (m - 1))^2,
$$

$$
s_i, t_i \in \mathbf{Z}, \ 0 \le s_i, t_i \le m - 1, \ \sum_1^m s_i = \sum_1^m t_i = \frac{m(m-1)}{2}.
$$

We next observe that if $x \le y$ are two integers then $x^2 + y^2 < (x - 2)^2 + (y + 2)^2$. We now maximize the expression

$$
\text{(3.26)} \qquad \sum_1^m (2s_i - m)^2 + (2t_i - (m - 1))^2.
$$

on all tournament matrices of an order $n = 2m$. Let $T = (t_{ij})_1^n$. Assume that $t_{ij} = 1$, $1 \le i \ne j \le m$. Then we may change the value of (3.26) by letting $t_{ij} = 0$. This is equivalent to replacing s_i, s_j by $s_i - 1$ and $s_j + 1$ and leaving all other s_k and t_k unchanged. Assume that T maximizes (3.26). We claim that after renaming the first m vertices if necessary we have the equality: $T_{11} = V$. In particular $s_i = m - i$, $i = 1, ..., m$. Assume that we renamed the indices so that $s_1 \ge s_2 \ge ... \ge s_m$. Suppose to the contrary that $s_m > 0$, that is, $t_{mj} = 1, 1 \le j < m$. Then replace s_m, s_j by $s_m - 1, s_j + 1$ to contradict the optimality of T. Thus, the last row of T_{11} is a zero row and the first $m - 1$ entries of the last column of T_{11} are equal to zero. It then follows that $s_i - 1$, $i = 1, ..., m - 1$ are the $i - th$ row sum of $(t_{ij})_1^{m-1}$. It now follows that $s_{m-1} - 1 = 0$. Thus, the last row of the matrix $(t_{ij})_1^{m-1}$ is equal to zero. Continuing in the same manner we deduce that $T_{11} = V$. Assume that we have renamed the indices $m + 1, ..., 2m$ so that $t_1 \ge ... \ge t_m$. Our arguments yield that $T_{22} = V$. In particular, Q maximize the expression (3.26).

Assume now that T is an almost regular matrix with $r_i(T) = m$, $i = 1, ..., m$, which maximizes (3.26). By renaming the indices in the sets $\{1, ..., m\}$ and $\{m + 1, ..., 2m\}$ if necessary we deduce from the arguments above that $T_{11} = T_{22} = V$. Lemma 3.18 yields that $T = Q$.

In what follows we assume that T is an almost regular tournament matrix. Using the standard formula

$$\sum_1^k i^2 = \frac{(2k+1)(k+1)k}{6}$$

we deduce that

$$\|Tb\|^2 \le \frac{(2m^2+1)m}{3}.$$

From (3.25), (3.4) and the above inequality we obtain the inequality (3.22) and the equality holds iff T is permutationally similar to the Brualdi-Li matrix. Theorem 2.20 yields the inequality (3.23). To deduce (3.24) let $r_n = \frac{n}{2} - \frac{3}{8n}$. A straightforward calculation shows that

$$\frac{r_n q_2(r_n)}{\delta - r_n} > \frac{n^2-4}{12}, \ \delta = \frac{n}{2}, \ \alpha_1^2 = \frac{1}{4}.$$

The arguments of the proof of Theorem 2.20 yield that

(3.27) $$\rho_3\left(\frac{1}{4}, \frac{n^2-4}{12}\right) < \frac{n}{2} - \frac{3}{8n}.$$

4. A generalization of Ostrowski's inequality and applications. In this section we let $h = (1, ..., 1)^T$.

THEOREM 4.1. *Let B be an $n \times n$ nonnegative irreducible matrix matrix. Set*

$$Bu = \rho(B)u, \ B^T v = \rho(B)v,$$

(4.2) $$u = (u_1, ..., u_n)^T, \ v = (v_1, ..., v_n)^T, \ 0 < u_i, v_i, \ i = 1, ..., n, \ \sum_1^n u_i v_i = 1.$$

Let $0 \le s \le \frac{1}{n}$, $0 \le t \le 1$ be given. Suppose that

(4.3) $$s \le u_i v_i, \ i = 1, ..., n.$$

Then

(4.4) $$\log \rho(B) \le s\left(\sum_1^n t \log r_i(B) + (1-t)\log c_i(B)\right)$$
$$+ (1-ns)\log(\max_{1 \le i \le n} t \log r_i(B) + (1-t)\log c_i(B)).$$

Assume that $0 < t < 1$ and B is fully indecomposable. (PBQ is irreducible for any permutation matrices P, Q.) Then the equality sign holds iff $\frac{B}{r_1(B)}$ is doubly stochastic.

Proof. The characterization of Friedland-Karlin [F-K] claims that

(4.5) $$\log \rho(B) = \min_{x=(x_1,...,x_n), 0 < x_i, i=1,...,n} \sum_1^n u_i v_i \log \frac{(Bx)_i}{x_i}.$$

By choosing $x = h$ we deduce that

$$\log \rho(B) \le \sum_{1}^{n} u_i v_i \log r_i(B).$$

Replace B by B^T to obtain

$$\log \rho(B) \le \sum_{1}^{n} u_i v_i \log c_i(B).$$

Multiply the first inequality by t the second inequality by $1 - t$ and use (4.3) to get (4.4). Suppose that B is fully indecomposable. Then the minimum in (4.5) is achieved only for the eigenvector $x = \alpha u, \alpha > 0$. (Contrary to the claim in [F-K], if B is irreducible but not fully indecomposable then the equality sign in (4.5) can be achieved for vectors x which may not be the eigenvector of B. See [Fri].) As $0 < t < 1$, the equality sign in (4.4) implies that h is the left and the right eigenvector of B. That is $\frac{B}{r_1(B)}$ is a doubly stochastic matrix. In that case we obviously have the equality sign in (4.4). □

Remark 4.6. We view (4.4) as a generalization of (3.9). Indeed, for any non-negative irreducible matrix B (4.4) yields the inequality

$$(4.7) \qquad \rho(B) \le \nu_t(B), \ \nu_t(B) = \max_{1 \le i \le n} r_i(B)^t c_i(B)^{1-t}, \ 0 \le t \le 1.$$

Use a continuity argument to deduce that the above inequality holds for any non-negative B. It is not hard to show that (3.9) implies (4.7). Of course, in the case B has a zero diagonal (3.9) is equivalent to (4.7).

Remark 4.8. Clearly, if B is reducible we might have the equality sign in the inequality $\rho(B) \le \nu_{\frac{1}{2}}(B)$ without the requirement that $\frac{A}{r_1(B)}$ is a doubly stochastic matrix. This also may be the case for an irreducible B which is not completely irreducible. Indeed, assume that C is a nonnegative cyclic matrix of even order - $n = 2m$. Suppose that all odd row sums of C and all even column sums of C are equal to $\alpha > 0$ and all even row sums and odd column row sums are equal to $\beta > 0$. As $C^n = \nu_{\frac{1}{2}}(B)^n I$ we deduce that we have the equality sign in $\rho(B) = \nu_{\frac{1}{2}}(B)$. Clearly, for $\alpha \ne \beta$ the matrix $\frac{B}{\alpha}$ is not doubly stochastic.

A generalized tournament matrix T is a nonnegative square matrix satisfying

$$(4.9) \qquad\qquad T + T^T = J - I.$$

See [M-P]. We now apply (4.4) to give an upper bound for the spectral radius of generalized tournament matrices. According to (3.1) any generalized tournament matrix is an almost skew symmetric matrix minus $\frac{I}{2}$. Thus, we can use the inequalities (3.6) − (3.7) for $\rho(T)$. As we pointed out before there is an intrinsic difficulty in applying (3.6) as $\rho_3(\alpha_1^2, \alpha_2^2)$ is a solution of a cubic equation. In what follows we shall deduce upper bounds for $\rho(T)$ using (4.4) provided that $\rho(T)$ is sufficiently big. This condition can be verified by using (3.7).

Let T be a generalized tournament matrix and assume that

(4.10) $\quad Tu = \rho(T)u, \ u = (u_1, ..., u_n)^T, \ u_i \geq 0, \ i = 1, ..., n, \ u^T u = 1, \ \alpha = \dfrac{\sum_1^n u_i}{n}.$

Combine (4.9) and (4.10) to deduce

(4.11) $\quad\quad\quad T^T u = Ju - (\rho(T) + 1)u = n\alpha h - (\rho(T) + 1)u.$

Multiply (4.11) by u^T to deduce:

(4.12) $\quad\quad\quad\quad\quad n^2\alpha^2 = 2\rho(T) + 1.$

It then follows

(4.13)
$$\epsilon = (\epsilon_1, ..., \epsilon_n)^T = u - \alpha h,$$
$$\epsilon^T\epsilon = u^T u - 2u^T\alpha h + \alpha^2 h^T h = 1 - n\alpha^2 = 1 - \frac{2\rho(T) + 1}{n}.$$

THEOREM 4.14. *Let T be an $n \times n$ generalized tournament matrix. Assume that*

(4.15) $$\rho(T) > \frac{n - 2}{2} + \frac{1}{2(n + 1)}.$$

Then T is an irreducible matrix. Furthermore

(4.16)
$$\log \rho(T) \leq \left(\frac{\sqrt{2\rho(T) + 1}}{n} - \sqrt{1 - \frac{2\rho(T) + 1}{n}} \right)^2 \frac{\sum_1^n \log r_i(T)c_i(T)}{2} +$$
$$\left(\left(1 - n\left(\frac{\sqrt{2\rho(T) + 1}}{n} - \sqrt{1 - \frac{2\rho(T) + 1}{n}} \right)^2 \right) \right) \log \nu_{\frac{1}{2}}(T).$$

Proof. In view of (4.12) and (4.13) we deduce that (4.15) is equivalent to the inequality $\alpha^2 > \epsilon^T\epsilon$. By the definition $\alpha = \frac{u^T h}{n} > 0$. Hence, the inequality (4.15) yields that $0 < u_i, \ i = 1, ..., n$. Let

(4.17) $\quad\quad T^T w = \rho(T)w, \ w = (w_1, ..., w_n)^T, \ 0 \leq w_i, \ i = 1, ..., n, \ w^T w = 1.$

Using the arguments for the eigenvector u we deduce:

(4.18) $\quad\quad \alpha = \frac{w^T h}{n}, \ \delta = (\delta_1, ..., \delta_n)^T = w - \alpha h, \ \delta^T\delta = 1 - \frac{2\rho(T) + 1}{n}.$

Thus, the assumption (4.15) yields that w is a vector with strictly positive coordinates. It is well known that if a nonnegative matrix with positive spectral radius has strictly positive left and right eigenvectors then T is a direct sum of irreducible matrices. The assumption (4.9) implies that this sum reduces only to one component, i.e. T is irreducible. We next note that $0 < w^T u \leq (w^T w u^T u)^{\frac{1}{2}} = 1$. Also, the vector v satisfying (4.2) is given by:

(4.19) $$v = \frac{1}{w^T u} w.$$

Therefore

(4.20)
$$u_i v_i \geq (\alpha - (\epsilon^T\epsilon)^{\frac{1}{2}})(\alpha - (\delta^T\delta)^{\frac{1}{2}})$$
$$= \left(\frac{\sqrt{2\rho(T) + 1}}{n} - \sqrt{1 - \frac{2\rho(T) + 1}{n}} \right)^2 = s, \ i = 1, ..., n.$$

Use (4.4) with $t = \frac{1}{2}$ and the above s to deduce (4.16). □

THEOREM 4.21. *Let T be an $n \times n$ generalized tournament matrix. Assume that*

$$(4.22) \qquad \sqrt{2\rho(T)+1} = \sqrt{n}\left(1 - \frac{\tau^2}{8n}\right), \quad 0 \le \tau \le 1.$$

Then

$$(4.23) \qquad \log \rho(T) \le (1-\tau)\log\left(\prod_1^n r_i(T)c_i(T)\right)^{\frac{1}{2n}} + \tau \log \nu_{\frac{1}{2}}(T).$$

Proof. The condition (4.22) implies:

$$(4.24) \qquad \frac{2\rho(T)+1}{n} = 1 - \frac{\tau^2}{4n} + \frac{\tau^4}{64n^2}.$$

Hence

$$(4.25) \qquad \sqrt{1 - \frac{2\rho(T)+1}{n}} \le \frac{\tau}{2\sqrt{n}}.$$

Therefore:

$$(4.26) \qquad \left(\sqrt{\frac{2\rho(T)+1}{n}} - \sqrt{1 - \frac{2\rho(T)+1}{n}}\right)^2 \ge \frac{1}{n}\left(1 - \frac{\tau}{2} - \frac{\tau^2}{8n}\right)^2 \ge \frac{1-\tau}{n}.$$

Use the above inequality in (4.16) to obtain (4.23). □

For an $n \times n$ generalized tournament matrix T let:

$$(4.27) \qquad \begin{aligned} &\bar{r}_i(T) = r_i(T) - \frac{n-1}{2}, \quad \bar{c}_i(T) = c_i(T) - \frac{n-1}{2} = -\bar{r}_i(T), \\ &|\bar{r}_i(T)| = |\bar{c}_i(T)| \le \frac{n-1}{2}, \quad i = 1, ..., n. \end{aligned}$$

Then:

$$(4.28) \qquad r_i(T)c_i(T) = \left(\frac{n-1}{2}\right)^2 - \bar{r}_i(T)^2 = \left(\frac{n-1}{2}\right)^2 - \bar{c}_i(T)^2.$$

Set:

$$(4.29) \qquad \omega(T) = \left(\prod_1^n r_i(T)c_i(T)\right)^{\frac{1}{2n}}.$$

Combine the standard inequality $\log(1-x) \le -x$, $x \le 1$ with (4.28) to deduce

$$(4.30) \qquad \log \omega(T) \le \log \frac{n-1}{2} - \frac{2}{n(n-1)^2}\sum_1^n \bar{r}_i(T)^2.$$

COROLLARY 4.31. *Let the assumptions of Theorem 4.21 hold. Then*

$$(4.32) \qquad \log \frac{2\rho(T)}{n-1} \le -(1-\tau)\frac{2}{n(n-1)^2}\sum_1^n \bar{r}_i(T)^2.$$

We stress again that the inequalities (4.16), (4.23) and (4.32) should be viewed as a refined version of (3.11). (Note that for generalized tournament matrices T $\mu_{\frac{1}{2}}(T) = \nu_{\frac{1}{2}}(T)$.) To apply these inequalities one should first use the lower bound (3.7) which in particular results in an upper bound for τ.

5. Remarks and open problems. Inequalities (3.17) and (3.21) yield that a matrix T which satisfy (3.15) also satisfy $\rho(T) < \rho^*$, i.e. T is not maximal. The inequality (3.21) implies

(5.1) $$\frac{n-1}{2} - \frac{1}{2n-1} < \rho_n^*.$$

The inequality (3.14) imply

(5.2) $$\rho_n^* < \frac{n-1}{2} - \frac{1}{4n}.$$

Note that for almost regular matrices (3.24) gives a better upper bound than (5.2). Brualdi-Li conjecture implies in particular that the inequality (3.23) holds for all tournament matrices of an even order n. We conjecture that for any tournament matrix of an even order n we have the inequality:

(5.3) $$\rho_3(\alpha_1^2, \alpha_2^2) \le \rho_3(\frac{1}{4}, \frac{n^2-4}{12}).$$

Here α_1^2, α_2^2 are defined by (3.4). Furthermore, the equality sign holds iff T is permutationally similar to Brualdi-Li matrix. The inequalities $(5.1) - (5.2)$ imply that

(5.4) $$\rho_n^* = \frac{n-1}{2} - \frac{K_n}{n}.$$

Roughly speaking $\frac{1}{4} \le K_n \le \frac{1}{2}$. Conjecture (5.3) yields the inequality $\frac{3}{8} \le K_n$. It would be interesting to find the asymptotic distribution of the spectral radius of Q_n given by (3.12). A straightforward computation shows that

(5.5) $$\rho_3\left(\frac{1}{4}, \frac{n^2-4}{12}\right) - \frac{1}{2} = \frac{n-1}{2} - \frac{3}{8n} + O\left(\frac{1}{n^2}\right).$$

Thus, it is possible to estimate the asymptotics of $\rho(Q_n)$ by finding the corresponding value of α_3^2 and evaluating the asymptotics of $\rho_4(\alpha_1^2, \alpha_2^2, \alpha_3^2)$ for the matrix Q_n. We decided to skip this computation here. We conjecture that

(5.6) $$\lim_{n \to \infty} K_n = \frac{3}{8}.$$

However, $\rho_3\left(\frac{1}{4}, \frac{n^2-4}{12}\right) - \frac{1}{2} > \rho(T)$ for any almost regular tournament matrix.

LEMMA 5.7. *Let $n = 2m$ and assume that T is a generalized tournament matrix such that $r_i(T) = m$, $i = 1, ..., m$, $r_i(T) = m - 1$, $i = m + 1, ..., 2m$. Suppose furthermore that $T = (T_{ij})_1^2$, $T_{ij} \in M_m(\mathbf{R})$, $T_{11} = T_{22} = V$ where V is given by (3.12). Then $\rho(T) \le \rho_3(\frac{1}{4}, \frac{n^2-4}{12}) - \frac{1}{2}$. Furthermore, $\rho(T) = \rho_3(\frac{1}{4}, \frac{n^2-4}{12}) - \frac{1}{2}$ iff the following condition hold:*

(5.8) $$\begin{aligned} (T - T^T)(w, -w)^T &= \tau(v, -v)^T, \\ v = (1, ..., 1), w = (w_1, ..., w_m) &\in \mathbf{Z}^m, w_i = m + 1 - 2i, \\ i = 1, ..., m, \tau &= -\frac{2(m^2 - 1)}{3}. \end{aligned}$$

There is no almost regular tournament matrix which satisfies $\rho(T) = \rho_3\left(\frac{1}{4}, \frac{n^2-4}{12}\right) - \frac{1}{2}$.

Proof. Let $2A = T - T^T$, $u = (v,v)^T$ and consider the cyclic subspace V generated by $u, Au, ..., A^{n-1}u$. As in the proof of Theorem (3.20), we deduce that for any generalized tournament matrix which satisfies the conditions of the theorem one has the the the equalities $\alpha_1^2 = \frac{1}{4}$, $\alpha_2^2 = \frac{n^2-4}{12}$. Theorem 2.30 implies the inequality $\rho(T) \leq \rho_3\left(\frac{1}{4}, \frac{n^2-4}{12}\right) - \frac{1}{2}$ and the equality sign holds iff V is three dimensional. Clearly, $u, b = (v, -v)^T \in V$. It is straightforward to show that Tb and hence Ab in view of (3.25) is spanned by $u, b, x = (w, -w)^T$. Furthermore, $u^T w = b^T w = 0$. Thus V is three dimensional iff Ax is a linear combination of u, b, w. As $x^T Ax = u^T Ax = 0$ it follows that V is three dimensional iff $2Ax = \tau b$. Moreover, $\tau = -\frac{x^T 2Ab}{b^T b}$. A straightforward calculation shows that $\tau = -\frac{2(m^2-1)}{3}$.

Assume that T is an almost regular tournament matrix which satisfies $\rho(T) = \rho_3\left(\frac{1}{4}, \frac{n^2-4}{12}\right) - \frac{1}{2}$. From the proof of Theorem 2.6 we obtain that $\text{spec}(T) \supset \text{spec}(F - \frac{I}{2})$, $F \in A_{n-3}(\mathbf{R})$. As $n - 3$ is odd we deduce that $0 \in \text{spec}(F)$. That is $-\frac{1}{2} \in \text{spec}(T)$. This is impossible, since all the eigenvalues of T are algebraic integers. $\square$

We close the paper by giving some numerical data. For $n = 4, 6, 8, 10, 12$ we computed the quantities

$$\rho_2\left(\frac{1}{4}\right) - \frac{1}{2} < \rho(Q_n) < \rho_3\left(\frac{1}{4}, \frac{n^2-4}{12}\right) - \frac{1}{2}, \ \delta = \frac{n}{2}$$

using MATLAB. The matrix Q_n is given by (3.12). Here are the results:

$$
\begin{array}{llll}
n = 4: & 1.3660, & 1.3953, & 1.3969; \\
n = 6: & 2.4142, & 2.4340, & 2.4350; \\
n = 8: & 3.4365, & 3.4513, & 3.4521; \\
n = 10: & 4.4495, & 4.4614, & 4.4620; \\
n = 12: & 5.4580, & 5.4680, & 5.4684.
\end{array}
$$

Acknowledgements. I would like to thank to R. Brualdi, S. Kirkland, H. Schneider and B. Shader for useful remarks.

REFERENCES

[Bru] R.A. BRUALDI, *Matrices, eigenvalues and directed graphs*, Linear and Multilinear Alg. 11 (143–165).

[B-L] R.A. BRUALDI AND Q. LI, *Problem 31*, Discrete Math. 43 (1983), pp. 329–330.

[Fri] S. FRIEDLAND, *Convex spectral functions*, Linear and Multilinear Alg. 9 (1981), pp. 299–316.

[F-K] S. FRIEDLAND AND S. KARLIN, *Some inequalities for the spectral radius of nonnegative matrices and applications*, Duke Math. J. 42 (1975), pp. 459–490.

[G-V] G.H. GOLUB AND C.F. VAN LOAN, *Matrix computations*, John Hopkins Univ. Press, 2nd edition, 1989.

[Kat] T. KATO, *Perturbation theory for linear operators*, Springer-Verlag, 1966.

[Kir] S. KIRKLAND, *Hypertournament matrices, score vectors and eigenvalues*, Linear and Multilinear Alg., to appear.

[M-P] J.S. MAYBEE AND N.J. PULLMAN, *Tournament matrices and their generalizations, I*, Linear and Multilinear Alg. 28 (1990), pp. 57–70.

[Moo] J.W. MOON, *Topics on tournaments*, Holt, Rinehart and Winston, 1968.

[Ost] A. OSTROWSKI, *Über das nichtverschwinden einer klasse von determinanten und die lokalisierung der charakteristischen wurzeln von matrizen*, Compositio Math. 9 (1951), pp. 209–226.

[O-S] A. OSTROWSKI AND H. SCHNEIDER, *Some theorems on the inertia of general matrices*, J. Math. Analysis and Appl. 4 (1962), pp. 72–84.

COMBINATORIAL ORTHOGONALITY*

LEROY B. BEASLEY**, RICHARD A. BRUALDI† AND BRYAN L. SHADER‡

Abstract. By considering a combinatorial property of orthogonality we determine the minimum number of nonzero entries in an orthogonal matrix of order n which cannot be decomposed into two smaller orthogonal matrices.

1. Introduction. Let $Q = [q_{ij}]$ be a real orthogonal (or complex unitary) matrix of order $n \geq 2$. At least how many nonzero entries must Q have? The short answer is n, since the identity matrix I_n is an orthogonal matrix. In order to make the question more interesting, let us assume that Q cannot be written as the direct sum of two (orthogonal) matrices of smaller order (no matter how the rows and columns are permuted). M. Fiedler conjectured that Q must have at least $4(n-1)$ nonzero entries. For each $n \geq 2$ there exists an orthogonal matrix of order n with exactly $4(n-1)$ nonzero entries which cannot be written as a direct sum. Examples of such matrices are

$$(1) \quad \begin{bmatrix} -\frac{1}{\sqrt{2}} & \frac{1}{2} & \frac{1}{2} & 0 & 0 \\ -\frac{1}{\sqrt{2}} & -\frac{1}{2} & -\frac{1}{2} & 0 & 0 \\ 0 & \frac{1}{2} & -\frac{1}{2} & \frac{1}{\sqrt{3}} & \frac{1}{\sqrt{6}} \\ 0 & \frac{1}{2} & -\frac{1}{2} & -\frac{1}{\sqrt{3}} & -\frac{1}{\sqrt{6}} \\ 0 & 0 & 0 & \frac{1}{\sqrt{3}} & -\frac{2}{\sqrt{6}} \end{bmatrix}$$

and

$$(2) \quad \begin{bmatrix} -\frac{1}{\sqrt{2}} & \frac{1}{2} & \frac{1}{2} & 0 & 0 & 0 \\ -\frac{1}{\sqrt{2}} & -\frac{1}{2} & -\frac{1}{2} & 0 & 0 & 0 \\ 0 & \frac{1}{2} & -\frac{1}{2} & \frac{1}{2} & \frac{1}{2} & 0 \\ 0 & \frac{1}{2} & -\frac{1}{2} & -\frac{1}{2} & -\frac{1}{2} & 0 \\ 0 & 0 & 0 & \frac{1}{2} & -\frac{1}{2} & \frac{1}{\sqrt{2}} \\ 0 & 0 & 0 & \frac{1}{2} & -\frac{1}{2} & -\frac{1}{\sqrt{2}} \end{bmatrix}.$$

* This paper was written while LBB and RAB were members of the Institute for Mathematics and Its Applications at the University of Minnesota and BLS was a postdoctoral fellow.
** Department of Mathematics, Utah State University, Logan, UT 84322-3900.
† Department of Mathematics, University of Wisconsin, Madison, WI 53706. Research partially supported by NSF Grant DMS-8901445 and NSA grant MDA904-89-H-2060.
‡ Department of Mathematics, University of Wyoming, Laramie, WY 82071. Research partially supported by a University of Wyoming Basic Research Grant.
private communication

We show that Fiedler's conjecture follows from a more general combinatorial theorem.

Let $x = (x_1, \ldots, x_n)$ and $y = (y_1, \ldots, y_n)$ be two vectors. We define x and y to be *combinatorially orthogonal* provided

$$|\{i : x_i y_i \neq 0\}| \neq 1.$$

Thus the combinatorial orthogonality of two vectors depends only on the positions of their nonzero coordinates. The vectors $z_1, \ldots, z_k$ are *combinatorially orthogonal* provided z_i and z_j are combinatorially orthogonal for all $i \neq j$.

A *combinatorially row-orthogonal matrix* is a square matrix whose rows are combinatorially orthogonal. Combinatorial column-orthogonality is defined analogously. A *combinatorially orthogonal matrix* is a square matrix whose rows are combinatorially orthogonal and whose columns are combinatorially orthogonal. Clearly, an orthogonal matrix is combinatorially orthogonal. The combinatorial orthogonality of a matrix depends only on the zero-nonzero pattern of the matrix. Hence in discussing the combinatorial orthogonality of a square matrix A we may assume that A is a $(0,1)$-matrix. A $(0,1)$-matrix A is combinatorially orthogonal if and only if for all $i \neq j$ the inner product of row i with row j and that of column i with column j is not one. A zero matrix as well as the matrix

(3)
$$\begin{bmatrix} 1 & 1 & 0 & 0 \\ 1 & 1 & 0 & 0 \\ 1 & 1 & 1 & 1 \\ 1 & 1 & 1 & 1 \end{bmatrix}$$

are examples of combinatorially orthogonal matrices for which there does not exist an orthogonal matrix with the same zero-nonzero pattern.

Recall that a matrix A of order $n \geq 2$ is *partly decomposable* provided there exist permutation matrices P_1 and P_2 such that

(4)
$$P_1 A P_2 = \begin{bmatrix} A_1 & O \\ X & A_2 \end{bmatrix}$$

where A_1 and A_2 are square (nonvacuous) matrices. Equivalently, A is partly decomposable if and only if A has an r by s zero submatrix for some positive integers r and s with $r + s = n$. A matrix of order $n \geq 2$ is *fully indecomposable* provided it is not partly decomposable. If A is an orthogonal matrix which satisfies (4) then X is a zero matrix. Hence a partly decomposable orthogonal matrix can be written as a direct sum (after row and column permutations). The matrix (3) is a partly decomposable matrix.

There are fully indecomposable, combinatorially orthogonal $(0,1)$-matrices for which there does not exist an orthogonal matrix with the same zero-nonzero pattern. Let P be the permutation matrix of order 11 with 1's in positions $(1,2), \ldots, (10,11)$, $(11,1)$ and let

$$A = I_{11} + P + P^3 + P^6 + P^{10}.$$

Then

(5)
$$AA^T = A^T A = 3I_{11} + 2J_{11}$$

where J_{11} denotes the all 1's matrix of order 11 (A is an incidence matrix of the (11,5,2)-design). Thus A is a fully indecomposable, combinatorially orthogonal (0,1)-matrix. Suppose $Q = [q_{ij}]$ is an orthogonal matrix with the same zero-nonzero pattern as A. Let R be the $(0,1,-1)$-matrix obtained from Q by replacing the positive entries with 1 and the negative entries with -1. It follows from (5) that $RR^T = 5I_{11}$ and hence $(\det R)^2 = 5^{11}$. We conclude that there is no orthogonal matrix with the same zero-nonzero pattern as A.

In the next section we prove that a fully indecomposable, combinatorially orthogonal matrix of order $n \geq 2$ has at least $4(n-1)$ nonzero entries thereby proving Fiedler's conjecture. For $n \geq 5$, we show that up to row and column permutations and transposition, there is a unique fully indecomposable combinatorially orthogonal (0,1)-matrix of order n with exactly $4(n-1)$ 1's. In the last section we show that a fully indecomposable, combinatorially row-orthogonal (0,1)-matrix of order $n \geq 4$ can have $3n$ but no fewer 1's if n is even and $3n + 1$ if n is odd. Those matrices for which equality holds are characterized.

2. Combinatorially orthogonal matrices. We begin by recursively defining a family of (0,1)-matrices of order $n \geq 2$. Let

$$B_2 = \begin{bmatrix} 1 & 1 \\ 1 & 1 \end{bmatrix}.$$

If n is odd, define

$$B_n = \left[\begin{array}{c|c} B_{n-1} & \begin{matrix} 0 \\ \vdots \\ 0 \\ 1 \\ 1 \end{matrix} \\ \hline \begin{matrix} 0 & \cdots & 0 & 1 \end{matrix} & 1 \end{array} \right].$$

If n is even, define

$$B_n = \left[\begin{array}{c|c} B_{n-1} & \begin{matrix} 0 \\ \vdots \\ 0 \\ 1 \end{matrix} \\ \hline \begin{matrix} 0 & \cdots & 0 & 1 & 1 \end{matrix} & 1 \end{array} \right].$$

For example,

$$B_3 = \begin{bmatrix} 1 & 1 & 1 \\ 1 & 1 & 1 \\ 0 & 1 & 1 \end{bmatrix}, \quad B_4 = \begin{bmatrix} 1 & 1 & 1 & 0 \\ 1 & 1 & 1 & 0 \\ 0 & 1 & 1 & 1 \\ 0 & 1 & 1 & 1 \end{bmatrix},$$

and

$$B_5 = \begin{bmatrix} 1 & 1 & 1 & 0 & 0 \\ 1 & 1 & 1 & 0 & 0 \\ 0 & 1 & 1 & 1 & 1 \\ 0 & 1 & 1 & 1 & 1 \\ 0 & 0 & 0 & 1 & 1 \end{bmatrix}.$$

We shall make use of the following facts concerning the matrices B_n. If n is odd, then B_n has a unique row and a unique column with exactly two 1's. If n is even and greater than 2, then B_n has exactly two columns with exactly two 1's and every row has at least three 1's. For $n \geq 3$, there are exactly four 1's in B_n each of which is in a row with exactly two 1's and a column with exactly three 1's or a column with exactly two 1's and a row with exactly three 1's. Two rows (or two columns) of B_n whose inner product is greater than 2 are identical. It is easy to verify that B_n is a fully indecomposable, combinatorially orthogonal matrix for each $n \geq 2$. The matrices B_5 and B_6 have the same zero-nonzero pattern as the orthogonal matrices (1) and (2), respectively. For each n a similar construction yields an orthogonal matrix with the same zero-nonzero pattern as B_n.

The following technical lemma is used in the inductive proof of the main result of this section. In general we use J to denote an all 1's matrix. The r by s matrix of all 1's is denoted by $J_{r,s}$; in case $r = s$ this is shortened to J_r.

LEMMA 2.1. *Let*

$$M = \left[\begin{array}{c|c|c} U & V & O \\ \hline W & X & Y \\ \hline O & O & Z \end{array} \right]$$

be a fully indecomposable, combinatorially orthogonal $(0,1)$-matrix of order n where U is k by k, X is ℓ by p, and $\ell \geq 2$. Assume each row of $[W\ X]$ contains a 1, and if $p = 1$ then the inner product of each pair of rows of Y is at least 1. Let

$$\overline{M} = \left[\begin{array}{cc} J_{\ell,p} & Y \\ O & Z \end{array} \right]$$

be the matrix of order $n - k$. Then $\overline{M}$ is a fully indecomposable, combinatorially orthogonal matrix.

Proof. A zero submatrix of $\overline{M}$ is either a zero submatrix of

$$\left[\begin{array}{cc} O & Z \end{array} \right] \quad \text{or} \quad \left[\begin{array}{c} Y \\ Z \end{array} \right].$$

Since U is a square matrix, the full indecomposability of M implies the full indecomposability of $\overline{M}$. The assumptions imply that each nonzero column of Y contains at least two 1's. This fact, the combinatorial orthogonality of M, and the assumption in the case $p = 1$ now imply that $\overline{M}$ is combinatorially orthogonal. $\square$

The number of nonzero entries of a matrix A is denoted by $\sigma(A)$. The number of zeros of a matrix A is denoted by $\sigma_0(A)$.

THEOREM 2.2. *Let $M = [m_{ij}]$ be a fully indecomposable, combinatorially orthogonal $(0,1)$-matrix of order $n \geq 2$. Then*

$$\sigma(M) \geq 4(n-1).$$

Equality holds if and only if there exist permutation matrices P_1 and P_2 such that $P_1 M P_2$ equals B_n or B_n^T, or $n = 4$ and $P_1 M P_2$ equals $J_4 - I_4$.

Proof. Let r_i and s_i denote the number of 1's in row i and column i of M, respectively. It is easy to verify the theorem for $n \leq 4$. We now assume that $n > 4$ and proceed by induction on n.

First suppose that there exist i and j such that $m_{ij} = 1$ and $r_i = 2$. Let the other 1 in row i be in column k. Combinatorial row-orthogonality of M implies that columns j and k are identical. Full indecomposability of M implies that $s_j \geq 3$. We conclude that there do not exist integers i and j such that $m_{ij} = 1$ and $r_i = s_j = 2$. We divide the remainder of the proof into several cases.

Case 1: There exist i and j such that $m_{ij} = 1$, $r_i = 2$ and $s_j = 3$ (or $r_i = 3$ and $s_j = 2$).

Without loss of generality we assume that

$$M = \left[\begin{array}{ccc|cc} 1 & 1 & 0 \cdots 0 \\ 1 & & & & \\ 1 & X & & Y & \\ \hline 0 & & & & \\ \vdots & T & & Z & \\ 0 & & & & \end{array} \right].$$

The combinatorial row-orthogonality of M implies that $X = J_{2,1}$ and $T = O$. The combinatorial column-orthogonality of M implies that the two rows of Y are identical. Since M is fully indecomposable, the two rows of Y have a nonzero inner product. Hence by Lemma 2.1 the matrix

$$\overline{M} = \left[\begin{array}{cc} J_{2,1} & Y \\ O & Z \end{array} \right]$$

is a fully indecomposable, combinatorially orthogonal matrix of order $n - 1$. We have $\sigma(M) = \sigma(\overline{M}) + 4$, and hence by induction $\sigma(M) \geq 4(n - 1)$. Suppose that $\sigma(M) = 4(n - 1)$ and $n \geq 6$. Then $\sigma(\overline{M}) = 4(n - 2)$ and by induction $\overline{M}$ can be permuted to B_{n-1} or B_{n-1}^T. It follows that M can be permuted to B_n or B_n^T. It can be verified that the same conclusion holds if $n = 5$.

Case 2: There exist i, j, p, q, u, v such that $m_{ij} = 0$, $m_{iu} = m_{iv} = m_{pj} = m_{qj} = 1$, $m_{pu} = 1$, and $r_i = s_j = 2$.

Combinatorial row- and column-orthogonality imply that $m_{pv} = m_{qu} = m_{qv} = 1$. Without loss of generality we assume that

$$M = \left[\begin{array}{cc|cc} 0 & 1 \quad 1 & 0 \cdots 0 \\ 1 & & & \\ 1 & & & \\ 0 & X & & Y \\ \vdots & & & \\ 0 & & & \\ \hline 0 & & & \\ \vdots & O & & Z \\ 0 & & & \end{array} \right]$$

where X has no zero rows. Combinatorial orthogonality implies that rows 2 and 3 of M are identical and columns 2 and 3 are also identical; in particular, we have $X = J$. The matrix $\overline{M}$ obtained by applying Lemma 2.1 is a fully indecomposable, combinatorially orthogonal matrix of order $n - 1$. Since $X = J$ we have $\sigma(M) = \sigma(\overline{M}) + 4$ and hence by induction $\sigma(M) \geq 4(n - 1)$. If $\overline{M}$ can be permuted to B_{n-1} or B_{n-1}^T, then it is easy to see that M is partly decomposable. Hence it follows by induction that when $n \geq 6$ we have $\sigma(M) > 4(n - 1)$. The same conclusion holds when $n = 5$.

Case 3: There exists a row with exactly two 1's and a row with exactly three 1's whose inner product is 2 (or there exist two columns with these properties).

We assume that the rows in Case 3 are rows 1 and 2, respectively, and that $m_{11} = m_{12} = m_{13} = m_{22} = m_{23} = 1$. Row-orthogonality implies that columns 2 and 3 of M are identical. Permuting rows, M has the form

$$\left[\begin{array}{c|c|c} 1 & 1 \; 1 & 0 \cdots 0 \\ \hline W & J & Y \\ \hline O & O & Z \end{array} \right].$$

The matrix $\overline{M}$ obtained by applying Lemma 2.1 is a fully indecomposable, combinatorially orthogonal matrix of order $n-1$. Since M is fully indecomposable, $\sigma(W) \geq 1$ and hence $\sigma(M) = \sigma(\overline{M}) + 3 + \sigma(W)$. By induction, $\sigma(M) \geq 4(n - 1)$ with equality if and only if $\sigma(W) = 1$. If $\sigma(W) = 1$ then Case 1 applies.

Case 4: There exists a row with exactly two 1's and a row with exactly four 1's whose inner product is 2 (or there exist two columns with these properties).

Without loss of generality we assume that

$$M = \left[\begin{array}{c|cc|c} J_2 & 0 \; 0 & & O \\ & 1 \; 1 & \\ \hline W & X & Y \\ \hline O & O & Z \end{array} \right]$$

where $[W\,X]$ has no zero rows. Row-orthogonality implies that each row of $[W\,X]$ is either 0 0 1 1 or is of the form 1 1 * *. We again apply Lemma 2.1 and obtain a matrix $\overline{M}$ of order $n - 2$ for which

$$\sigma(M) = \sigma(\overline{M}) + 6 + \sigma(W) - \sigma_0(X).$$

Considering inner products of columns 3 and 4 with column 1 of M, we see that either some row of $[W\,X]$ equals 1 1 1 1, or both 1 1 1 0 and 1 1 0 1 occur as rows of $[W\,X]$. Hence $\sigma(W) - \sigma_0(X) \geq 2$, and by induction $\sigma(M) \geq 4(n - 1)$. Suppose that $\sigma(M) = 4(n - 1)$. Then $\sigma(\overline{M}) = 4(n - 3)$, and by induction $\overline{M}$ can be permuted to B_{n-2} or B_{n-2}^T, or to $J_4 - I_4$ if $n = 6$. It is easy to verify that the latter cannot occur. The definition of B_{n-2} implies that there is a 1 in Z such that the row of $\overline{M}$ which contains it has exactly two 1's and the column which contains it has exactly three 1's, or the other way around. The row and column of M that contains this 1 have the same number of 1's as the corresponding row and column of $\overline{M}$. This reduces the case of equality to that of Case 1.

Case 5: There exist two rows with exactly two 1's and another row whose inner product with each of these two rows is nonzero (or there exist columns with these properties).

The combinatorial row-orthogonality of M allow us to assume that

$$
M = \left[
\begin{array}{cc|cc|c}
1 & 1 & 0 & 0 & \multirow{2}{*}{O} \\
0 & 0 & 1 & 1 & \\
\hline
\multicolumn{2}{c|}{W} & \multicolumn{2}{c|}{X} & Y \\
\hline
\multicolumn{2}{c|}{O} & \multicolumn{2}{c|}{O} & Z
\end{array}
\right]
$$

where $[W\ X]$ has no zero rows and has at least one row equal to 1 1 1 1. By Lemma 2.1 the matrix $\overline{M}$ of order $n-2$ satisfies

$$\sigma(M) = \sigma(\overline{M}) + 4 + \sigma(W) - \sigma_0(X).$$

The combinatorial row-orthogonality of M implies that each row of $[W\ X]$ is one of 0 0 1 1, 1 1 0 0, and 1 1 1 1. The combinatorial orthogonality of columns 1 and 3 implies that at least two rows of $[W\ X]$ equal 1 1 1 1. Hence by induction, $\sigma(M) \geq 4(n-1)$. As above we reduce the case of equality to that of Case 1.

Case 6: There exist two rows, one of which has exactly three 1's, whose inner product is 3 (or there exist two columns with these properties).

Without loss of generality we assume that

$$
M = \left[
\begin{array}{c|cc|c}
1 & 1 & 1 & O \\
\hline
W & \multicolumn{2}{c|}{X} & Y \\
\hline
O & \multicolumn{2}{c|}{O} & Z
\end{array}
\right],
$$

where $[W\ X]$ has no zero rows and has at least one row equal to 1 1 1. The combinatorial row-orthogonality of M implies that each row of $[W\ X]$ contains at least two 1's. The matrix $\overline{M}$ obtained by applying Lemma 2.1 satisfies

$$\sigma(M) = \sigma(\overline{M}) + 3 + \sigma(W) - \sigma_0(X).$$

Since $[W\ X]$ has at least one row equal to 1 1 1, it follows by induction that $\sigma(M) \geq 4(n-1)$. Again the case of equality can be reduced to that of Case 1.

Case 7: There exist i and j such that $m_{ij} = 1$ and $r_i = s_j = 3$.

Without loss of generality we assume that $i = j = 1$ and that

$$
M = \left[
\begin{array}{c|cc|c}
1 & 1 & 1 & O \\
\hline
W & \multicolumn{2}{c|}{X} & Y \\
\hline
O & \multicolumn{2}{c|}{O} & Z
\end{array}
\right],
$$

where $[W\ X]$ has no zero rows. If $[W\ X]$ has a row equal to 1 1 1, then Case 6 applies. Otherwise, we may assume that $m_{21} = m_{31} = m_{22} = m_{33} = 1$ and that $m_{32} = m_{23} = 0$. By combinatorial row-orthogonality of M each row of $[W\ X]$ other than rows 1 and 2 equals 0 1 1. We may apply a similar argument to M^T and hence may assume that

$$M = \begin{bmatrix} 1 & J_{1,2} & O & O \\ J_{2,1} & I_2 & J_{2,u} & O \\ O & J_{v,2} & Y_1 & Y_2 \\ O & O & Z_1 & Z_2 \end{bmatrix}.$$

where u and v are positive integers. By applying Lemma 2.1 to M we obtain the fully indecomposable, combinatorially orthogonal matrix

$$\overline{M} = \begin{bmatrix} J_2 & J_{2,u} & O \\ J_{v,2} & Y_1 & Y_2 \\ O & Z_1 & Z_2 \end{bmatrix}.$$

Considering inner products of rows of M with row 2 we see that Y_1 has no zero rows. Similarly, Y_1 has no zero columns. One easily verifies that the matrix $\tilde{M}$ obtained from $\overline{M}$ by changing the 1 in its first row and second column to 0 is also fully indecomposable and combinatorially orthogonal. We have $\sigma(M) = \sigma(\overline{M}) + 3 = \sigma(\tilde{M}) + 4$. Applying induction to $\tilde{M}$ we obtain $\sigma(M) \geq 4(n-1)$.

Suppose $\sigma(M) = 4(n-1)$. Then $\sigma(\tilde{M}) = 4(n-2)$ and by induction $\tilde{M}$ can be permuted to B_{n-1} or B_{n-1}^T, or to $J_4 - I_4$ if $n = 5$. It is easy to verify that the latter cannot occur. By the full indecomposability of M we have $u \geq 1$ and $v \geq 1$. If $u \geq 2$, then rows 1 and 2 of $\tilde{M}$ have inner product 3 or more and are not identical, a property that does not hold for any two rows of B_{n-1} or B_{n-1}^T. We conclude that $u = v = 1$. Considering the inner product of row 2 and 4 of M we have $m_{44} = 1$ and hence the inner product of rows 2 and 3 of $\tilde{M}$ equals 3. Therefore rows 2 and 3 of $\tilde{M}$ are identical. Since $n \geq 5$ M is partly decomposable, contrary to assumption. Hence equality cannot occur. (Note that if in the above analysis we had $n = 4$ then we obtain $M = J_4 - I_4$.)

Case 8: There exist two rows each with exactly three 1's whose inner product is 2 (or there exist two columns with these properties).

We assume that the rows in Case 8 are rows 1 and 2. Without loss of generality we assume that

$$\begin{bmatrix} J_2 & I_2 & O \\ \hline W & X & Y \\ \hline O & O & Z \end{bmatrix},$$

where $[W \ X]$ has no zero rows. Assume that Case 6 does not apply. Then considering inner products of rows 1 and 2 with the other rows we see that each row of $[W \ X]$ is one of 1 1 0 0, 1 0 1 1, and 0 1 1 1. By considering inner products of columns 1 and 2 with columns 3 and 4, we conclude that both 1 0 1 1 and 0 1 1 1 occur as a row of $[W \ X]$. The matrix $\overline{M}$ of order $n-2$ obtained by applying Lemma 2.1 satisfies

$$\sigma(M) \geq \sigma(\overline{M}) + 8,$$

and hence by induction $\sigma(M) \geq 4(n-1)$. The case of equality once again can be reduced to that of Case 1.

We now assume that none of Cases 1 to 8 applies. Then every row with two or three 1's has an inner product of 0 with every other row with two or three 1's. A similar statement holds for the columns of M. We now permute the rows and columns so that all rows with exactly two 1's and the columns containing their 1's come first. Then all the rows with three 1's and the columns containing their 1's come next. The columns containing exactly two 1's and the rows containing their 1's can be taken to be last (since Cases 1 and 7 do not occur) preceded by the columns containing exactly three 1's and the rows containing their 1's. Without loss of generality we may assume that the number of rows of M with exactly three 1's is at least as large as the number of columns with exactly three 1's. We may further permute rows and columns to bring M to the form

$$
\begin{bmatrix}
C & O & O & O & O & O & O \\
O & F & O & O & O & O & O \\
O & O & G & O & O & O & O \\
 & & & & & O & O \\
 & & & & & O & O \\
 & & & & & H & O \\
S & O & T & O & U & O & D
\end{bmatrix},
$$

where each row of C and each column of D has exactly two 1's, each row of F and G and each column of H has exactly three 1's, each column of U has at least one 1, and no row of G has a inner product zero with all the rows of T. Since Case 2 does not apply we have $S = O$. Since Case 4 does not apply each column of M containing a column of U has at least five 1's.

Consider a row α of F or G and the three columns of M containing its 1's. Each of these columns contains at least four 1's. Since Case 6 does not apply, the inner products of the row of M containing α with the other rows of M are 0 or 2. This implies that the three specified columns of M have a total of at least thirteen 1's. If α is a row of G then since Cases 3 and 4 do not apply, the three columns of M have at least fifteen 1's.

Let the number of columns of C, F, G, U, H and D be c, f, g, u, h and d, respectively. Note that the number of rows of M with exactly three 1's is $(f + g)/3$ and hence $h \leq (f + g)/3$. Each of the last d columns of M has a nonzero inner product with at leat three other columns of M. Moreover, since Case 5 does not apply, a column of M cannot have a nonzero inner product with two of the last d columns of M. This implies that $g + u \geq 3d$.

Computing the number of 1's in each of the seven groups of columns of M we obtain

$$
\begin{aligned}
\sigma(M) \;\geq\; & 4c + \frac{13f}{3} + \frac{15g}{3} \\
& + 4(n - c - f - g - u - h - d) + 5u + 3h + 2d \\
=\; & 4n + g + u + \frac{f}{3} - h - 2d \\
=\; & 4n + \frac{f + g}{3} - h + \frac{2}{3}(g + u) - 2d + \frac{u}{3} \\
\geq\; & 4n + \frac{u}{3}.
\end{aligned}
$$

Hence if none of Cases 1 to 8 apply the number of 1's of M is at least $4n$, and the proof is complete. $\qquad\square$

3. Combinatorially row-orthogonal matrices. A real matrix whose rows are mutually orthogonal, unit vectors is an orthogonal matrix and hence has mutually orthogonal columns. But there exist combinatorially row-orthogonal $(0,1)$-matrices with the same number of 1's in each row which are not combinatorially column-orthogonal. For example, the matrix

$$\begin{bmatrix} 1 & 1 & 1 & 0 \\ 1 & 1 & 0 & 1 \\ 0 & 1 & 1 & 1 \\ 0 & 1 & 1 & 1 \end{bmatrix}$$

is a combinatorially row-orthogonal but not column-orthogonal matrix.

Let n be a positive even integer and let

$$R_n = \begin{bmatrix} J_2 & O & O & \cdots & O & X \\ X & J_2 & O & \cdots & O & O \\ O & X & J_2 & \cdots & O & O \\ \vdots & \vdots & \vdots & \ddots & \vdots & \vdots \\ O & O & O & \cdots & J_2 & O \\ O & O & O & \cdots & X & J_2 \end{bmatrix},$$

be a matrix of order n where

$$X = \begin{bmatrix} 1 & 1 \\ 0 & 0 \end{bmatrix}.$$

Now let $n > 1$ be an odd integer. We define matrices S_n and T_n by

$$S_n = \left[\begin{array}{ccccc|c} & & & & & 0 \\ & & R_{n-1} & & & \vdots \\ & & & & & 0 \\ & & & & & 1 \\ \hline 0 & \cdots & 0 & 1 & 1 & 1 \end{array}\right],$$

and

$$T_n = \left[\begin{array}{ccccc|c} & & & & & 0 \\ & & R_{n-1} & & & \vdots \\ & & & & & 0 \\ & & & & & 1 \\ & & & & & 0 \\ \hline 0 & \cdots & 0 & 1 & 1 & 1 \end{array}\right].$$

Then R_n, S_n and T_n are fully-indecomposable, combinatorially row-orthogonal matrices. If $n \geq 4$ is an even integer, then $\sigma(R_n) = 3n$. If $n \geq 5$ is an odd integer then $\sigma(S_n) = \sigma(T_n) = 3n + 1$.

THEOREM 3.1. *Let* $M = [m_{ij}]$ *be a fully-indecomposable, combinatorially row-orthogonal* $(0,1)$*-matrix of order* $n \geq 5$. *Then*

(6)
$$\sigma(M) \geq \begin{cases} 3n & \text{if } n \text{ is even} \\ 3n+1 & \text{if } n \text{ is odd.} \end{cases}$$

Proof. We prove by induction on n that (6) holds and that if n is even and $\sigma(M) = 3n$, then each column of M has exactly three 1's. This can be verified for $n = 5$ and $n = 6$. If $\sigma(M) \geq 3n + 1$, there is nothing to prove. We now assume that $n \geq 7$ and that $\sigma(M) \leq 3n$.

Let r_i and s_i denote the number of 1's in row i and column i, respectively. First assume that there do not exist integers i and j with $m_{ij} = 1, r_i = 2$ and $s_j = 3$. If each column of M has at least three 1's, then since $\sigma(M) \leq 3n$ each row and column of M has exactly three 1's. Since a fully indecomposable, combinatorially row-orthogonal $(0,1)$-matrix with exactly three 1's in each row and column has order less than 5, some column of M has exactly two 1's. After row and column permutations,

$$M = \begin{bmatrix} 1 & u \\ 1 & v \\ O & Z \end{bmatrix}.$$

Let

$$\overline{M} = \begin{bmatrix} u \oplus v \\ Z \end{bmatrix}$$

where $u \oplus v$ is the $(0,1)$-vector of the same length as u and v with 1's in those positions in which u or v has a 1. Then $\overline{M}$ is a fully indecomposable, combinatorially row-orthogonal matrix of order $n - 1$. The combinatorial row-orthogonality of rows 1 and 2 of M implies that $\sigma(M) \geq \sigma(\overline{M}) + 3$. If n is even, then by induction $\sigma(M) \geq 3n+1$. Suppose n is odd. Since $\sigma(M) \leq 3n$, it follows from the inductive assumption that $\sigma(M) = 3n$, $\sigma(\overline{M}) = 3(n - 1)$, $uv^T = 1$ and each column of $\overline{M}$ has exactly three 1's. This implies that M has a column with exactly two 1's, a column with exactly four 1's and the remaining columns each have exactly three 1's. Since no 1 lies in a row with two 1's and a column with three 1's, each row of M has exactly three 1's. The full indecomposability and combinatorial row-orthogonality of M now lead to the contradiction that $n = 4$.

Now assume that there exist integers i and j with $m_{ij} = 1, r_i = 2$ and $s_j = 3$. Without loss of generality we assume that $i = j = 1$. The combinatorial row-orthogonality of M implies that M can be permuted to the form

$$M = \begin{bmatrix} 1 & 1 & 0 \cdots 0 \\ 1 & 1 & w \\ 1 & 1 & x \\ O & & Z \end{bmatrix}.$$

Then

$$\overline{M} = \begin{bmatrix} w \oplus x \\ Z \end{bmatrix},$$

is a fully indecomposable, combinatorially row-orthogonal matrix of order $n-2$ with $\sigma(M) \geq \sigma(\overline{M})+6$. Hence, by induction (6) holds. Suppose n is even and $\sigma(M) = 3n$. Then $\sigma(\overline{M}) = 3n-6$ and $wx^T = 0$. By induction each column of $\overline{M}$ has exactly three 1's and hence each column of M has exactly three 1's. $\qquad\square$

With a more careful inductive argument it is possible to characterize those matrices M for which equality holds in (6).

THEOREM 3.2. *Let M be a fully indecomposable, combinatorially row-orthogonal $(0,1)$-matrix of order $n \geq 6$. If n is even, then $\sigma(M) = 3n$ if and only if M can be permuted to R_n. If n is odd, then $\sigma(M) = 3n+1$ if and only if M can be permuted to S_n or T_n.*

The inequality (6) and Theorem 3.2 can be proved together by showing inductively that if M is a fully indecomposable, combinatorially row-orthogonal $(0,1)$-matrix of order $n \geq 6$ then

    (i) (6) holds with equality if and only if M is as given in Theorem 3.2;
    (ii) if n is even and $\sigma(M) = 3n+1$ then each row of M has at most four 1's.

This involves some tedious checking whose details we omit.

THE SYMMETRIC GROUP AS A POLYNOMIAL SPACE

MARSTON CONDER* AND CHRIS D. GODSIL†

Abstract. A design in a polynomial space is a natural generalisation of the concepts of design and orthogonal array in design theory. In this paper we further develop the second author's theory of polynomial spaces. As a consequence we prove that a subgroup of the symmetric group is a t-design if and only it is t-transitive.

1. Polynomial spaces. This section gives the definition of a polynomial space. These were first introduced in [3], although the axioms there are somewhat more restrictive than those we present now. For a more detailed exposition based on the new axioms, see [4].

A *polynomial space* consists of a set Ω, a real function ρ on $\Omega \times \Omega$ and an inner product space of functions on Ω. We call ρ the *distance function* of the polynomial space. Four axioms are needed to describe a polynomial space, of which the first is:

I If $x, y \in \Omega$ then $\rho(x, y) = \rho(y, x)$.

For each point a in Ω define the function ρ_a on Ω by

$$\rho_a(x) = \rho(a, x).$$

Given ρ_a, any polynomial f in one real variable gives rise by composition to a function $f \circ \rho_a$ on Ω. We call such functions *zonal polynomials* with respect to the point a, and we define $\mathrm{Zon}(\Omega, r)$ to be the space spanned by the functions $f \circ \rho_a$, as a varies over the points in Ω and f over the polynomials with degree at most r.

The product gh of two zonal polynomials g and h is defined by

$$(gh)(x) = g(x)h(x),$$

for all x in Ω. The product of two zonal polynomials will be a function on Ω but, unless they are zonal with respect to the same point, will not be a zonal polynomial in general. Define $\mathrm{Pol}(\Omega, 0)$ to be the space formed by the constant functions on Ω, and let $\mathrm{Pol}(\Omega, 1)$ be $\mathrm{Zon}(\Omega, 1)$. Define $\mathrm{Pol}(\Omega, r + 1)$ inductively to be the span of the set

$$\{gh : g \in \mathrm{Pol}(\Omega, r), \ h \in \mathrm{Pol}(\Omega, 1)\}.$$

On occasion we will have to consider situations where we have more than one polynomial space on the same underlying set. In this case we will write $\mathrm{Pol}_\rho(\Omega, r)$ rather than $\mathrm{Pol}(\Omega, r)$. The union of all the spaces $\mathrm{Pol}(\Omega, r)$ will be denoted by $\mathrm{Pol}(\Omega)$. We will often refer to the elements of $\mathrm{Pol}(\Omega)$ as *polynomials* on Ω, and to the elements of $\mathrm{Pol}(\Omega, r) \setminus \mathrm{Pol}(\Omega, r - 1)$ as *polynomials of degree r*. We can now state our second axiom:

*Department of Mathematics and Statistics, University of Auckland, Private Bag 92019, Auckland, New Zealand. Support from the N. Z. Lottery grants Board and the Auckland University Research Committee is gratefully acknowledged.

†Department of Combinatorics and Optimization, University of Waterloo, Waterloo Ont. N2L 3G1, Canada. Support from grant OGP0093041 of the National Sciences and Engineering Research Council of Canada is gratefully acknowledged.

II The dimension of the vector space $\text{Pol}(\Omega, 1)$ is finite.

This axiom has no content if Ω is finite, since in that case $\text{Pol}(\Omega)$ is contained in the space of all real-valued functions on Ω, which has dimension equal to $|\Omega|$.

To complete the description of a polynomial space, we must provide an inner product $\langle\,,\,\rangle$ on $\text{Pol}(\Omega)$. This is required to satisfy the remaining two axioms:

III For all polynomials f and g on Ω,

$$\langle f, g \rangle = \langle 1, fg \rangle.$$

IV If f is a non-negative polynomial on Ω then $\langle 1, f \rangle \geq 0$, with equality if and only if f is identically zero.

Axiom IV is vacuous in many cases, in particular when $\text{Pol}(\Omega)$ is the set of all functions on Ω. (The reader may treat this as an exercise.) For all the finite polynomial spaces we consider, the inner product of polynomials f and g will be given by

$$\langle f, g \rangle = \frac{1}{|\Omega|} \sum_{x \in \Omega} f(x)g(x).$$

Suppose we are given two distance functions ρ and σ on the same underlying set Ω. It follows from our definitions that, if there are real numbers a and b such that

$$\rho(x, y) = a\sigma(x, y) + b$$

for all x and y in Ω, the spaces $\text{Pol}_\rho(\Omega, r)$ and $\text{Pol}_\sigma(\Omega, r)$ are equal for all r. It is important to note that we have equality, and not just isomorphism. Hence if (Ω, ρ) is a polynomial space with respect to some inner product then (Ω, σ) is a polynomial space with respect to the same inner product. We will say that these two spaces are *affinely equivalent*. We will not normally need to distinguish between affinely equivalent polynomial spaces.

A polynomial space (Ω, ρ) is *spherical* if there is an injection, τ say, of Ω into a sphere with centre at the origin in some real vector space such that:

(a) the image $\tau(\Omega)$ is closed,

(b) for any x and y in Ω we have $\rho(x, y) = \langle \tau(x), \tau(y) \rangle$,

(c) if f and g are polynomials on Ω and μ is the standard measure on the sphere, then

$$\langle f, g \rangle = \int f(\tau(x))g(\tau(x)) \, d\mu.$$

The *dimension* of a polynomial space (Ω, ρ) is defined to be $\dim \text{Pol}(\Omega, 1) - 1$. A spherical polynomial space with dimension d can always be embedded as above in a sphere in $\mathbf{R}^d$. (This follows from [3: Lemma 5.7].) We will see in the next section that the symmetric group is spherical.

2. Two examples and two applications. To give some idea of what can be done with polynomial spaces, we briefly discuss two examples and then state two of the main results from the theory. The notation we introduce to present these results will be re-used in the following sections.

Our first example is the *Johnson scheme* $J(v, k)$. In this case Ω is the set of all k-subsets of a fixed set with v elements and if x and y are elements of Ω define

$$\rho(x, y) = |x \cap y|.$$

The second example is the *symmetric group* $\mathrm{Sym}(n)$. Here Ω is the set of all permutations on n points, and for two permutations α and β we define $\rho(\alpha, \beta)$ to be the number of points fixed by $\beta \alpha^{-1}$. Note that the mapping τ which takes a permutation π to the corresponding $n \times n$ permutation matrix is a spherical embedding of $\mathrm{Sym}(n)$ into $\mathbf{R}^{n^2}$. (However we will see in Section 4, following the proof of Theorem 4.3, that $\mathrm{Sym}(n)$ actually has dimension $1 + (n-1)^2$.)

Let (Ω, ρ) be a polynomial space and let Φ be a subset of Ω. The *degree* of Φ is the cardinality of the set

$$\{\rho(x, y) : x, y \in \Phi, \ x \neq y\}.$$

In all cases of interest to us, the degree will be finite. A subset Φ of Ω is *orderable* if there is a linear ordering of Φ, denoted by '$<$', such that for each element a of Φ,

$$\rho(a, a) \notin \{\rho(a, x) : x < a\}.$$

If Ω is orderable then any subset of it is orderable, and in this case we will say that the polynomial space itself is orderable. It is easy to verify that both the Johnson scheme and the symmetric group are orderable.

2.1 THEOREM [4]. *Let (Ω, ρ) be a polynomial space and let Φ be an orderable subset of Ω with degree s. Then*

$$|\Phi| \leq \dim \mathrm{Pol}(\Omega, s). \quad \square$$

For the Johnson scheme it can be shown that if $r \leq k$ then

$$\dim \mathrm{Pol}(\Omega, r) = \binom{v}{r}.$$

Given this we see that, for $J(v, k)$, Theorem 2.1 already occurs as in [8]. Determining $\dim \mathrm{Pol}(\Omega, r)$ for $\mathrm{Sym}(n)$ is more difficult, and will be carried out for the first time in Section 4 of this paper.

A *t-design* in a polynomial space (Ω, ρ) is a finite subset Φ of Ω such that, for all f in $\mathrm{Pol}(\Omega, t)$,

$$(2.1) \qquad \langle 1, f \rangle = \frac{1}{|\Phi|} \sum_{x \in \Phi} f(x).$$

The right side of (2.1) will be denoted by $\langle 1, f \rangle_\Phi$. The *strength* of a subset Φ of a polynomial space is the largest value of t such that, for all f in $\mathrm{Pol}(\Omega, t)$,

$$\langle 1, f \rangle_\Phi = \langle 1, f \rangle.$$

A *t-design* in $J(v, k)$ is a *t-design* in the usual sense. (See [3].) For information on *t*-designs in $\mathrm{Sym}(n)$, see the next section.

2.2 THEOREM [3]. *Let* (Ω, ρ) *be a polynomial space and let* Φ *be a t-design in* Ω. *Then*

$$|\Phi| \geq \dim \mathrm{Pol}(\Omega, \lfloor \tfrac{t}{2} \rfloor). \quad \square$$

For the Johnson scheme this implies both Fisher's inequality and the inequalities of Ray-Chaudhuri and Wilson [8].

3. $\mathrm{Pol}(\Omega, r)$ and $\mathrm{Zon}(\Omega, r)$. It is clear that $\mathrm{Zon}(\Omega, r)$ is a subspace of $\mathrm{Pol}(\Omega, r)$; we are interested in determining classes of polynomial spaces for which these spaces coincide for all r. Our first lemma provides one important family. Some further notation is needed for the proof: we will denote the Kronecker product of r copies of A by $A^{\otimes r}$.

3.1 THEOREM. *If* (Ω, ρ) *is a spherical polynomial space then* $\mathrm{Zon}(\Omega, r) = \mathrm{Pol}(\Omega, r)$ *for all non-negative integers* r.

Proof. If (Ω, ρ) is spherical then there is no loss in identifying it with a subset of the unit sphere in $\mathbf{R}^n$, for some n, and thus $\rho(a, b) = a^T b$. Let $\Omega(s)$ denote the set of all vectors

$$w_1 \otimes \cdots \otimes w_s, \quad \text{where } w_1, \dots, w_s \in \Omega.$$

For each element $w_1 \otimes \cdots \otimes w_s$ of $\Omega(s)$ the function on Ω defined by

$$x \mapsto \prod_{i=1}^{s} w_i^T x$$

lies in $\mathrm{Pol}(\Omega, s)$. If V_s denotes the space spanned by all these functions then it is not hard to see that $\mathrm{Pol}(\Omega, r)$ is the join of $V_0, \dots, V_r$. Let U_s denote the span of the functions

$$x \mapsto (w^T x)^s, \quad \text{for } w \in \Omega.$$

Then $\mathrm{Zon}(\Omega, r)$ is the join of the spaces $U_0, \dots, U_r$ and so, since U_s is a subspace of V_s, we may complete the proof by showing that $\dim V_s \leq \dim U_s$ for all s.

Assume that γ is a function on Ω with finite support such that for all $b \in \Omega$,

$$(3.1) \qquad \sum_{a \in \Omega} \gamma(a)\rho_b(a)^s = 0.$$

Then

$$\sum_{a,b \in \Omega} \gamma(b)\gamma(a)\rho_b(a)^s = 0$$

and, as $\rho_b(a) = b^T a$, it follows that

$$\sum_{a \in \Omega} \gamma(a) a^{\otimes s} = 0.$$

Hence, for any sequence $b_1, \ldots, b_s$ of elements of Ω,

$$\sum_{a \in \Omega} \gamma(a)(b_1 \otimes \cdots b_s)^T a^{\otimes s} = 0,$$

yielding in turn that

$$(3.2) \qquad \sum_{a \in \Omega} \gamma(a)\rho_{b_1}(a) \cdots \rho_{b_s}(a) = 0.$$

If U_s has dimension d then, for every set of $d+1$ points Φ from Ω, there is a function γ supported by Φ such that (3.1) holds. Since (3.1) implies (3.2), we deduce that $\dim V_s \leq d$. $\square$

We do not really need (Ω, ρ) to be spherical in the above argument. It suffices that there be an embedding τ of Ω into $\mathbf{R}^n$, for some n, such that

$$\rho(a, b) = \tau(a)^T \tau(b).$$

For example, the conclusion of Theorem 3.1 holds for the polynomial space formed by the power set of finite set, where $\rho(a, b)$ is defined as $|a \cap b|$.

4. The symmetric group. The main result of this section is the following.

4.1 THEOREM. *Let (Ω, ρ) be* $\mathrm{Sym}(n)$, *viewed as a polynomial space. If Φ is a subgroup of* $\mathrm{Sym}(n)$ *then Φ is a t-design if and only if it is t-transitive.*

Proof. Let $u_{(r)}$ be defined inductively by setting $u_{(0)}$ identically equal to one and, if $r > 0$,

$$u_{(r)} = (u - r + 1)u_{(r-1)}.$$

Assume that $\mathrm{Sym}(n)$ acts on the set $X = \{1, \ldots, n\}$ and denote its identity element by 1. Suppose Φ is a subgroup of Ω. If $a \in \Omega$ then $\rho(a, x)_{(s)}$ is equal to the number of ordered s-tuples σ of distinct elements of X such that $\sigma a = \sigma x$. We observe that, by "Burnside's lemma", the number of orbits of Φ on the ordered s-tuples of distinct elements of X is equal to

$$(4.1) \qquad \langle 1, (\rho_1)_{(s)} \rangle_\Phi.$$

Hence we see that Φ is s-transitive if and only if the inner product in (4.1) is equal to 1. If Φ is a t-design and $s \leq t$ then

$$\langle 1, (\rho_1)_{(s)} \rangle_\Phi = \langle 1, (\rho_1)_{(s)} \rangle = 1$$

which implies that Φ is t-transitive.

Thus we may now assume that Φ is a t-transitive subgroup of Ω, and seek to prove that Φ is a t-design. Our first step is to show that, for any element b of Ω,

(4.2) $$\langle 1, (\rho_b)_{(t)} \rangle = 1.$$

Now

$$|\Phi| \langle 1, (\rho_b)_{(t)} \rangle = \sum_{x \in \Phi} \rho(b, x)_{(t)}$$

and the sum here is equal to the number of ordered pairs (x, α) such that $x \in \Phi$, α is an ordered t-tuple of distinct elements of X and $\alpha x = \alpha b$. This equals

(4.3) $$\sum_\alpha |\{x \in \Phi : \alpha x = \alpha b\}|.$$

Since Φ is t-transitive,

$$|\{x \in \Phi : \alpha x = \alpha b\}| = |\{x \in \Phi : \alpha x = \alpha\}|,$$

from which it follows that the sum in (4.3) is equal to $|\Phi|$, independent of the choice of b from Ω.

Hence, for any integer s such that $0 \leq s \leq t$ and any element a of Ω, we have

$$\langle 1, (\rho_b)_{(s)} \rangle_\Phi = 1 = \langle 1, (\rho_a)_{(s)} \rangle.$$

Consequently

$$\langle 1, f \rangle_\Phi = \langle 1, f \rangle$$

for any f in $\mathrm{Zon}(\Omega, t)$. By Lemma 3.1, $\mathrm{Zon}(\Omega, t) = \mathrm{Pol}(\Omega, t)$ and thus we deduce that Φ is a t-design. $\square$

The rest of this section will be devoted to deriving an expression for $\dim \mathrm{Pol}(\Omega, t)$. This will require a knowledge of the basics of the theory of group representations. As before, we denote the identity element of a group G by 1.

If φ is a function on G, let $M(\varphi)$ be the matrix with rows and columns indexed by the elements of G, and with

$$(M(\varphi))_{xy} = \varphi(x^{-1}y).$$

If φ is a character on G then $M(\varphi)$ is a hermitian matrix. If φ and ψ are two irreducible characters on G then, by [2: Theorem 2.5] we have:

$$M(\varphi)M(\psi) = \begin{cases} \frac{|G|}{\varphi(1)} M(\varphi) & \text{if } \varphi = \psi \\ 0 & \text{otherwise}. \end{cases}$$

This implies that all eigenvalues of $M(\varphi)$ are equal to $|G|/\varphi(1)$ or 0. Since $\mathrm{tr}\, M(\varphi) = |G|\varphi(1)$ it follows that exactly $\varphi(1)^2$ of the eigenvalues of $M(\varphi)$ are non-zero and therefore $M(\varphi)$ has rank $\varphi(1)^2$. More generally:

4.2 LEMMA. *Let ψ be a character of the finite group G, and suppose that $\psi = \sum_\varphi c_\varphi \varphi$ is the unique representation of ψ as a linear combination of irreducible characters. Then*

$$\operatorname{rank}(M) = \sum_{\varphi : c_\varphi \neq 0} \varphi(1)^2. \quad \square$$

We are now going to apply the above to the symmetric group. Let ψ be the character which maps an element of $\operatorname{Sym}(n)$ to the number of points it fixes, and let 1 denote the trivial character. Then $\psi - 1$ is an irreducible character. We denote the usual inner product on characters by $(\, , \,)$ and define the *depth* of an irreducible character φ of $\operatorname{Sym}(n)$ to be the least integer d such that (φ, ψ^d) is non-zero. (For a little more on this see, e.g., [5: pp. 99, 234]. In [1] it is referred to as the *dimension* of the character.)

4.3 THEOREM. *If (Ω, ρ) is the symmetric group then $\dim \operatorname{Pol}(\Omega, r)$ is equal to the sum of the squares of the degrees of the irreducible characters of the symmetric group with depth at most r.*

Proof. First, $\operatorname{Pol}(\Omega, r)$ and $\operatorname{Zon}(\Omega, r)$ are the same, so we will compute the dimension of the latter space. This dimension is equal to the dimension of the join of the row spaces of the matrices $M(\psi^s)$ for $s = 0, \ldots, r$. But, by the previous lemma, each of these matrices is a linear combination of the matrices $M(\varphi)$, where φ ranges over the characters of $\operatorname{Sym}(n)$ of depth at most r. Since these matrices are pairwise orthogonal, the lemma follows. $\square$

Combining Theorem 4.3 and Theorem 2.1 we obtain an upper bound on the cardinality of a subset of $\operatorname{Sym}(n)$ with degree s. A related, but somewhat stronger, bound occurs as Theorem 2.5 in [1]. (It can be derived by using [4: Theorem 2.2] in place of Theorem 2.1.) For the symmetric group we find that $\dim \operatorname{Pol}(\Omega, 1) = 1 + (n-1)^2$. (See [5: Chapter 2].) Thus we deduce that a subset of degree one in the symmetric group has cardinality at most $n^2 - 2n + 2$. Subsets of degree one in $\operatorname{Sym}(n)$ are known as *equidistant permutation* arrays. The *index* λ of such an array is the value of $\rho(a, b)$ for any pair of distinct permutations a and b in it. When $\lambda = 1$ the best upper bound is due to Mathon [6], and is of order $n^2 - 4n$, while if $\lambda > 1$ and n is large enough the best bound [9], due to van Rees and Vanstone, is

$$(n - \lambda)^2 - 5(n - \lambda) + 7.$$

The proof of the latter is complicated, and so it is interesting that our general approach yields such a good bound.

5. Coordinate rings. Let S be a subset of $\mathbf{R}^n$. The *coordinate ring* of S is the ring formed by the restrictions to S of the real polynomials in n variables. The elements of the coordinate ring are usually referred to as *polynomials on* S.

5.1 THEOREM. *Let (Ω, ρ) be a spherical polynomial space with dimension n. If Ω is viewed as a subset of a sphere centred at the origin in $\mathbf{R}^n$, then $\operatorname{Pol}(\Omega, r)$ is the set of all polynomials with degree at most r in n variables, restricted to Ω.*

Proof. There is a spherical embedding τ of Ω into $\mathbf{R}^n$ such that $\tau(\Omega)$ spans $\mathbf{R}^n$ and there is no loss in identifying Ω with its image $\tau(\Omega)$. Let $e_1, \dots, e_n$ denote the standard basis vectors for $\mathbf{R}^n$. Since Ω spans $\mathbf{R}^n$ and $\rho(a, b)$ is linear in each coordinate, for $i = 1, \dots, n$ there are points $a_1, \dots, a_m$ and scalars $\lambda_1, \dots, \lambda_m$ such that

$$e_i^T y = \sum_{r=1}^{m} \lambda_r \rho(a_r, y).$$

Thus $\mathrm{Pol}(\Omega, 1)$ contains the linear coordinate functions on Ω, and it follows that $\mathrm{Pol}(\Omega, 1)$ is spanned by these functions. The theorem follows now from the definition of $\mathrm{Pol}(\Omega, r)$. □

Now let (Ω, ρ) be the symmetric group, viewed as a polynomial space, and define ψ by

$$\psi(x, y) = \rho(x, y) - 1.$$

Then (Ω, ψ) is a polynomial space, affinely equivalent to (Ω, ρ). If 1 is the identity in $\mathrm{Sym}(n)$ then the function $\psi(1, y)$ is the character of an irreducible representation of $\mathrm{Sym}(n)$ with degree $n - 1$. This representation gives rise to an embedding, $\hat{\psi}$ say, of $\mathrm{Sym}(n)$ in $O(n-1)$. From Theorem 4.3 and the remarks which follow it, $\mathrm{Sym}(n)$ has dimension $1 + (n-1)^2$ as a polynomial space. Hence $\hat{\psi}$ is a spherical embedding of $\mathrm{Sym}(n)$, and therefore we may identify $\mathrm{Pol}(\Omega)$ in a natural way with the coordinate ring of $\hat{\psi}(\mathrm{Sym}(n))$.

If A and B are two $n \times n$ orthogonal matrices, define $\rho(A, B)$ to be $\operatorname{tr} A^T B$. If Ω is the group of all $m \times m$ orthogonal matrices $O(m)$ then (Ω, ρ) is a polynomial space. (We take the inner product of functions f and g to be

$$\int fg \, d\mu$$

where μ is the Haar measure on $O(m)$, although this will not be needed in what follows.) Define the coordinate functions f_{ij} on $O(m)$ by

$$f_{ij}(A) = (A)_{ij}.$$

Any finite subgroup Γ of $O(m)$ also gives rise to a polynomial space with the same function ρ, and we denote this by (Γ, ρ). A theorem due to Burnside (see, e.g., [7: Theorem IX.2]) tells us that if Γ is irreducible then the matrices in it span $\mathbf{R}^{m^2}$. The proof of Theorem 5.1 then shows that $\mathrm{Pol}(\Gamma, r)$ is precisely the set of polynomials of degree at most r in the coordinate ring of Γ, viewed as a subset of $\mathbf{R}^{m^2}$.

The coordinate ring of Γ is the quotient of the ring of all polynomials in m variables with respect to the ideal formed by the polynomials in m variables which vanish on Γ. The space spanned by this ideal, together with the constant functions, is the ring of invariants of Γ. Thus we have yet another view of $\mathrm{Pol}(\Gamma, r)$.

REFERENCES

[1] P. J. CAMERON, M. DEZA AND P. FRANKL, *Sharp sets of permutations*, J. Algebra, 111 (1987), pp. 220–247.

[2] M. J. COLLINS, *Representations and Characters of Finite Groups*, Cambridge U. P. (Cambridge) 1990.

[3] C. D. GODSIL, *Polynomial spaces*, Discrete Math. 73 (1988/89), pp. 71–88.

[4] C. D. GODSIL, *Q-Polynomial spaces*, submitted.

[5] G. JAMES AND A. KERBER, *The Representation Theory of the Symmetric Group*, Addison-Wesley (Reading, Massachusetts) 1981.

[6] R. MATHON, *Bounds for equidistant permutation arrays of index one*, in Combinatorial Mathematics: Proceedings of the the Third International Conference. Eds. G. S. Bloom, R. L. Graham, J. Malkevitch, Annals N. Y. Acad. Sci. Vol. 555 (1989).

[7] M. NEWMAN, *Integral Matrices*, Academic Press (New York) 1972.

[8] D. K. RAY-CHAUDHURI AND R. M. WILSON, *On t-designs*, Osaka J. Math. 12 (1975), pp. 737–744.

[9] G. H. J. VAN REES AND S. A. VANSTONE, *Equidistant permutation arrays: a bound*, J. Austral. Math. Soc. (Series A) 33 (1982), pp. 262–274.

COMPLETELY POSITIVE GRAPHS*

ABRAHAM BERMAN**

Abstract. The paper describes a qualitative study of completely positive matrices. The main result is

THEOREM. *A graph is completely positive if and only if it does not contain an odd cycle of length greater than 4.*

This theorem is a result of works with Danny Hershkowitz, Bob Grone and Nataly Kogan. The paper is divided into three parts: definitions, a sketch of the proof, and questions.

Definitions. A matrix A is completely positive if it can be decomposed as $A = BB^T$ where B is a (not necessarily square) nonnegative matrix. This is not to be confused with the matrix being totally positive, which means that all the minors are nonnegative. However, symmetric totally positive matrices are completely positive, [9].

Example 1. If $a_{11} \neq 0$,

$$A = \begin{pmatrix} a_{11} & a_{12} \\ a_{12} & a_{22} \end{pmatrix} = \begin{pmatrix} \sqrt{a_{11}} & 0 \\ \frac{a_{12}}{\sqrt{a_{11}}} & \sqrt{\frac{\det A}{a_{11}}} \end{pmatrix} \begin{pmatrix} \sqrt{a_{11}} & \frac{a_{12}}{\sqrt{a_{11}}} \\ 0 & \sqrt{\frac{\det A}{a_{11}}} \end{pmatrix}$$

Applications. Block designs, [3], statistics and "a proposed mathematical model of energy demand for certain sectors of the U.S. economy" [4]; Geometric interpretation [4]: Given vectors $x_1, \ldots x_n$ in R^m, do there exist a number k and an isometry $T : R^m \rightarrow R^k$ so that $T(x_1), \ldots, T(x_n)$ lie in R_+^k? Such an isometry exists if and only if the Gram matrix $A = (\langle x_i, x_j \rangle)$ is completely positive. Another interesting geometric observation is the following:

A real symmetrix matrix A is copositive if

$$x \geq 0 \Rightarrow x^T A x \geq 0.$$

Consider the vector space S of $n \times n$ real symmetric matrices. Let CP and C denote, respectively, the sets of completely positive and copositive matrices in S. Then, CP and C are closed convex cones and each set is the dual of the other, [6].

Clearly if A is completely positive, then it is positive semidefinite and element-wise nonnegative. We shall refer to being nonnegative in these two senses as being doubly nonnegative, and say that a graph G is completely positive if every doubly nonnegative matrix realization of G is completely positive.

---

*Presented in the workshop on "Combinatorial and Graph Theoretic Problems in Linear Algebra", Institute of Mathematics and its Applications, The University of Minnesota, November 1991. Research supported by the Henry Guthwright Research Fund & by the New York Metropolitan Research Fund.
**Department of Mathematics, Technion–Israel Institute of Technology, Haifa 32000, ISRAEL.

Example 2. Doubly stochastic tridiagonal matrices are completely positive so a graph which is a path or a union of distinct paths in completely positive.

Example 3. If A is a doubly nonnegative matrix of order less than 5, then there exists a square nonnegative matrix such that $A = BB^T$, [11].

Thus small graphs, $V(G) \leq 4$, are completely positive.

Example 4. The Alfred Horn matrix

$$A = \begin{pmatrix} 4 & 0 & 0 & 2 & 2 \\ 0 & 4 & 3 & 0 & 2 \\ 0 & 3 & 4 & 2 & 0 \\ 2 & 0 & 2 & 4 & 0 \\ 2 & 2 & 0 & 0 & 4 \end{pmatrix}$$

is doubly nonnegative but not completely positive.

Example 5.

$$A = \begin{pmatrix} 2 & 0 & 0 & 1 & 1 \\ 0 & 2 & 0 & 1 & 1 \\ 0 & 0 & 2 & 1 & 1 \\ 1 & 1 & 1 & 3 & 0 \\ 1 & 1 & 1 & 0 & 3 \end{pmatrix} = \begin{pmatrix} 1 & 1 & 0 & 0 & 0 & 0 \\ 0 & 0 & 1 & 1 & 0 & 0 \\ 0 & 0 & 0 & 0 & 1 & 1 \\ 0 & 1 & 0 & 1 & 0 & 1 \\ 1 & 0 & 1 & 0 & 1 & 0 \end{pmatrix} \begin{pmatrix} 1 & 0 & 0 & 0 & 1 \\ 1 & 0 & 0 & 1 & 0 \\ 0 & 1 & 0 & 0 & 1 \\ 0 & 1 & 0 & 1 & 0 \\ 0 & 0 & 1 & 0 & 1 \\ 0 & 0 & 1 & 1 & 0 \end{pmatrix}$$

but there is *no square* nonnegative matrix B such that $A = BB^T$, [7].

A Sketch of the Proof. We divide the proof into five lemmas and give references to the original proofs.

LEMMA 1. [2]. *Let G be a graph with $V(G) = \langle n \rangle$, and let $\ell \in \langle n \rangle$. If A is a doubly nonnegative matrix realization of G and if*

there exist positive numbers d_j, $j \in N(\ell)$, such that

(*) $$\sum_{j \in N(\ell)} a_{\ell j}^2 / d_j \leq a_{\ell \ell} \text{ and the matrix } H = A(\ell) - \sum_{j \in N(\ell)} d_j E_{jj}$$

is completely positive.

then A is completely positive.

If ℓ does not lie on a triangle in G then for every doubly nonnegative matrix realization A of G, A is completely positive if and only if (*) holds.

Example 6. Let k be an odd number, $k \geq 5$ and let B be the $k \times k$ matrix

$$B = \begin{pmatrix} 1 & 1 & 0 & \cdots & 0 & 1 \\ 1 & 2 & \ddots & \ddots & & 0 \\ 0 & \ddots & \ddots & & & \vdots \\ 0 & & & & & 0 \\ 0 & & & & 2 & 1 \\ 1 & 0 & \cdots & 0 & 1 & k-2 \end{pmatrix}$$

Then the first $k - 1$ leading principal minors of B are all equal to 1 and $\det B = 0$ so B is positive semidefinite and thus doubly nonnegative. But B is not completely positive, for if it was, then by Lemma 1 with $l = k$ there would exist positive numbers d and e such that

$$\frac{1}{d} + \frac{1}{e} \leq k - 2$$

and

$$H = \begin{pmatrix} 1 - d & 1 & 0 & \cdots & & 0 \\ 1 & 2 & 1 & & & \vdots \\ 0 & 1 & 2 & & & 0 \\ \vdots & & \ddots & \ddots & 2 & 1 \\ 0 & & \cdots & 0 & 1 & 2 - e \end{pmatrix}$$

is completely positive.

But then

$$\det H[\langle k - 2\rangle | \langle k - 2\rangle] = 1 - (k - 2)d \geq 0$$

So $\frac{1}{d} \geq k - 2$.

LEMMA 2. [2]. *A graph which contains an odd cycle of length greater than 4 is not completely positive.*

LEMMA 3. [3]. *Bipartite graphs are completely positive.*

LEMMA 4. [9]. *Let G_1 and G_2 be two subgraphs of a graph G, connected only by a cutpoint k, i.e., $G = G_1 \cup G_2$ and $G_1 \cap G_2 = \{k\}$. Then, if both G_1 and G_2 are completely positive, then G also has this property.*

For example, in the butterfly (Figure 1), the vertex 3 is a cutpoint.

Figure 1

LEMMA 5. [9]. *A graph T_n consisting of n triangles with a common base (Figure 2) is completely positive.*

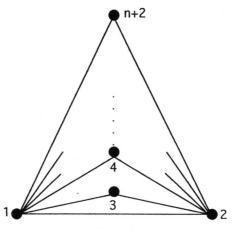

Figure 2

With these lemmas the proof is completed by observing that if G does not contain an odd cycle of length greater than 4 and if B is a block of G, i.e., a maximal connected subgraph which does not contain a cutpoint, then B must be bipartite or K_4 (the complete graph on four vertices) or of the form T_n.

A proof of Lemmas 4 and 5, using Schur complements, was suggested by Ando. See the very nice survey, [1].

Questions.

Question 1 Find quantitative conditions for complete positivity of matrices.

For example, [8], diagonally dominant nonnegative matrices are completely positive. In [12] it is observed that this result can be derived from Lemma 1.

Question 2 The factorization index, $\varphi(A)$, of a completely positive matrix A is the smallest number of columns in a nonnegative matrix B such that $A = BB^T$. Given that A is completely positive, what is its factorization index?

Clearly $\varphi(A) \geq$ rank A. (In Example 5, rank $A = 4$, the nonnegative rank is 5 and the factorization index is 6). For acyclic matrices $\varphi(A) =$ rank A. Let $2N$ be the maximal number of zero entries in a nonsingular principal submatrix of A. Then it is shown in [7] that

$$\varphi(A) \leq \frac{(\text{rank } A)((\text{ rank } A) + 1)}{2} - N.$$

Let d be the number of components of $G(A)$ and let ν be the nullity of A. If A is diagonally dominant then $\varphi(A) \leq E(G(A)) + d$ [8]. If $G(A)$ is bipartite, then, [3],

$$|E(G(A))| \leq \varphi(A) \leq |E(G(A))| + d - \nu$$

Example 7.

$$A = \begin{pmatrix} 2 & 1 \\ 1 & 1 \end{pmatrix}, \qquad \varphi(A) = 2 = 1 + |E(G(A))|$$

$$A = \begin{pmatrix} 3 & 0 & 1 & 1 \\ 0 & 3 & 1 & 1 \\ 1 & 1 & 3 & 0 \\ 1 & 1 & 0 & 3 \end{pmatrix}, \quad \varphi(A) = \text{rank } A = |E(G(A))|$$

Question 3 What is the set $I(G)$ of factorization indices of all completely positive matrix realizations of a graph G?

REFERENCES

[1]   T. ANDO, *Completely positive matrices*, Lecture Notes Sapporo (1991).
[2]   A. BERMAN AND D. HERSHKOWITZ, *Combinatorial results on completely positive matrices*, Lin. Alg. Appl. 95 (1987), pp. 111–125.
[3]   A. BERMAN AND R. GRONE, *Bipartite completely positive matrices*, Proc. Cambridge Philos. Soc. 103 (1988), pp. 269–276.
[4]   L.J. GRAY AND D.G. WILSON, *Nonnegative factorization of positive semidefinite nonnegative matrices*, Lin. Alg. Appl. 31 (1980), pp. 119–127.
[5]   M. HALL JR., *Combinatorial Theory*, Blaisdell, Lexington, 1967, 2nd ed 1986.
[6]   M. HALL JR. AND M. NEWMANN, *Copositive and completely positive quadratic forms*, Proc. Cambridge Philos. soc. 59 (1963), pp. 329–339.
[7]   J. HANNA AND T.J. LAFFEY, *Nonnegative factorization of completely positive matrices*, Lin. Alg. Appl. 55 (1983), pp. 1–9.
[8]   M. KAYKOBAD, *On nonnegative factorization of matrices*, Lin. Alg. Appl. 96 (1987), pp. 27–33.
[9]   N. KOGAN AND A. BERMAN, *Characterization of completely positive graphs*, to appear in Discrete Math..
[10] T.L. MARKHAM, *Factorization of completely positive matrices*, Proc. Cambridge Philos. Soc. 69 (1971), pp. 53–58.
[11] J.E. MAXFIELD AND H. MINC, *On the equation $X'X = A$*, Proc. Edinburgh Math. Soc. 13 (1962), pp. 125–129.
[12] L. SALCE AND P. ZANARDO, *Completely positive matrices, associated graphs and least square solutions*, preprint.

HADAMARD MATRICES

W. D. WALLIS†

Abstract. In this survey the ideas of Hadamard matrices are briefly outlined. We mention the existence problem and constructions, relations to block designs, and special types of Hadamard matrices.

0. Introduction. In 1893, Jacques Hadamard published a paper [2] which considered the following optimization problem:

What is the maximum determinant possible for an $n \times n$ matrix whose entries are all bounded in norm by a constant B?

Clearly it is sufficient to take $B = 1$. Hadamard completely solved this problem – the maximum is $n^{n/2}$, and it can always be attained.

Hadamard then considered the harder case when his problem is restricted to the reals. He proved that the stated maximum can only be attained by the determinants of real $(1, -1)$ matrices H satisfying

$$HH^T = nI.$$

He proved that (except for the trivial cases $n = 1$ and 2) there are solutions only when 4 divides n. He found solutions for orders up to 20. Because of this connection, square $(1, -1)$ matrices whose rows are pairwise orthogonal have been called *Hadamard matrices*.

These matrices were not first discussed by Hadamard, however. In 1867 Sylvester [10] discussed these matrices and orthogonal matrices in general. He proved the existence of what would later be called Hadamard matrices, for all orders a power of 2.

It has been conjectured that Hadamard matrices exist for every order divisible by 4. No counterexample exists, but no proof exists either.

Hadamard matrices arise in statistics. The most obvious way is in weighing experiments. Given a chemical balance and n objects to be weighed, a weighing consists of placing some objects in the left-hand pan, some in the right, and then measuring the weight difference using standard weights. The unknown weights are estimated by comparing these measurements; clearly at least n weighings must be made. A weighing can be represented by a vector of length n; entry i is 1, -1 or 0 according as object i is placed in the left pan, the right pan, or omitted from the weighing. The matrix with these vectors as rows is the *weighing matrix* of the experiment. It can be shown that the most efficient weighing design for n weighings of n objects (in the usual statistical sense of minimizing expected experimental error) is given by a Hadamard matrix, if one exists.

†Department of Mathematics, Southern Illinois University, Carbondale, Illinois 62901-4408

There are other relationships between Hadamard matrices and statistical designs. In section 2 we shall discuss Hadamard matrices and balanced incomplete block designs. Further relations between designs and Hadamard matrices, and other applications of Hadamard matrices, are discussed in [3]. In particular Hadamard matrices are used in constructing error-correcting codes; for a practical example, see [7].

1. Preliminaries. A *Hadamard matrix* H of order or side n is an $n \times n$ matrix with all its entries chosen from $\{1, -1\}$, in which the rows are pairwise orthogonal: that is,

$$HH^T = nI.$$

Observe that such a matrix has an inverse – $H^{-1} = n^{-1}H^T$ – and

$$H^T H = nH^{-1}H = nI,$$

so H^T is Hadamard if and only if H is. Consequently we could say the columns are pairwise orthogonal, rather than the rows, in the definition.

It is obvious that a Hadamard matrix retains its defining properties if rows or columns are negated or permuted. We shall say two Hadamard matrices are *equivalent* if one can be obtained from the other in this fashion. In particular, we shall define a Hadamard matrix to be *standardized* if its first row contains only positive entries, and *normalized* if both the first row and first column have this property. Clearly any Hadamard matrix is equivalent to a standardized (or a normalized) Hadamard matrix.

The first question which comes to mind is, for what n does there exist a Hadamard matrix of order n? The following theorem provides a necessary condition.

THEOREM 1.1 [2]. *If there is a Hadamard matrix of order n, then $n = 1$ or $n = 2$ or n is divisible by 4.*

Proof. Without loss of generality we may assume $n \geq 3$. If there is a Hadamard matrix of side n, there is a normalized Hadamard matrix of order n; call it H. We look at the first three entries in each row: say H has

α	rows of form	1	1	1	...
β	rows of form	1	1	−	...
γ	rows of form	1	−	1	...
δ	rows of form	1	−	−	...

(here and on many occasions subsequently we simply write - for -1). Then

$$\begin{aligned}
\alpha + \beta + \gamma + \delta &= n \quad \text{(count rows)} \\
\alpha + \beta - \gamma - \delta &= 0 \quad \text{(column 1} \cdot \text{column 2)} \\
\alpha - \beta + \gamma - \delta &= 0 \quad \text{(column 1} \cdot \text{column 3)} \\
\alpha - \beta - \gamma + \delta &= 0 \quad \text{(column 2} \cdot \text{column 3)}
\end{aligned}$$

and adding we have

$$4\alpha = n,$$

so n is divisible by 4. □

Of course this only answers one half of the question. For which of these orders is there a Hadamard matrix? The two exceptional small orders are easily handled:

$$1 \qquad \begin{array}{cc} 1 & 1 \\ 1 & - \end{array}$$

and it has been conjectured that such matrices exist for all orders divisible by 4. However this is a very difficult question. Many constructions have been found. We shall introduce a few of the methods in Section 3, but a proper discussion goes well beyond the scope of an elementary survey – the reader should consult [1], [9] and [13]. We can currently construct Hadamard matrices of order $4t$ for all $4t$ less than 428; there are eight unknown cases under 1000, and twelve further cases under 2000, namely

$$428, 652, 668, 716, 764,$$
$$852, 892, 956, 1004, 1132,$$
$$1244, 1388, 1436, 1676, 1772,$$
$$1852, 1912, 1916, 1948, 1964.$$

(These are taken from the extensive listing in [9].)

2. The relationship with block designs. A *balanced incomplete block design* with parameters (v, k, λ), or $(v, k, \lambda)-bibd$, is a way of selecting k-sets (called *blocks*) from a set of v objects so that any two objects occur together in precisely λ subsets. To avoid triviality we shall always assume that $v > k$.

Suppose element x belongs to r of the blocks. If we count all the occurrences of other objects in those r blocks we get $r(k-1)$ occurrences (as there are $k-1$ further objects in each block). On the other hand, each of the $v-1$ remaining objects belongs to λ blocks which contain x. So

$$(2.1) \qquad\qquad r(k-1) = \lambda(v-1)$$

If we count all entries in all blocks, we get bk entries. But each object must have been counted r times. So

$$(2.2) \qquad\qquad bk = vr.$$

We define the incidence matrix of a (v, k, λ)-bibd as follows. First select an ordering of the elements and an ordering of the blocks. Then define

$$b_{ij} \begin{array}{l} = \quad 1 \text{ if object } i \text{ lies in block } j \\ = \qquad 0 \text{ otherwise.} \end{array}$$

Then the $v \times b$ matrix $B = (b_{ij})$ is the *incidence matrix* of the design. It is easy to see that

$$BB^T = (r - \lambda)I + \lambda J.$$

Then BB^T has determinant $(r - \lambda)^{v-1}(r + (v - 1)\lambda)$, which is non-zero (if $r = \lambda$ then $k = v$, and that case was outlawed). Since BB^T is $v \times v$, we have

(2.3) $$v \leq b$$

(which is called *Fisher's inequality*).

Given a Hadamard matrix H of order $4n$, let us define a matrix B as follows. First, normalize H (by negating rows and columns). Then change every entry -1 to 0. Finally, erase the first row and column. Then B is the incidence matrix of a $(4n - 1, 2n - 1, n - 1) - $ bibd. The construction is reversible, so we have

THEOREM 2.1. *There is a Hadamard matrix of order $4n$ if and only if there is a $(4n - 1, 2n - 1, n - 1) - $ bibd.*

(For further relationships between Hadamard matrices and block designs of various types, see [3].)

As an example of Theorem 2.1, consider the Hadamard matrix

$$
\begin{array}{cccccccc}
1 & 1 & 1 & 1 & 1 & 1 & 1 & 1 \\
1 & 1 & 1 & 1 & - & - & - & - \\
1 & 1 & - & - & 1 & 1 & - & - \\
1 & 1 & - & - & - & - & 1 & 1 \\
1 & - & 1 & - & 1 & - & 1 & - \\
1 & - & 1 & - & - & 1 & - & 1 \\
1 & - & - & 1 & 1 & - & - & 1 \\
1 & - & - & 1 & - & 1 & 1 & - \\
\end{array}
$$

The corresponding incidence matrix is

$$
\begin{array}{ccccccc}
1 & 1 & 1 & 0 & 0 & 0 & 0 \\
1 & 0 & 0 & 1 & 1 & 0 & 0 \\
1 & 0 & 0 & 0 & 0 & 1 & 1 \\
0 & 1 & 0 & 1 & 0 & 1 & 0 \\
0 & 1 & 0 & 0 & 1 & 0 & 1 \\
0 & 0 & 1 & 1 & 0 & 0 & 1 \\
0 & 0 & 1 & 0 & 1 & 1 & 0 \\
\end{array}
$$

which is the incidence matrix of a $(7, 3, 1) - $ bibd with blocks

$$
\begin{array}{cccc}
123 & 145 & 167 & 246 \\
257 & 347 & 356 &
\end{array}
$$

The design corresponding to a Hadamard matrix has the special property that $b = v$ (and $r = k$). A design of this special kind is called *symmetric*. Symmetric designs are particularly interesting for a number of reasons, one of which is because they are the extreme case in Fisher's inequality. The equations (2.1) and (2.2) boil down to

(2.4) $$k(k - 1) = \lambda(v - 1)$$

in the symmetric case.

It is easy to see that if B is the incidence matrix of a symmetric (v, k, λ)-bibd then B^T is also the incidence matrix of a symmetric (v, k, λ)-bibd. In that case B satisfies

(i) $\quad BB^T = (k - \lambda)I + \lambda J, \qquad B^T B = (k - \lambda)I + \lambda J,$

(ii) $\quad AJ = kJ, \qquad\qquad\qquad JA = kJ.$

THEOREM 2.2 [8]. *Any non-singular $v \times v$ $(0, 1)$-matrix which satisfies one of the equations (i) and one of the equations (ii) must satisfy all four equations, and be the incidence matrix of a $(v, k, \lambda) - $ bibd (which will be symmetric).*

3. Some constructions. In this section we briefly survey some of the more important easy constructions for Hadamard matrices. As was pointed out, this is only a short outline, and the interested reader should consult [1], [9] or [13].

3.1 Kronecker product construction. Given matrices A and B, define the Kronecker product $A \otimes B$ to be the block matrix whose blocks are the same size as B, with (i, j) block

$$a_{ij}B.$$

THEOREM 3.1 [2]. *If H and K are Hadamard then so is $H \otimes K$.*

COROLLARY. *If there are Hadamard matrices of order h and k, there is one of order hk.*

The following particular case was in fact discussed by Sylvester [9]. Suppose

$$A = \begin{matrix} 1 & 1 \\ -1 & 1 \end{matrix}$$

Then

$$A \otimes B = \begin{matrix} B & B \\ -B & B \end{matrix}$$

and in particular $A \otimes A$ is

$$\begin{matrix} 1 & 1 & 1 & 1 \\ -1 & 1 & -1 & 1 \\ -1 & -1 & 1 & 1 \\ -1 & 1 & 1 & -1 \end{matrix}$$

So there is a Hadamard matrix of order 2^n for every positive integer n.

3.2 Paley's construction. Paley [6] observed the following. Say q is a prime power congruent to $3 \pmod 4$; say $q = 4n - 1$. Label the elements of the finite field $GF[q]$ as $x_1, x_2, \ldots$ in some order. Write $b_{ij} = 0$ if $x_i - x_j$ is a quadratic element (that is, if $z^2 = x_i - x_j$ has a solution in $GF[q]$), and 1 otherwise. Then $B = (b_{ij})$ is the incidence matrix of a $(4n - 1, 2n - 1, n - 1)$-bibd. From Theorem 2.1 we have

THEOREM 3.2. *There is a Hadamard matrix of order $4n$ whenever $4n - 1$ is a prime power.*

As an example consider the case $4n - 1 = 7$. The quadratic elements modulo 7 are 0, 1, 2 and 4, because:

$$
\begin{aligned}
0 &= 0^2 \\
1 &= 1^2 = 6^2 \\
2 &= 3^2 = 4^2 \\
4 &= 2^2 = 5^2
\end{aligned}
$$

The block design has incidence matrix

$$
\begin{array}{ccccccc}
0 & 0 & 0 & 1 & 0 & 1 & 1 \\
1 & 0 & 0 & 0 & 1 & 0 & 1 \\
1 & 1 & 0 & 0 & 0 & 1 & 0 \\
0 & 1 & 1 & 0 & 0 & 0 & 1 \\
1 & 0 & 1 & 1 & 0 & 0 & 0 \\
0 & 1 & 0 & 1 & 1 & 0 & 0 \\
0 & 0 & 1 & 0 & 1 & 1 & 0
\end{array}
$$

and the Hadamard matrix is constructed as in the preceding section.

3.3 Williamson's method. Williamson [14] used the array

$$
\begin{array}{cccc}
A & B & C & D \\
-B & A & -D & C \\
-C & D & A & -B \\
-D & -C & B & A
\end{array}
$$

The rows of this matrix are formally orthogonal. If A, B, C and D are replaced by $n \times n$ $(1, -1)$ matrices which satisfy

$$ AA^T + BB^T + CC^T + DD^T = 4nI $$

and which satisfy

$$ XY^T = YX^T $$

whenever X and Y are in $\{A, B, C, D\}$, then the result is Hadamard. This technique has been extensively generalized.

In his paper Williamson made the following further assumptions: A, B, C and D are circulant and symmetric. Then $XY^T = YX^T$ becomes $XY = YX$, and this is always satisfied because the matrices are all polynomials in the circulant matrix

$$
\begin{array}{ccccccc}
0 & 1 & 0 & 0 & 0 & 0 & \ldots \\
0 & 0 & 1 & 0 & 0 & 0 & \ldots \\
0 & 0 & 0 & 1 & 0 & 0 & \ldots
\end{array}
$$

$$ \ldots $$

As an example, in the case $n = 3$, one may take

$$
A = \begin{array}{ccc} 1 & 1 & 1 \\ 1 & 1 & 1 \\ 1 & 1 & 1 \end{array}
\qquad
B = C = D = \begin{array}{ccc} 1 & - & - \\ - & 1 & - \\ - & - & 1 \end{array}
$$

4. Special Hadamard matrices. We conclude with a discussion of some types of Hadamard matrices with special defining properties, which have been of particular interest for various reasons.

4.1 Regular Hadamard matrices. A Hadamard matrix is called *regular* if every row contains the same number of entries +1. Say H has order $4n$ and has h 1's per row, so that it has constant row sum $h - (4n - h) = 2h - 4n$. Consider the matrix $A = (H + J)/2$. A satisfies

$$
\begin{aligned}
AA^T &= (HH^T + HJ + (HJ)^T + J^2)/4 \\
&= (4nI + 2(2h - 4n)J) + 4nJ)/4 \\
&= nI + (h - n)J.
\end{aligned}
$$

(4.1)

Also

$$AJ = hJ$$

From Theorem 2.2, A is the incidence matrix of a $(4n, h, h - n)$-bibd, which is symmetric. Equation (2.4) becomes

$$h(h - 1) = (h - n)(4n - 1)$$

so $h = 2n \pm \sqrt{n}$, and n must be a perfect square.

THEOREM 4.1. *Regular Hadamard matrices can only exist if the order is a perfect square.*

There are several theorems for constructing regular Hadamard matrices (see [9], [13]). The regularity property is preserved by Kronecker product, so infinite classes are easily constructed.

4.2 Circulants. A *circulant* matrix is one obtained by circulating its first row: for example,

$$
\begin{array}{cccc}
a & b & c & d \\
d & a & b & c \\
c & d & a & b \\
b & c & d & a
\end{array}
$$

There is a circulant Hadamard matrix of order 4:

$$
\begin{array}{cccc}
1 & - & 1 & 1 \\
1 & 1 & - & 1 \\
1 & 1 & 1 & - \\
- & 1 & 1 & 1
\end{array}
$$

Are there any others? If so they would be of considerable interest in communications theory and in the theory of difference sets. Turyn [11, 12] has shown that such matrices must be rare, and recently Jedwab and Lloyd [4] showed that no such matrix can exist of order less than 10^8, with 6 possible exceptions. (The latter paper has an up-to-date bibliography of the problem.)

The existence or otherwise of circulant Hadamard matrices has attracted a number of researchers, probably because the problem is so easy to state, and many false "proofs" exist. Turyn alone reports that he has caught at least eight false proofs in the refereeing process, and a number of others have seen print. (See, for example, [5].)

4.3 Graphical Hadamard matrices. Symmetric Hadamard matrices with constant diagonal have been discussed because they have an obvious graph-theoretic interpretation, and the graphs arising are of special interest. In fact, such matrices have been called *graphical*.

If H is any symmetric Hadamard matrix of order n, it satisfies

$$H^2 = nI,$$

so the minimum polynomial of H divides $x^2 - n$. Therefore the eigenvalues of H are the roots of this equation – say $\sqrt{n}$ occurs α times and $-\sqrt{n}$ β times. H is a real symmetric matrix, so the sum of the eigenvalues equals the trace:

$$(\alpha - \beta)\sqrt{n} = tr(A).$$

In particular, if H is graphical, its trace is n. So

$$\begin{aligned}(\alpha - \beta)\sqrt{n} &= n, \\ (\alpha - \beta) &= \sqrt{n},\end{aligned}$$

and n must be a perfect square.

THEOREM 4.2. *Graphical Hadamard matrices can only exist if the order is a perfect square.*

Again, Kronecker product preserves the graphical property – for example, the existence of a graphical Hadamard matrix of order 4 ensures that there is one of order every even power of 2.

Observe the similarity to the case of regular Hadamard matrices – in fact, regular symmetric Hadamard matrices with constant diagonal constitute an interesting class of matrices, and the special case of regular graphical Hadamard matrices has been very useful in the construction of regular Hadamard matrices.

4.4 Skew-Hadamard matrices. Obviously a Hadamard matrix cannot be skew, because the diagonal of a skew matrix must be zero. But there is a closely related idea. We say a Hadamard matrix H is *skew-type* of its diagonal entries are all 1's and it is skew otherwise:

$$(H - I)^T = -(H - I).$$

These matrices are also called *skew-Hadamard* matrices.

If H is skew-type, then $H - I$ is the incidence matrix of a special type of tournament (called a *doubly-regular* tournament).

Unlike the situation for symmetric Hadamard matrices, it seems likely that a skew-Hadamard matrix exists whenever a Hadamard matrix exists. For example, the matrices from Paley's construction are all skew-Hadamard. If H is skew-Hadamard, then, writing $H = S + I$,

$$\begin{array}{cc} S+I & S+I \\ S-I & -S+I \end{array}$$

is also skew-Hadamard.

Skew-Hadamard matrices have been very useful in the construction of Hadamard matrices.

REFERENCES

[1] S. S. AGAIAN, *Hadamard Matrices and their Applications*, (LNiM 1168), Springer-Verlag, Heidelberg (1985).

[2] J. HADAMARD, *Résolution d'une question relative aux déterminants*, Bull. Sci. Math 17 (1893), pp. 240–246.

[3] A. HEDAYAT AND W. D. WALLIS, *Hadamard matrices and their applications*, Ann. Statist. 6 (1978), pp. 1184–1238.

[4] J. JEDWAB AND S. LLOYD, *On the nonexistence of Barker sequences*, Designs, Codes and Cryptography (to appear).

[5] C. LIN AND W. D. WALLIS, *On the circulant Hadamard matrix conjecture*, Proc. Marshall Hall Memorial Conference (to appear).

[6] REAC PALEY, *On orthogonal matrices*, J. Math. and Phys. 12 (1933), pp. 311–320.

[7] E. C. POSNER, *Combinatorial structures in planetary reconnaissance*, Error Correcting Codes (H. B. Mann, ed.), Wiley, New York (1968), pp. 15–46.

[8] H. J. RYSER, *Matrices with integer elements in combinatorial investigations*, Amer. J. Math. 74 (1952), pp. 769–773.

[9] J. R. SEBERRY AND M. YAMADA, *Hadamard matrices, sequences and block designs*, in Contemporary Design Theory – A Collection of Surveys (J. H. Dinitz and D. R. Stinson, eds.), Wiley, New York (1992).

[10] J. J. SYLVESTER, *Thoughts on universe orthogonal matrices, simultaneous sign successions, and tessellated pavements in two or more colours, with applications to Newton's rule, ornamental tile-work, and the theory of numbers*, Phil. Mag. 34 (1867), pp. 461–475.

[11] R. J. TURYN, *Character sums and difference sets*, Pacific J. Math 15 (1965), pp. 319–346.

[12] R. J. TURYN, *Sequences with small correlations*, in Error Correcting Codes (H. B. Mann, ed), Wiley, New York (1968), pp. 195–228.

[13] W. D. WALLIS, A. P. STREET AND J. S. WALLIS, *Combinatorics: Room Squares, Sum-Free Sets, Hadamard Matrices*, (LNiM292), Springer-Verlag, Heidelberg (1972).

[14] J. WILLIAMSON, *Hadamard's determinant theorem and the sum of four squares*, Duke Math. J. 11 (1944), pp. 65–81.

SELF-INVERSE SIGN PATTERNS

CAROLYN A. ESCHENBACH*†, FRANK J. HALL* AND
CHARLES R. JOHNSON**‡

Abstract. We first show that if A is a self-inverse sign pattern matrix, then every principal submatrix of A that is not combinatorially singular (does not require singularity) is also a self-inverse sign pattern. Next we characterize the class of all n-by-n irreducible self-inverse sign pattern matrices. We then discuss reducible self-inverse patterns, assumed to be in Frobenius normal form, in which each irreducible diagonal block is a self-inverse sign pattern matrix. Finally we present an implicit form for determining the sign patterns of the off-diagonal blocks (unspecified block matrices) so that a reducible matrix is self-inverse.

0. Introduction. A matrix whose entries consist of the symbols $+$, $-$, and 0 is called a *sign pattern matrix*. For a real matrix B, by sgn B we mean the sign pattern matrix in which each positive (respectively, negative, zero) entry is replaced by $+$ (respectively, $-$, 0). If $A = (a_{ij})$ is an n-by-n sign pattern matrix, then the *sign pattern class of A* is defined by

$$Q(A) = \{B \in M_n(R)|\, \text{sgn}\, B = A\}.$$

A sign pattern matrix A is said to be combinatorially singular if every $B \in Q(A)$ is singular.

We shall be interested in the cycles in a sign pattern matrix, since every real matrix associated with it has the same qualitative cycle structure. If $A = (a_{ij})$ is an n-by-n sign pattern matrix, then a product of the form $\gamma = a_{i_1 i_2} a_{i_2 i_3} \ldots a_{i_k i_1}$, in which the index set $\{i_1, i_2, \ldots, i_k\}$ consists of distinct indices is called a *simple cycle of length k*. A cycle is said to be *negative* (respectively, *positive*) if it contains an *odd* (respectively, *even*) number of negative entries and no entries equal to zero. In the remainder of this paper, when we say simple cycle, we mean a nonzero simple cycle.

A *matching of size k* in an n-by-n matrix $A = (a_{ij})$ corresponds to k entries in the matrix among whose collective initial indices there are no repetitions, and among whose collective terminal indices there are no repetitions. We call a matching *principal* if the set of initial indices is the same as the set of terminal indices. A principal matching in an n-by-n matrix is called a *complete* matching if the cardinality of the set of initial (terminal) indices is n. The product of entries in a principal matching of size k is either a simple k-cycle or a product of simple cycles whose total length is k and whose index sets are mutually disjoint. Henceforth

---

*Department of Mathematics and Computer Science, Georgia State University, Atlanta, GA 30303.

†The work of this author was supported in part by the National Science Foundation, grant number DMS-9109130.

**Department of Mathematics, College of William and Mary, Williamsburg, VA 23187-8795.

‡The work of this author was supported in part by the National Science Foundation, grant number DMS-90-00839 and by the Office of Naval Research contract N00014-90-J-1739.

when we use the term principal matching, we mean the product of entries in the matching. If γ is a principal matching, sgn γ will refer to the sign that the indices of γ takes as a permutation.

Suppose P is a property a real matrix may or may not have. Then A is said to *require* P if *every* real matrix in $Q(A)$ has property P, or to *allow* P if *some* real matrix in $Q(A)$ has property P. Let A be an n-by-n given sign pattern matrix, and let P be the property "B in $Q(A)$ implies B^{-1} exists and sgn $B^{-1} = A$." We call such a sign pattern *self-inverse*. In particular, B and B^{-1} have the same zero pattern. The question of how to characterize the self-inverse sign patterns was raised in [J]. Our purpose here is to give a characterization.

Let A be an n-by-n sign pattern matrix, and let $A(\alpha, \beta)$ denote the submatrix of A obtained by deleting the rows in the index set α and the columns in the index set β. Let $A[\alpha, \beta]$ be the submatrix that lies in the rows of A indexed by α and the columns indexed by β. If $\alpha = \beta$, then $A[\alpha, \alpha]$, shortened to $A[\alpha]$, and $A(\alpha, \alpha)$, shortened to $A(\alpha)$, are principal submatrices of A. To further simplify notation, we shorten $A(\{i\}, \{j\})$ and $A(\{i\})$ to $A(i, j)$ and $A(i)$, respectively.

The *undirected graph* $G(A)$ of an n-by-n sign pattern matrix A is the undirected graph on n vertices $1, 2, \ldots, n$ such that there is an undirected edge in $G(A)$ from i to j, denoted by the unordered pair $\{i, j\}$, if and only if $a_{ij} \neq 0$ or $a_{ji} \neq 0$. The *directed graph* $D(A)$ of A is the directed graph on n vertices $1, 2, \ldots, n$, such that there is a directed edge in $D(A)$, denoted by (i, j), if and only if $a_{ij} \neq 0$. The set of all vertices is called the *vertex set* V, and the set of all edges is called the *edge set* E. An *undirected path in* $G(A)$ (respectively, directed path in $D(A)$) from i to j is a sequence of undirected (respectively, directed) edges $\{i, i_1\}, \{i_1, i_2\}, \ldots, \{i_{k-1}, j\}$ $((i, i_1), (i_1, i_2), \ldots, (i_{k-1}, j))$. Here the number k of edges is called the length of the undirected (respectively, directed) path. A directed path $(i, i_1), (i_1, i_2), \ldots, (i_{k-1}, j)$ is called a *simple* path if the vertices $i, i_1, \ldots, i_{k-1}, j$ are distinct.

We call a graph *bipartite* if it is possible to partition the vertex set V into two subsets V_1 and V_2 such that every element in the edge set E joins a vertex of V_1 to a vertex of V_2 or a vertex of V_2 to a vertex of V_1. Recall that the lengths of the cycles in a bipartite graph are even.

We say that a sign pattern matrix $A = (a_{ij})$ is a tree sign pattern, $(t.s.p.)$ matrix (see [JJ]) if A is combinatorially symmetric, that is, $a_{ij} \neq 0$ if and only if $a_{ji} \neq 0$; and its undirected graph is connected, but acyclic, that is, the graph is a tree.

If A is a self-inverse sign pattern, then A is sign nonsingular, that is, every $B \in Q(A)$ is nonsingular. Sign nonsingular matrices have been heavily studied, and it is well known that they have unambiguously signed determinants; that is, there is at least one nonzero term in the determinant, and all nonzero terms have the same sign.

1. Results. Let **A** be the class of all self-inverse sign pattern matrices. Since the equivalences in our first lemma are clear, it is stated without proof.

1.1 LEMMA. *The class* **A** *is closed under the following operations:*

(i) *signature similarity (that is, similarity by a diagonal matrix, each of whose diagonal entries is nonzero)*

(ii) *permutation similarity;*

(iii) *negation; and*

(iv) *transposition.*

1.2 LEMMA. *If $A \in \mathbf{A}$ is a matrix of order n, α is a subset of $\{1, 2, \ldots, n\}$, and $A[\alpha]$ is not combinatorially singular, then $A[\alpha]$ and $A(\alpha)$ are sign nonsingular.*

Proof. Let $A \in \mathbf{A}$ be of order n and assume $A[\alpha]$ is not combinatorially singular. Then there is some $B \in Q(A)$ such that $B[\alpha]$ is nonsingular. Consequently $\det B[\alpha] \neq 0$, and since $A \in \mathbf{A}$, $\det B \neq 0$. Use of Jacobi's identity gives

$$\det B^{-1}(\alpha) = \frac{\det B[\alpha]}{\det B} \neq 0,$$

that is, $B^{-1}(\alpha)$ is nonsingular. Since the sign pattern A is self-inverse, $\operatorname{sgn}(B^{-1}(\alpha)) = A(\alpha)$ and it follows that $A(\alpha)$ is also not combinatorially singular. The sign nonsingularity of A now implies that both $A[\alpha]$ and $A(\alpha)$ are sign nonsingular. □

1.3 LEMMA. *If $A \in \mathbf{A}$ is a matrix of order n and $\alpha \subseteq \{1, \ldots, n\}$ is such that $A[\alpha]$ is not combinatorially singular, then $A[\alpha]$ and $A(\alpha)$ are in class $\mathbf{A}$.*

Proof. Let $A = (a_{ij}) \in \mathbf{A}$, and $\alpha \subseteq \{1, 2, \ldots, n\}$. Assume $A[\alpha]$ is not combinatorially singular. From lemma 1.2 we know $A[\alpha]$ and $A(\alpha)$ are sign nonsingular. For contradiction, assume $A[\alpha] \notin \mathbf{A}$, and let $C \in Q(A[\alpha])$ be such that $C^{-1} \notin Q(A[\alpha])$. Let D be a fixed matrix in $Q(A(\alpha))$, and let $B = C \oplus D$. For $\in > 0$, define $B_\in \in Q(A)$ as follows: $B_\in[\alpha] = C$; $B_\in(\alpha) = D$; and all other entries of $B_\in$ are chosen to be 0, respectively, $\pm \in$, if a_{ij} is 0, respectively, $\pm$. For sufficiently small $\in$, we regard $B_\in$ as a perturbation of B. Since

$$\left(\lim_{\in \to 0} B_\in \right)^{-1} = (B_\in[\alpha])^{-1} \oplus (B_\in(\alpha))^{-1} = C^{-1} \oplus D^{-1},$$

we have, for sufficiently small ε, that $A[\alpha] = \operatorname{sgn}(B_\in^{-1}[\alpha]) = \operatorname{sgn} C^{-1}$, which contradicts the assumption that $C^{-1} \notin Q(A[\alpha])$. □

If a sign pattern matrix $A = (a_{ij})$ contains a simple k-cycle, say, $\gamma = a_{i_1 i_2} a_{i_2 i_3} \cdots a_{i_k i_1}$, then the product $\bar{\gamma} = a_{i_1 i_k} a_{i_k i_{k-1}} \cdots a_{i_2 i_1}$ is called the *complementary cycle* of γ. If A contains a simple k-cycle, say, the γ given above, and if $\alpha = \{i_1, i_2, \ldots, i_k\}$, then $A[\alpha]$ is not combinatorially singular. From lemma 1.3, we know $A[\alpha]$ is a k-by-k matrix in class $\mathbf{A}$ that also contains a simple k-cycle. The nature of lemmas 1.4, 1.7 and 1.8 allows us to use this fact, and to assume, without loss of generality, that if $A \in \mathbf{A}$ contains a simple k-cycle, then A is a k-by-k matrix.

1.4 LEMMA. *If $A \in \mathbf{A}$ contains a simple k-cycle ($k \geq 2$), then it contains the complementary k-cycle.*

Proof. Without loss of generality, suppose $A = (a_{ij}) \in \mathbf{A}$ is a k-by-k sign pattern matrix that contains a simple k-cycle γ. By the equivalence in lemma 1.1, we may take $\gamma = a_{12}a_{23}\ldots a_{k1}$. For any $0 \neq a_{ij}$ lying on γ, the complementary block $A(i,j)$ contains a complete matching, namely, $a_{12}\cdots a_{i-1,j-1}\,a_{i+1,j+1}\cdots a_{k1}$, and thus, is not combinatorially singular. Further, if $A(i,j)$ is not sign nonsingular, then there is some $B_1 \in Q(A)$ and $B_2 \in Q(A)$ such that $\det B_1(i,j) = 0$ and $\det B_2(i,j) \neq 0$. Consequently

$$(B_1)^{-1}_{ji} = \frac{(-1)^{i+j}\det B_1(i,j)}{\det B_1} = 0$$

and

$$(B_2)^{-1}_{ji} = \frac{(-1)^{i+j}\det B_2(i,j)}{\det B_2} \neq 0,$$

which contradicts the assumption that $A \in \mathbf{A}$. Thus $A(i,j)$ is sign nonsingular, and since $A \in \mathbf{A}$, this implies that $a_{ji} \neq 0$ for any a_{ij} lying on the cycle γ. We conclude that the complementary cycle $\bar{\gamma}$ is nonzero. $\square$

1.5 LEMMA. *If $A = (a_{ij}) \in \mathbf{A}$ is a matrix of order n, then $a_{ij}a_{ji} \geq 0$, for all i and j in $\{1,2,\ldots,n\}$.*

Proof. Suppose $A \in \mathbf{A}$, and assume $a_{ij} \neq 0$. Then if we expand about the ith row of A to obtain $\det A$, we know $a_{ij}(-1)^{i+j}\det A(i,j)$ is weakly of the same sign as $\det A$. Consequently

$$(1.5a) \qquad a_{ji} = \operatorname{sgn}(A^{-1})_{ji} = \frac{(-1)^{i+j}\det A(i,j)}{\det A}$$

implies that

$$a_{ij}a_{ji} = \frac{a_{ij}(-1)^{i+j}\det A(i,j)}{\det A} \geq 0$$

for all i and j in $\{1,2,\ldots,n\}$. $\square$

1.6 LEMMA. *If $A \in \mathbf{A}$ is a matrix of order n, $\alpha \subseteq \{1,2,\ldots,n\}$, and $A[\alpha]$ is irreducible, then $A[\alpha]$ is sign symmetric.*

Proof. Let $A = (a_{ij}) \in \mathbf{A}$, and $\alpha \subseteq \{1,2,\ldots,n\}$. Assume $A[\alpha]$ is irreducible, and $0 \neq a_{ij}$ is an entry of $A[\alpha]$, $(i \neq j)$. Then $A[\alpha]$ irreducible implies there is a simple path in the directed graph of $A[\alpha]$ from j to i, say $j \rightarrow i_1 \rightarrow i_2 \rightarrow \cdots \rightarrow i_k \rightarrow i$, whose vertices are in the index set α. Thus $a_{ij}a_{ji_1}\ldots a_{i_k i}$ is a nonzero simple cycle in A containing a_{ij}. Now from lemma 1.4, we know $a_{ji} \neq 0$; and from lemma 1.5 we know that $a_{ij}a_{ji} \geq 0$. Thus $a_{ij}a_{ji} > 0$, and it follows that $a_{ij} = a_{ji}$, that is, $A[\alpha]$ is sign symmetric. $\square$

1.7 LEMMA. *If $A \in \mathbf{A}$, then A contains no simple odd cycles of length $\ell \geq 3$.*

Proof. Let $A = (a_{ij}) \in \mathbf{A}$.
Case (i). Let $\ell = 3$. Using equivalence 1.1 (ii), we may assume, without loss of generality, that A contains the simple 3-cycle $\gamma = a_{12}a_{23}a_{31}$. Further we may assume that $\gamma = (+)(+)(*)$, where $*$ is $+$, respectively, $-$, depending upon whether

γ is a positive, respectively, negative 3-cycle. Let $\alpha = \{1, 2, 3\}$. From lemmas 1.4 and 1.6 we know $A[\alpha]$ contains $\bar{\gamma}$, and $\bar{\gamma} = a_{13}a_{32}a_{21} = (*)(+)(+)$. Since $A[\alpha]$ is not combinatorially singular, lemma 1.2 implies that $A[\alpha]$ and $A(\alpha)$ are sign nonsingular. We want to determine if a_{11} is zero or nonzero. To this end, consider the complementary block of the principal submatrix (a_{11}), namely, $A(1)$. Since $A(\alpha)$ is an $(n-3)$-by-$(n-3)$ sign nonsingular matrix, it contains a complete matching, say, δ, which is also a principal matching in $A(1)$. Further $a_{23}a_{32} = \beta$ is a simple 2-cycle contained in A as well as the principal submatrix $A(1)$. Consequently $\beta \cdot \delta$ is a complete matching in $A(1)$, and it follows that $A(1)$ is not combinatorially singular. From lemma 1.2, we know the complementary submatrix (a_{11}) is sign nonsingular, that is, $a_{11} \neq 0$. Similarly it can be shown that $a_{22} \neq 0$ and $a_{33} \neq 0$. Thus $A[\alpha]$ is an entrywise nonzero matrix, and it can easily be checked that $A[\alpha]$ cannot be sign nonsingular. However, from lemma 1.2, we know $A[\alpha]$ is sign nonsingular, and we have a contradiction. We conclude that A contains no simple 3-cycles.

Case (ii). Let $A = (a_{ij})$ be a k-by-k (k odd) sign pattern matrix in $\mathbf{A}$, $k \geq 5$. Without loss of generality, assume A contains the simple k-cycle $\gamma = a_{12}a_{23} \ldots a_{k1}$. By lemma 1.4, A contains the complementary cycle $\bar{\gamma}$. The submatrix $C = A(2, k)$ is not combinatorially singular since it contains a complete matching, namely,

$$c_{12}c_{24}c_{46} \cdots c_{k-3,k-1}c_{k-1,1}c_{33}c_{55} \cdots c_{k-2,k-2} =$$

$$a_{12}a_{34}a_{56} \cdots a_{k-2,k-1}a_{k1}a_{43}a_{65} \cdots a_{k-1,k-2}.$$

Thus there is a matrix B in $Q(A)$ such that $\det B(2, k) \neq 0$, and

$$(B^{-1})_{k2} = \frac{-\det B(2, k)}{\det B}.$$

Since $A \in \mathbf{A}$, we know $\operatorname{sgn}(B^{-1})_{k2} = \operatorname{sgn}(b_{k2}) = a_{k2} \neq 0$. Thus A contains the simple 3 cycle $a_{1k}a_{k2}a_{21}$, which contradicts the assumption that $A \in \mathbf{A}$. $\square$

1.8 LEMMA. *If $A \in \mathbf{A}$, then A contains no simple even cycles of length $\ell \geq 6$.*

Proof.

Case (i). Let $A = (a_{ij})$ be a 6-by-6 matrix in $\mathbf{A}$. Without loss of generality, assume A contains the simple 6-cycle $\gamma = a_{12}a_{23} \ldots a_{61} = (+)(+) \ldots (+)(*)$. By lemma 1.4, A contains the complementary cycle $\bar{\gamma} = a_{16}a_{65} \ldots a_{21} = (*)(+) \ldots (+)$. Since $\operatorname{sgn}(a_{12}a_{21}) \operatorname{sgn}(a_{34}a_{43}) \operatorname{sgn}(a_{56}a_{65}) = -$ is a term in $\det A$, and A is sign nonsingular, it follows that $* = +$. We first show that $a_{25} = -$. To this end, consider the complementary block $C = A(2, 5)$, which contains a complete matching, namely, $a_{16}a_{61}a_{32}a_{43}a_{54}$. Then $a_{16}a_{61}a_{32}a_{43}a_{54} = c_{15}c_{51}c_{22}c_{33}c_{44}$, and $-(c_{15}c_{51})c_{22}c_{33}c_{44} = -$ is a term in $\det A(2, 5)$. Thus $\mathbf{a}_{52} = (\mathbf{A}^{-1})_{52} = \frac{(-1)^7 \det \mathbf{A}(2,5)}{\det A} = -$. Further $a_{25} = a_{52} = -$, since A is sign symmetric. Similar arguments show that $a_{14} = a_{41} = -$ and $a_{36} = a_{63} = -$. Consequently $\operatorname{sgn}(a_{36}a_{63}) \operatorname{sgn}(a_{25}a_{54}a_{41}a_{12}) = +$ is a term in $\det A$, which contradicts the sign nonsingularity of A. We conclude that A contains no simple 6-cycles.

Case (ii). Let $A = (a_{ij})$ be a k-by-k (k even) sign pattern matrix in $\mathbf{A}$, and assume $k \geq 8$. Without loss of generality, assume A contains the simple k-cycle

$\gamma = a_{12}a_{23}\ldots a_{k1}$. By lemma 1.4, A contains the complementary cycle $\bar{\gamma}$. Further, using induction, assume A has no simple m-cycles for any even integer m such that $6 \leq m \leq k-2$. If $m = k$, then $C = A(2,5)$ is not combinatorially singular since it contains a complete matching, namely, $c_{12}c_{24}c_{45}c_{56}\ldots c_{k-2,k-1}c_{k-1,1}c_{33} = a_{12}a_{34}a_{56}a_{67}\ldots a_{k-1,k}a_{k1}a_{43}$. Thus there is a matrix B in $Q(A)$ such that $\det B(2,5) \neq 0$, and by the same argument used in the proof of lemma 1.7, it follows that $a_{52} \neq 0$. However this implies that A contains the simple $(k-2)$-cycle $a_{21}a_{1k}a_{k,k-1}\ldots a_{65}a_{52}$, which contradicts the induction hypothesis. We conclude that A contains no simple even cycles of length $\ell \geq 6$. ∎

For any $A \in \mathbf{A}$, using lemmas 1.7 and 1.8, we conclude that the only possible simple cycles in A are 1-cycles, 2-cycles and/or 4-cycles. Consequently the graphs of matrices in class $\mathbf{A}$ are bipartite graphs, in addition to possible loops at the vertices. The bipartite nature of the graph of a matrix in class $\mathbf{A}$ is an important ingredient needed to prove the major result in this section.

1.9 LEMMA. *If A is an irreducible n-by-n sign pattern matrix in $\mathbf{A}$, and A has no simple 4 cycles, then $n = 1$ or $n = 2$.*

Proof. Suppose A is an irreducible sign pattern matrix in $\mathbf{A}$, and has no simple 4 cycles. By lemmas 1.7 and 1.8, it follows that A has no simple cycles of length $\ell \geq 3$. From lemma 1.6, we know that A is sign symmetric, and, therefore is a t.s.p. matrix. For contradiction, assume $n \geq 3$.

Case (i). Suppose the bipartite graph of A (relabeled by permutation similarity, if necessary) contains a vertex that is adjacent to two or more vertices, one of which is adjacent to a second vertex. Without loss of generality, assume the first vertex is adjacent to $q(q \geq 2)$ vertices, say, $2, 3, \ldots, q$, and the qth vertex is adjacent to the $(q+1)^{\text{st}}$ vertex. Then part of the graph of A is

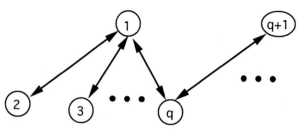

For $B \in Q(A)$, let $C = B^{-1}$. Then the $(1, q+1)$ off-diagonal entry in the product BC, is given by $(BC)_{1,q+1} = 0 + \cdots + 0 + b_{1q}c_{q,q+1} + 0 + \cdots + 0 = b_{1q}\, c_{q,q+1} \neq 0$, which contradicts the assumption that A is in $\mathbf{A}$.

Case (ii). Suppose the bipartite graph of A contains a vertex that is adjacent to two or more vertices, none of which is adjacent to a second vertex. Then part of the graph of A is given by

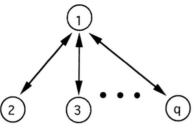

Consequently for any $B \in Q(A)$, with $C = B^{-1}$, we have $(BC)_{23} = \sum_{k=1}^{q} b_{2k}c_{k3} = b_{21}c_{13} \neq 0$, which contradicts the assumption that A is in class **A**. ☐

1.10 LEMMA. *If A is an irreducible n-by-n sign pattern matrix in* **A**, *and A contains a simple 4-cycle, then $n = 4$.*

Proof. Without loss of generality, suppose A is an n-by-n irreducible sign pattern matrix in **A** that contains the simple 4-cycle $\gamma = a_{12}a_{23}a_{34}a_{41} = (+)(+)(+)(*_1)$. From lemma 1.4, we know A contains the complementary cycle $\bar{\gamma}$. Let $V = \{1, 2, 3, 4\}$, and assume $n \geq 5$. Then there is some $j \leq n$ such that $j \notin V$. Further A irreducible implies there is a simple path from i to j and from j to i, for each $i \in V$. It suffices to choose any vertex in V, since the arguments for all vertices are the same. Henceforth let $i = 1$.

Case (i). Suppose there is a path of length 1 from vertex 1 to j, say, $a_{1j} = *_2 \neq 0$. Since A is irreducible, we know from lemma 1.6, that $a_{j1} = *_2$. Let $\alpha = \{1, 2, 3, j\}$. Then $A[\alpha]$ is not combinatorially singular, and, thus, is in **A**. Let $C = A[\alpha - \{3\}, \alpha - \{j\}]$. Then C contains a complete matching, namely, $c_{12}c_{23}c_{31} = a_{12}a_{23}a_{j1}$ which implies that $a_{3j} \neq 0$. Now $a_{3j} \neq 0$ implies that $\gamma_1 = a_{j1}a_{12}a_{23}a_{3j} \neq 0$ in $A[\alpha]$. Since $\gamma_2 = a_{23}a_{32}a_{j1}a_{1j} \neq 0$ in $A[\alpha]$, and $A[\alpha]$ is sign nonsingular, it follows that $(\text{sgn}\,\gamma_1)\gamma_1 = (\text{sgn}\,\gamma_2)\gamma_2$. Consequently $a_{3j} = -a_{j1}$. Next let $\alpha' = \{1, 3, 4, j\}$. Then $A[\alpha']$ is not combinatorially singular, and, thus, is in **A**. Further $\gamma_3 = a_{14}a_{43}a_{3j}a_{j1} = -*_1*_2^2$ and $\gamma_4 = a_{1j}a_{j1}a_{34}a_{43} = *_2^2$ are complete matchings in $A[\alpha']$. By the sign nonsingularity of $A[\alpha']$, we conclude that $(\text{sgn}\,\gamma_3)\gamma_3 = (\text{sgn}\,\gamma_4)\gamma_4$, that is, $*_1*_2^2 = *_2^2$, or $*_1 = +$. Now, $A[V]$ is not combinatorially singular, and, hence, is sign nonsingular. However $(\text{sgn}\,\gamma)\gamma = -$, and if $\gamma' = a_{12}a_{21}a_{34}a_{43}$, then $(\text{sgn}\,\gamma')\gamma' = +$, which contradicts the sign nonsingularity of $A[V]$.

Case (ii). Assume there is a path of even length $p \geq 2$ from vertex 1, to vertex j, $a_{1k_1}a_{k_1k_2} \ldots a_{k_{p-1}j}$. Assume $k_i \notin V$, for all $k_i = k_1, k_2, \ldots, k_{p-1}$. Since A is irreducible, we know from lemma 1.6 that $a_{mi} \neq 0$ for each a_{im} in the path from 1 to j. Let $\alpha = \{1, 2, k_1, k_2, \ldots, k_{p-1}, j\}$. Then $A[\alpha]$ is not combinatorially singular, and, thus, is in **A**. Let $C = A[\alpha - \{2\}, \alpha - \{j\}]$. Then C contains a complete matching, namely, $c_{12}c_{21}c_{33} \ldots c_{p+1,p+1} = a_{12}a_{k_1,1}a_{k_2k_1} \ldots a_{j,k_{p-1}}$, which implies that $a_{j2} \neq 0$. If $p = 2$, then A contains the 6-cycle $a_{1k_1}a_{k_1j}a_{j2}a_{23}a_{34}a_{41}$; if $p > 2$, then A contains an even cycle of length $(p+2) \geq 6$, namely, $a_{1k_1}a_{k_1k_2}\ldots a_{k_{p-1},j}a_{j2}a_{21}$. In both cases, we contradict lemma 1.8.

Case (iii). Suppose there is a path of odd length $p \geq 3$, from vertex 1 to j, say, $a_{1k_1}a_{k_1k_2}\ldots a_{k_{p-1},j}$. Assume $k_i \notin V$, for all $k_i = k_1, k_2, \ldots, k_{p-1}$. From lemma 1.6, it follows that $a_{mi} \neq 0$ for each a_{im} in the path 1 to j. Let $\alpha = \{1, 2, 3, k_1, \ldots, k_{p-1}, j\}$. Then $A[\alpha]$ is not combinatorially singular, and is, therefore, in $\mathbf{A}$. Let $C = A[\alpha - \{3\}, \alpha - \{j\}]$. Then C contains the complete principal matching $c_{12}c_{23}c_{31}c_{44}\ldots c_{p+2,p+2} = a_{12}a_{23}a_{k_11}a_{k_2k_1}\ldots a_{jk_{p-1}}$, which implies that $a_{j3} \neq 0$. Consequently A contains the even cycle $a_{1k_1}a_{k_1k_2}\ldots a_{k_{p-1},j}a_{j3}a_{32}a_{21}$, whose length $\ell \geq 6$, contradicting, lemma 1.8. From cases i – iii, it follows that there is no path from vertex 1 to vertex j, which contradicts the assumption that A is irreducible. We conclude that $n = 4$. □

1.11 THEOREM. *If A is an irreducible n-by-n sign pattern matrix in $\mathbf{A}$, then up to equivalence (as stated in lemma 1.1), A must be one of the following:*

$$
(+); \begin{pmatrix} 0 & + \\ + & 0 \end{pmatrix}; \quad \begin{pmatrix} + & + \\ + & - \end{pmatrix} \text{ or } \begin{pmatrix} 0 & + & 0 & - \\ + & 0 & + & 0 \\ 0 & + & 0 & + \\ - & 0 & + & 0 \end{pmatrix}.
$$

Proof. Suppose A is an n-by-n irreducible sign pattern matrix in $\mathbf{A}$. From lemmas 1.9 and 1.10, we know $n = 1$, $n = 2$ or $n = 4$. If $n = 1$, then it is clear that A must be nonzero. If $n = 2$, then up to equivalence, the only sign symmetric sign pattern class (meeting the requirement of lemma 1.2) not listed in the theorem is

$$
\begin{pmatrix} + & + \\ + & + \end{pmatrix}.
$$

However this matrix is not sign nonsingular, and, consequently, is not in $\mathbf{A}$. Simple algebraic calculations using formula 1.5a show that the two 2-by-2 t.s.p. matrices given in the theorem are in $\mathbf{A}$. The only other 4-by-4 matrices, up to equivalences, that contain no 3-cycles, are irreducible and sign symmetric are represented by the following

$$
\hat{A} = \begin{pmatrix} \square_1 & *_1 & 0 & + \\ *_1 & \square_2 & *_2 & 0 \\ 0 & *_2 & \square_3 & *_3 \\ + & 0 & *_3 & \square_4 \end{pmatrix}
$$

where the $\square$'s and $*$'s are entries in $\{+, -, 0\}$. If $*_i = +$, for all $i = 1, 2, 3$, then the sign pattern of $\hat{A}$ does not have a signed determinant, therefore, is not in $\mathbf{A}$. If $*_1 = *_2 = -$ and $*_3 = +$, or if $*_1 = +$ and $*_2 = *_3 = -$, then the determinant of $\hat{A}$ is not signed, and, therefore, not in $\mathbf{A}$. If $*_1 = -$, for all $i = 1, 2, 3$, and if $\square_1 \neq 0$, formula 1.5a implies $\square_2 = -\square_1$ or $\square_4 = -\square_1$. In either case, $\det \hat{A}$ is not signed. Similar arguments show that if any diagonal entry is nonzero, then $\det \hat{A}$ is not signed. We conclude that $\hat{A}$ has a zero diagonal and is the negation of the 4-by-4 matrix given in the theorem.

Now suppose the sign pattern of $\hat{A}$ contains exactly one $*_i = -$, for some $i \in \{1, 2, 3\}$. For example, let $*_1 = -$ and let

$$\hat{A} = \begin{pmatrix} \square_1 & - & 0 & + \\ - & \square_2 & + & 0 \\ 0 & + & \square_3 & + \\ + & 0 & + & \square_4 \end{pmatrix}.$$

First we note that if any diagonal entry is nonzero, say, $\square_1 \neq 0$, then formula 1.5a implies that $\square_2 = -\square_1$ or $\square_4 = -\square_1$. In either case, $\det \hat{A}$ is not signed, and we conclude $\hat{A}$ has a zero diagonal. However the resulting $\hat{A}$ matrix is equivalent by signature similarity and permutation similarity to the 4-by-4 matrix given in the theorem. Since similar arguments hold if $*_2 = -$ or $*_3 = -$, we omit these cases. Thus the only sign patterns for $\hat{A}$ that are in $\mathbf{A}$, are equivalent to the matrix A in the theorem. We conclude the matrices listed in the theorem, and their equivalences as stated in lemma 1.1, are the only irreducible matrices in $\mathbf{A}$. □

2. Reducible sign patterns in A. Recall that a reducible matrix is permutation similar to a matrix in Frobenius normal form. If A is in $\mathbf{A}$, then from lemma 1.1, we know every permutation similarity of A is in $\mathbf{A}$. Consequently we may assume, without loss of generality, that a reducible matrix in $\mathbf{A}$ is in Frobenius normal form. Further it is clear that if $A = (A_{ij})$ is in $\mathbf{A}$, then each irreducible block A_{ii} of A is in class $\mathbf{A}$.

For reducible matrices in class $\mathbf{A}$, the key question is: What may the off-diagonal blocks look like given the irreducible diagonal blocks, which must be as in 1.11? To answer this question, is to find all possible solutions to a particular qualitative matrix completion problem. We wish to find the sign patterns of the off-diagonal blocks (unspecified block matrices) so that A is in class $\mathbf{A}$. Clearly, any of the matrices given in theorem 1.11 (up to the equivalences allowed by lemma 1.1) can occur as direct summands. A more interesting question occurs when the off-diagonal blocks are nonzero.

We begin by determining the possible off-diagonal blocks lying on the first superdiagonal, that is, we determine the possible sign patterns of the blocks $A_{i,i+1}$. Because of the way in which block triangular matrices multiply, requirements on these off-diagonal blocks depend only upon A_{ii} and $A_{i+1,i+1}$, and we may assume, then, that A has the form

$$(2.0) \qquad \begin{pmatrix} A_{ii} & A_{i,i+1} \\ 0 & A_{i+1,i+1} \end{pmatrix}.$$

2.1 THEOREM. *Let A be a reducible sign pattern matrix in $\mathbf{A}$ of the form (2.0). Then, for each irreducible pair A_{ii} and $A_{i+1,i+1}$ (up to negation) respectively, $A_{i,i+1}$*

must be of the following form:

	A_{ii}	$A_{i+1,i+1}$	$A_{i,i+1}$
i)	$(+)$	$(-)$	$(*)$
ii)	$\begin{pmatrix} 0 & + \\ + & 0 \end{pmatrix}$	$(-)$	$\begin{pmatrix} * \\ * \\ * \end{pmatrix}$
iii)	$(-)$	$\begin{pmatrix} 0 & + \\ + & 0 \end{pmatrix}$	$(*\ *)$
iv)	$\begin{pmatrix} 0 & + \\ + & 0 \end{pmatrix}$	$\begin{pmatrix} 0 & + \\ + & 0 \end{pmatrix}$	$\begin{pmatrix} *_1 & *_2 \\ -*_2 & -*_1 \end{pmatrix}$
v)	$\begin{pmatrix} 0 & + \\ + & 0 \end{pmatrix}$	$\begin{pmatrix} 0 & - \\ - & 0 \end{pmatrix}$	$\begin{pmatrix} *_1 & *_2 \\ *_2 & *_1 \end{pmatrix}$

where $*$, $*_1$, *and* $*_2$ *are any entries in* $\{+,-,0\}$. *All other irreducible* A_{ii} *and* $A_{i+1,i+1}$ *block combinations preclude* A *from being in* **A**, *unless* $A_{i,i+1} = 0$. *In particular,*

$$\begin{pmatrix} + & + \\ + & - \end{pmatrix} \quad \text{and} \quad \begin{pmatrix} 0 & + & 0 & - \\ + & 0 & + & 0 \\ 0 & + & 0 & + \\ - & 0 & + & 0 \end{pmatrix}$$

can occur as diagonal blocks only as direct summands.

Proof. Without loss of generality, and, to simplify notation, let $A_{ii} = A_{11}$, $A_{i+1,i+1} = A_{22}$, and $A_{i,i+1} = X$. Take $B \in Q(A)$, and let $C = B^{-1}$. Partition B and C conformably to A as follows:

$$B = \begin{pmatrix} B_{11} & B_{12} \\ 0 & B_{22} \end{pmatrix}, \quad C = \begin{pmatrix} C_{11} & C_{12} \\ 0 & C_{22} \end{pmatrix}.$$

Since $BC = I$, it follows that

$$B_{11}C_{12} + B_{12}C_{22} = 0;$$

or

$$C_{12} = -B_{11}^{-1}B_{12}C_{22};$$

or

(2.1a) $$C_{12} = -C_{11}B_{12}C_{12}.$$

Since (2.1a) holds for all $B \in Q(A)$, we get

(2.1b) $$\operatorname{sgn} X = -\operatorname{sgn} A_{11}XA_{22},$$

which must hold *unambiguously*.

Conversely, suppose X satisfies (2.1b) unambigously. Choose $C_{11} \in Q(A_{11})$, $C_{22} \in Q(A_{22})$, $B_{12} \in Q(X)$, and let $C_{12} = -C_{11}B_{12}C_{22}$. Then $C_{11}^{-1}C_{12} + B_{12}C_{22} = 0$, so that $X = A_{i,i+1}$.

Consequently, (2.1b) becomes the operative equation for determining $X = A_{i,i+1}$. The proof of the theorem then becomes routine. For example, with $A_{11} = \begin{pmatrix} + & + \\ + & - \end{pmatrix}$, each column of X must have a zero entry since $A_{11}X$ is unambiguous, say, $X = \begin{pmatrix} + & 0 \\ 0 & - \end{pmatrix}$. Then $A_{11}X = \begin{pmatrix} + & - \\ + & + \end{pmatrix}$. However, with any allowable choice for A_{22}, $\operatorname{sgn} X \neq - \operatorname{sgn} A_{11}XA_{22}$, unambiguously, unless $X = 0$.

With $A_{11} = \begin{pmatrix} 0 & + & 0 & - \\ + & 0 & + & 0 \\ 0 & + & 0 & + \\ - & 0 & + & 0 \end{pmatrix}$, and using the second and fourth rows of A_{11}, either the second or fourth entry in any column of X must be zero. Similarly, the first or third entry in any column of X must be zero, say, $\begin{pmatrix} *_1 \\ *_2 \\ 0 \\ 0 \end{pmatrix}$ is a column of X.

Then $\begin{pmatrix} *_2 \\ *_1 \\ *_2 \\ -*_1 \end{pmatrix}$ is a column of $A_{11}X$; and, again, (2.1b) will not hold unless $X = 0$. We omit the details for other allowable choices of an arbitrary column in X since the arguments are similar to the one given above.

Now let $A_{11} = (+)$ and $A_{22} = \begin{pmatrix} 0 & - \\ - & 0 \end{pmatrix}$. Then with $X = (*_1 \ *_2)$, we have $\operatorname{sgn} A_{11}XA_{22} = (- *_2 \ -*_1)$. In order for $\operatorname{sgn} X = - \operatorname{sgn} A_{11}XA_{22}$ to hold, $*_1 = *_2$, that is, $X = (* \ *)$. We omit the details of the remaining cases since the arguments are similar to the one above. □

Using (2.1b) and induction, one may prove our main theorem of this section. In Theorem 2.2, Q_{ij} denotes the set of increasing sequences from i to j. If $p = i, k_2, \ldots, k_{t-1}, j$ is such a sequence, then $|p| = t$. For simplicity, we present the solution only in an implicit form, since a description as explicit as theorem 2.1 would be prohibitive. Further the implicit form of the solution makes it clear how to check a given example.

2.2 THEOREM. *Let A be a reducible sign pattern matrix in $\mathbf{A}$, and assume A is in Frobenius normal form. Then, up to equivalence, the irreducible diagonal blocks are as given in Theorem 1.11, the first superdiagonal blocks are as indicated in Theorem 2.1; and, in general, for all $i < j$,*

$$A_{ij} = \sum_{p \in Q_{ij}} (-1)^{|p|} A_{ii} A_{ik_2} A_{k_2 k_2} A_{k_2 k_3} \ldots A_{k_{t-1}, j} A_{jj}$$

with the sum on the right unambiguous.

Proof. Sketching part of the proof, we take $B = (B_{ij}) \in Q(A)$, and let $C = (C_{ij}) = B^{-1}$, where B and C are partitioned conformally to A. Then for $i < j$, we have

(2.1c) $\qquad C_{ij} = -C_{ii}(B_{i,i+1}C_{i+1,j} + \cdots + B_{i,j-1}C_{j-1,j} + B_{ij}C_{jj})$.

Using induction, we have $C_{i+1,j}, \ldots, C_{j-1,j}$, each expressed as an appropriate sum. Substitute these sums into (2.1c) and simplify. Since (2.1c) holds for all $B \in Q(A)$, we get the restrictions on the signs in A_{ij}. $\Box$

It should be emphasized that each term in the sum for A_{ij} must be unambiguous. Furthermore all conditions of theorem 2.2 must hold. In particular, for *all* $i < j$, A_{ij} must unambiguously be the above sum. A subtle consequence of this is that there may be completions for certain blocks lying on, say, the kth superdiagonal that lead to ambiguities for blocks lying on a higher order superdiagonal. Clearly such completions preclude the matrix from being in class **A**. We illustrate this phenomenon with the following example.

2.3 Example. Let A be the partial block matrix

$$
\begin{pmatrix}
0 & + & | & - & 0 & 0 & 0 & 0 & | & & A_{14} & \\
+ & 0 & | & - & 0 & 0 & 0 & 0 & | & & & \\
\hline
& & & - & 0 & 0 & 0 & 0 & | & - & - & \\
\hline
& & & & 0 & + & 0 & - & | & 0 & 0 & \\
& & & & + & 0 & + & 0 & | & 0 & 0 & \\
& 0 & & & 0 & + & 0 & + & | & 0 & 0 & \\
& & & & - & 0 & + & 0 & | & 0 & 0 & \\
\hline
& & & & & & & & | & 0 & + & \\
& & & & & & & & | & + & 0 &
\end{pmatrix}
$$

Although the first and second superdiagonals were completed using theorems 2.1 and 2.2, there is no completion for A so that A is in **A**. However if $A_{12} = 0$ or $A_{24} = 0$, then

$$
A_{14} = \begin{pmatrix} *_1 & *_2 \\ -*_2 & -*_1 \end{pmatrix},
$$

where $*_1, *_2 \in \{0, +, -\}$, completes A so that A is in class **A**.

In [BF] the authors consider the graph whose vertices are sign pattern matrices, and whose edges connect those that are possible sign patterns of a matrix and its inverse. Consequently, self-inverse sign patterns we have characterized are the isolated vertices of the graph considered in [BF].

REFERENCES

[BMQ] BASSETT, L., JOHN MAYBEE AND JAMES QUIRK, *Qualitative Economics and the Scope of the Correspondence Principle*, Econometrica 35 (1967), pp. 544–563.

[BF] BERGER, M.A. AND A. FELZENBAUM, *Sign Patterns of Matrices and Their Inverses*, Linear Algebra and Its Applications, 86 (1987), pp. 161–177.

[JJ] JEFFRIES, CLARK AND C.R. JOHNSON, *Some Sign Patterns Which Preclude Matrix Stability*, SIAM J. on Matrix Analysis and Applications, 9 (1988), pp. 19–25.

[J] JOHNSON, C.R., *Sign Patterns of Inverse Nonnegative Matrices*, Linear Algebra and Its Applications, 24 (1979), pp. 75–83.

OPEN PROBLEMS

The following problems were presented at an open problems session at the work-shop.

1. (V. Klee) A real matrix A of order n is *potentially stable* provided there is a matrix $\tilde{A}$ such that $\tilde{A}$ has the same sign pattern as A and $\tilde{A}$ is stable (that is, all of its eigenvalues have negative real part). Now consider the decision problem

POTSTAB:

  Instance: A $(0, 1, -1)$-matrix A.

  Question: Is A potentially stable?

Conjecture 1: POTSTAB $\in \mathcal{NP}$.

Conjecture 2: POTSTAB is $\mathcal{NP}$-hard.

Remark: The two conjectures are stated separately because it is not obvious whether POTSTAB $\in \mathcal{NP}$.

2. (V. Klee) Let N^+ denote the set of all positive integers. For each function $f : N^+ \to N^+$. Consider the decision problem

P_f:

  Instance: An $(n + f(n))$ by n $(0, 1, -1)$-matrix A.

  Question: Does the sign pattern of A imply that A is of rank n?

General question: How does the complexity of P_f depend on f?

Remark 1: When $f \equiv 0$, P_f is polynomially equivalent to the infamous even cycle problem.

Remark 2: When (for a fixed $k > 0$) $f(n) = \lfloor n^{1/k} \rfloor$, the problem P_f is NP-complete [1].

Special question: Is there a k such that P_f is NP-complete when $f \equiv k$?

[1] V. Klee, R. Ladner and R. Manber, Signsolvability revisited, *Linear Algebra and its Applications*, 59: 131-157 (1984).

3. (V. Klee) Suppose that F is an ordered field, C is a bounded, closed convex subset of the product space F^n, and $p \in F^n \setminus C$. Must there exist a linear functional $\phi : F^n \to F$ such that $\phi(c) < \phi(p)$ for all $c \in C$?

Remark. The answer is 'yes' when the field is archimedean.

4. (M. Boyle) Let A be a nonnegative matrix of order n with nonzero spectrum $\delta = (\lambda_1, \ldots, \lambda_k)$ with $\lambda_1 = |\lambda_1| \geq |\lambda_j|$ for $j \geq 2$. Ashley's inequality [1] states

$$\lambda^n - (\lambda_1)^n \geq \lambda^{n-k} \prod_{i=1}^{k} |\lambda - \lambda_i| \quad (\lambda > \lambda_1);$$

equivalently,

$$\lambda^k - \left(\frac{\lambda_1}{\lambda}\right)^{n-k} (\lambda_1)^k \geq \prod_{i=1}^{k} |\lambda - \lambda_1| \quad (\lambda > \lambda_1).$$

This implies

$$\lambda^k > \prod_{i=1}^{k} |\lambda - \lambda_i| \quad (\lambda > \lambda_1) \qquad (*).$$

Problem: Is $(*)$ implied by the necessary conditions [2] on the nonzero spectrum of a primitive matrix ?

The necessary conditions are that $\lambda_1 = |\lambda_1| \geq |\lambda_j|$ ($j \geq 2$), the coefficients of the polynomial $\prod_{i=1}^{k}(t - \lambda_i)$ are real, and for all positive integers n and ℓ, $\sum_{i=1}^{k} \lambda_i^n \geq 0$ and $\sum_{i=1}^{k} \lambda_i^n > 0$ implies $\sum_{i=1}^{k} \lambda^{n\ell} > 0$.

[1] J. Ashley, On the Perron-Frobenius eigenvector for non-negative integral matrices whose largest eigenvalue is integral, *Linear Algebra and its Applications*, 94: 103-108 (1987).

[2] Mike Boyle and David Handelman, The spectra of nonnegative matrices via symbolic dynamics, *Annals of Mathematics*, 133: 249-316 (1991).

5. (W. Barrett) Let $A_n = [a_{ij}]$ be the matrix of order n defined by

$$a_{ij} = \begin{cases} 1 & \text{if } i|j \text{ or if } j = 1, \\ 0 & \text{otherwise.} \end{cases}$$

For example,

$$A_6 = \begin{bmatrix} 1 & 1 & 1 & 1 & 1 & 1 \\ 1 & 1 & 0 & 1 & 0 & 1 \\ 1 & 0 & 1 & 0 & 0 & 1 \\ 1 & 0 & 0 & 1 & 0 & 0 \\ 1 & 0 & 0 & 0 & 1 & 0 \\ 1 & 0 & 0 & 0 & 0 & 1 \end{bmatrix}.$$

Problem. Find the rate of growth of $|\det A_n|$ as $n \to \infty$.

Remark: The matrix A_n is of number-theoretical interest because

$$|\det A_n| = O(n^{\frac{1}{2}+\epsilon}) \quad \text{(for all } \epsilon > 0\text{)}$$

if and only if the Riemann hypothesis is true. The connection is a result of the fact, proved by Redheffer when he first introduced A_n [1], that

$$\det A_n = \sum_{k=1}^{n} \mu(k)$$

where $\mu : Z^+ \to \{-1, 0-, 1\}$ is the Möbius function.

Quite a lot is now known about the eigenvalues of A_n, see e.g. [2], but no interesting bound for $|\det A_n|$ has yet been obtained using matrix theory, not even $|\det A_n| \leq n$. A proof that $|\det A_n| = o(n)$ would constitute a new proof of the prime number theorem.

[1] Ray M. Redheffer, Eine explizit lösbare Optimierungsaufgabe, *Internat. Schriftenreihe Numer. Math.* 36: 213-216 (1977).

[2] Wayne W. Barrett and Tyler J. Jarvis, Spectral properties of a matrix of Redheffer, *Linear Algebra and its Applications*, 162-164: 673-683 (1992).

6. (Peter Gibson) Let Ω_n be the polytope of doubly stochastic matrices of order n. If $A, B \in \Omega_n$ and

$$\text{per}(rA + (1-r)B) = \text{per}A \quad (0 \leq r \leq 1),$$

then A and B form a *permanental pair* and B is a *permanental mate* of A. Let

$$\mathcal{M}(A) = \{B \in \Omega_n : B \text{ is a permanental mate of } A\}.$$

A *permanental polytope* is a subpolytope Γ of Ω_n such that $\text{per}A = \text{per}B$ for all $A, B \in \Gamma$. Let

$$\mathcal{M}_n = \{\mathcal{M}(A) : A \in \Omega_n, \mathcal{M}(A) \text{ is convex}\}$$

and

$$\mathcal{P}_n = \{\Gamma : \Gamma \text{ is a permanental polytope of } \Omega_n\}.$$

Let

$$\gamma_n = \max\{\dim \Gamma : \Gamma \in \mathcal{M}_n\} \text{ and } \delta_n = \max\{\dim \Gamma : \Gamma \in \mathcal{P}_n\}.$$

Problem 1: For $n \geq 4$, determine

(a) whether $\gamma_n = \delta_n$,

(b) the value of γ_n,

(c) the values of k for which there exists $\Gamma \in \mathcal{M}_n$ with $\dim \Gamma = k$.

$\mathcal{M}(A)$ has a *finite convex decomposition* provided that for some positive integer k there exist maximal permanental polytopes $\Gamma_1, \ldots, \Gamma_k$ of Ω_n such that

$$\mathcal{M}(A) = \cup_{i=1}^{k} \Gamma_i,$$

and

$$\{A\} = \Gamma_i \cap \Gamma_j \text{ whenever } i \neq j.$$

Problem 2: Let $n \geq 3$.

(a) Does $\mathcal{M}(A)$ have a finite convex decomposition for all $A \in \Omega_n$?

(b) When $A \in \Omega_n$ is such that $\mathcal{M}(A)$ has a finite convex decomposition, must there exist $B_1, \ldots, B_k \in \Omega_n$ such that $\mathcal{M}(B_i)$ is convex for all i, and

$$\mathcal{M}(A) = \cup_{i=1}^{k}\mathcal{M}(B_i), \{A\} = \mathcal{M}(B_i) \cap \mathcal{M}(B_j) \quad (i \neq j)?$$

7. (John Maybee) Let A be an irreducible, sign nonsingular matrix of order n which is normalized so that no element on the main diagonal equals 0. A zero submatrix of A of size p by q is said to have *size* $p + q$.

Conjecture: If A has more than $4(n-1)$ nonzero entries, then A has a zero submatrix of size $n - 1$.

8. (Richard A. Brualdi and Jennifer J. Q. Massey) Let A be an m by n $(0,1)$-matrix with maximum row sum α and maximum column sum β.

Conjecture: There exist $\alpha\beta$ permutation submatrices $\{P_i : 1 \leq i \leq \alpha\beta\}$ of A which together contain all the 1's of A. (Note that it is important that the P_i be submatrices of A, implying that A has 0's in all positions corresponding to the 0's of the P_i. By a theorem of König, there always exists $k = \max\{\alpha, \beta\}$ subpermutation matrices $Q_1, \ldots, Q_k$ such that $A = Q_1 + \cdots + Q_k$.)

This conjecture has been proved in [1] in case the length of each cycle of the bipartite graph associated with A is divisible by 4.

[1] R.A. Brualdi and J.J.Q. Massey, Incidence and strong edge colorings of graphs, *Discrete Applied Mathematics*, to be published.